The
MORRIS MOTOR CAR
1913 • 1983

The MORRIS MOTOR CAR

1913·1983

Harry Edwards

ROADMASTER PUBLISHING
CHATHAM · KENT

Dedicated to Babs, who has absolutely
no interest in old cars but has tolerated
my fanaticism uncomplainingly

First published in 1983 in Great Britain by
Moorland Publishing Co. Ltd.
Moor Farm Road
Airfield Estate
Ashbourne, Derbyshire DE6 1HD

This revised edition published in 1997 by
Roadmaster Publishing
PO Box 176
Chatham
Kent ME5 9AQ

ISBN 1 871814 01 4

British Library Cataloguing in Publication Data

A catalogue record of this book is available from the British Library

Printed and bound in Great Britain by
Bookcraft (Bath) Ltd, Midsomer Norton, Avon

Contents

Introduction

Despite the prolific output of books published for old-car enthusiasts, surprisingly little has been written in depth on the Morris motor car. *The Bullnose and Flatnose Morris* by Lytton P. Jarman and Robin I. Barraclough, and Paul Skilleter's *Morris Minor* are two of the few. True, there have been many general histories of the motor car but these tend to take a superficial look at the Morris and, often as not, get the model and dating wrong.

In writing this book I have drawn upon my twenty years' experience as honorary historian of the Morris Register to decide what information to include; useful to present day owners of a vintage Morris and yet not too lacking in marque history for those more generally interested in the Morris. Because I appreciate the value of a mediocre period photograph as a guide to originality, against a modern professional composition depicting a newly restored car, I make no apologies for including many early photographs.

I owe a debt of gratitude to innumerable members of the Morris Register who, over the years, have posed questions about their cars. Questions which, in order to give satisfactory answers, have sometimes forced me to delve into obscure aspects of the Morris car and as a result increased my knowledge of the marque. A high percentage of the many letters I receive are from enthusiasts who are not quite sure of the exact model or year of the Morris that they have just acquired. Or the query may be on the subject of the correct engine, or original colour scheme. Within these pages I hope I have provided most of the answers, but colours pose something of a problem for those cars made before the 1960s. I have only been able to quote the colours as described in contemporary catalogues, etc, and not the actual shades. 'Red' can be anything just beyond pink at one extreme, to a shade almost maroon at the other. One of the difficulties of recording colour schemes is that words without a supporting colour chart are meaningless. British Standards colour charts do not appear to have existed before the war. The paint finish used from the late 'twenties on Morris cars and Morris commercials was a cellulose finish called 'Bripal' made by the British Paint & Lacquer Co, of Cowley — a firm which is now part of Imperial Chemical Industries. It should be added that by 1947 some Morris cars were being turned out with a new synthetic 'Synobel' material. Regardless of the body colour, the wheels, mudguards, aprons, etc, on all Morris cars from the early 'twenties to the Series II coupé models were black stove enamelled finish — all, that is, except the 1934-5 Ten-Six Special mudguards and the colourful wheels and brake drums on the early Isis. Commercial vehicles were, in many instances, sold in primer finish only so the present day restorer has less of a problem when considering the original colour scheme. For the more modern Morris, the traditional listing of colour schemes in catalogues ceased about 1952 and as a consequence it has not been possible to detail the colour options for these later models.

Early 'Hotchkiss' type 11.9hp engines appear to have always been painted red. When the larger bore 13.9hp version was introduced in the early 'twenties, these were painted blue to distinguish, it was said, the special cast block used to allow for the larger bore of 75mm. Research suggests that all other engines until the late 'thirties were black, with reconditioned 'UB' 8hp units finished green and later gold for the 'Gold Seal' replacements. Series E Eight engines, and those of the Series III overhead-valve range, were bluish-grey.

It is important to recognise the difference between the 'model year' and the calendar year. Unlike the latter, the model year ran from Motor Show to Motor Show (approximately September-August) and so it is quite possible for a 1934 model Morris Minor to have been registered in September 1933. In this book, all my references are to the model year, or Series.

Inevitably, during the years covered by this book some of the persons mentioned received honours, resulting in a change of name. Their name applying at the period about which I am writing has been used. Similarly organisations and companies have been referred to by the name applicable at the time. For instance, the manufacturer of the Morris car has changed over the years from W.R.M. Motors Ltd, to Morris Motors (W.R. Morris) Ltd and, subsequently, Morris Motors Ltd, Morris Motors (1926) Ltd, reversion to Morris Motors Ltd as part of the British Motor Corporation, British Leyland (Austin-Morris) Ltd and more recently BL Ltd.

In adopting a strictly model-by-model approach in this book there is unavoidable repetition of some of the background details relating to features or components which were common to several models. It is hoped that the reader understands the reason for this, as this approach is the most useful for enthusiasts wishing to know about a particular model.

During my years as honorary historian of the Morris Register many people have helped me with my researches, culminating in the appearance of this book. Thus I find the hardest task of all is to list my acknowledgements; to mention everyone who has assisted me over the years with information and photographs would require a list almost as long as the book itself. To all those people I record my grateful thanks. But I must single out for special gratitude one or two people who have been particularly helpful with data, photographs and other research material. Peter Burdon of Kenilworth, ex-editor of *High Road* magazine; the late Denis Lowe of BL Photographic Department; Lytton P. Jarman who's assistance with information on the early Bullnose cars was invaluable, and Anders Ditlev Clausager of BL Heritage Ltd who provided me with production figures and data on the little-known export wide-track Cowleys. Above all I must express my gratitude to motoring historian Michael Sedgwick who went through my manuscript, making corrections and offering valuable criticism — appropriately, for it was he, and William Boddy of *Motor Sport,* who had unknowingly given me the confidence to complete this work.

Harry Edwards

The Bullnose
Oxford 1913-26

His long standing idea of making a car of his own had to remain a pipe dream for William Richard Morris until 1910 when, assured of sufficient finance and new enlarged premises at Longwall, he was able to go ahead with plans for his new car. Morris knew what he wanted and, unlike Carl Benz, he was willing to learn from the mistakes and successes of others. His experiences while handling the Oxford agency for established makes, such as Arrol-Johnston, Humber, Singer, Standard, Wolseley, Hupmobile, and Belsize, had given him an insight into features to be avoided, as well as designs worth emulating. His vehicle was to be a complete car and not one of the unsuccessful cycle-cars which his contemporaries produced to cater for the lower price market.

By the end of 1912 the design work had been completed, mostly on the drawing boards of the Coventry based firm of White & Poppe who would supply the engines for the two-seater, to be called the 'Morris Oxford Light Car'. While it was not going to be the cheapest car on the market, the stated policy was to provide a machine in which freedom from trouble and simplicity of control when driving was of first importance. Perhaps a hectic drive delivering a customer's Lacoste et Battman car from Paris to Stirling (when parts in the gearbox locked, the propeller shaft seized twice, and the bevel gear in the back axle stripped teeth on three occasions) was responsible for the meticulous attention given to the design of the transmission, which was totally enclosed from the starting handle to the rear wheels. The American Hupmobile cars had a similar feature and Morris did not hesitate to adopt a good idea. His gearbox design was said to be influenced by the unit used in the Belsize car.

Morris had hoped to have the Oxford ready for the Motor Show of 1912, but because of delays in the supply of components this was not to be. In particular, the White & Poppe prototype engine was not available, mainly because the Rugby foundry of Willans & Robinson, who were to supply the castings for the cylinder block and head, produced the first one half-scale! If he could not put his car on display Morris decided he would show the blue-prints and announce the specification. At least one trader, a Mr Gordon Stewart, was so impressed that on the strength of the design alone he placed a firm order with Morris for 400 cars. This faith paid off when his company, Stewart & Ardern Ltd, were later appointed sole distributors in London.

The formation of a new company, W.R.M. Motors Ltd, in August 1912 and the acquisition of an old military training college at Temple Cowley set the scene for the first Oxfords. Several months of preparation passed before the first production models saw the road but, paradoxically, before this event a Morris Oxford was exhibited at Manchester's 'North of England Motor Show' held in the Exhibition Hall, Rusholme, during 14-22 February 1913. However, all was not what it seemed for the delay in the supply of castings dictated that the smart little vehicle on display was 'powered' by a dummy engine made of wood! Meanwhile, Stewart & Ardern were already advertising the Morris Oxford in *The Autocar* with a specification which included a capability of 45-50mpg. This proved to be the last ever motor show to be held in the large, mainly timber constructed, exhibition building in Old Hall Lane. Ten months later the Suffragette movement had burnt the building to the ground.

At the end of March 1913 the first Oxford was ready and Gordon Stewart travelled from London to take delivery. Unfortunately he had only driven a few hundred yards when the cast-iron universal joint snapped! A new one only lasted until High Wycombe before it too shattered. Phosphor bronze joints subsequently made for the Oxfords solved the problems of shock loads.

In 1935 an article in *The Morris Owner* by W. Gordon Aston describes an instance in 1934 when the then owner of 'No 1 Morris Oxford' made enquiries at Cowley for a spare part. Sir William Morris (as he then was), learning that the elusive fox had been run to earth, offered its owner a brand new saloon of the latest type as a straight exchange. The owner replied, rather indignantly, that he wanted a spare part, that he was not interested in a new car, and that he would be obliged if the organisation would get on with its legitimate business!

Then, as now, in any new design there are teething troubles and the Oxford was no exception — the universal joint mentioned provides an example. The multi-plate clutch, a sandwich of thirty-six steel and bronze plates running in oil, refused to free when the oil was cold and thick. A reduction in the number of plates to thirty-four helped, but the problem was not solved until a deliberate buckle was put in the centre of each plate. Steering proved to be a major problem on a design incorporating a worm and sector box operating a draglink in an unconventional position under the front axle. The Oxford

1913 Morris Oxford against a background of the Hodal Rest House in the Southern Punjab, India.
(R.R. Harrison)

In 1958 a competition to find the earliest surviving Morris Oxford was sponsored by the Daily Express. *Lord Nuffield enjoys a joke while inspecting one of the entrants' cars, a Morris Oxford originally purchased by a Mr F. G. Brown in May 1913 (chassis No 111).*

exhibited a wanderlust because of the combination of the cobbles and tramlines of 1913, and the narrow, 700mm × 80mm, Dunlop bedded edge tyres. A measure of stability was obtained when, after some 150 cars had been made, the track dimension was increased by 2in.

With production underway the cars were being distributed by Messrs H.W Cranshaw of Deansgate, Manchester, for the north; W.H.M. Burgess in the south, while London area sales were handled by Stewart & Ardern Ltd. No effort was made to conceal the fact that the new Morris Oxford was an assembly of proprietary parts, indeed it became a sales feature. E.G. Wrigley & Co of Birmingham supplied the front and rear axles; the hollow pressed steel wheels were made by the firm of Sankey (the use of these wheels making the Morris Oxford one of the first, if not the first, English car to adopt the steel wheel); Doherty Motor Components Ltd of Coventry made the radiators; from White & Poppe came the engine, gearbox, clutch and carburetter; tyres were

William Morris and his accountant, Mr Varney, in a 1913 Oxford outside the original premises of Stewart & Ardern Ltd, 18 Woodstock Street, Off Bond Street, in London.
(Stewart & Ardern Ltd)

*The twelfth SMMT Motor Show at Olympia, London, in
1913, and the first for W.R.M. Motors Ltd. Included on
stand 9 is the Oxford de luxe model for the 1914 season.*
(Stewart & Ardern Ltd)

made by Dunlop; and lamps were obtained from Powell & Hanmer — which, if
contemporary photographs are any guide, were the self-contained carbide type
No 500 with Mangin reflectors. For the two-seater torpedo body Morris turned
to the old established Oxford coachbuilding firm of Charles Raworth & Sons,
while the actual chassis frames were imported from Belgium. The ignition
system used the reliable Bosch ZF4 magneto and Bosch sparking plugs. The
suppliers of parts were of good repute and the public immediately accepted the
car. 'Morris is the only Light Car', said the Stewart & Ardern advertisement of
1913, 'which embodies the joint production of the greatest British experts';
conveniently forgetting the use of German and Belgian components.

The Motor Show held at Olympia in 1913 must have given William Morris
much pleasure. At last he was able to show his models, albeit on a very
crowded stand 9 (erected by Harrods Ltd, which may explain the fact that this
well known London store were offering the Morris Oxford for sale that year).
Not only was there the established Oxford Light Car (now referred to as the
'Standard Model'), but also two-seater and coupé examples of the improved
Oxford Light Car to be known as the de luxe model which, due to the wider

track of 45in (as opposed to the standard model's 42in, after the modification mentioned earlier) and revised steering arrangement, had now lost the wandering tendency inherent in the earlier car. This new steering design, shown in large diagrams, dominated the background to the display. Other de luxe differences included a longer wheelbase, larger radiator capacity and modifications to the rear axle ratio, which altered the final drive ratio from 3.5 or 4.2:1 to 4.6:1.

The full list of the new (1914) models offered the de luxe chassis with a choice of torpedo two-seater, coupé, sports or delivery van body. The chassis used on the standard two-seater was also available for £5 less with a narrower two-seater body and poorer quality finish, called the 'commercial' model. For the same sum, £5, the buyer of the de luxe was given nickel fittings as an alternative to brass.

Useful publicity and a reputation for reliability was gained by the entry of Morris Oxford cars in various reliability and hill-climb trials by William Morris himself. As early as June 1913 a Morris Oxford with its originator at the wheel took part in the Caerphilly hill climb, to be placed fifth out of the six entrants in a class for four-cylinder cars not exceeding 70mm bore. Not the result Morris had hoped for, but success came in the Dutch six-day reliability trial in the same year, a Morris Oxford gaining the first prize with full marks — a success repeated when Morris cars came first, second and third in the Oxfordshire Motor Club annual hill climb in 1913. An event on a much larger scale was the London-Edinburgh Trial organised by the Motor Cycling Club during the Whitsun holiday 1914. Six Morris Oxfords took part. W.R. Morris with his passenger Henry Galpin entered a drop-head coupé (registered FC1711), emphasizing the advantage of a closed vehicle by wearing lounge suits, in contrast to the overalls worn by most other competitors! The other Morris entries were driven by Gordon Stewart of Stewart & Ardern Ltd, L.P. Kent, E. Channon, S.J. Prevost, and Harry Bashall. Bashall was a well-known motor cyclist of the time having, two years before, won the Junior TT race. The reliability of the Morris car was publicly demonstrated (despite the failure of Bashall's magneto) with the award of gold medals to all Morris entries and a special trophy, 'The Light Car Cup', collected by William Morris himself. In other 1914 events, two Oxfords gained gold medals in the London-Exeter trial for completing the course in the scheduled time, and in July a Mrs R. Wikey entered her Oxford in the Lady Driver's Handicap Race at Saltburn speed trials and came second to a 12/16 Sunbeam. W.R. Morris entered in the Coventry Hill Climb which took place at Irondown Hill, near Deddington, about the same time. His success in reaching the top was marred by the subsequent discovery that one piston had split in two with only the gudgeon pin holding the halves together! Even Brooklands saw the Morris Oxford competing when in the Three Lap Cycle Car Handicap Race held around August 1914, a Morris Oxford was placed third to a Morgan and a GWK.

An event involving another make of car, a Gräf und Stift, at Sarajevo on 28 June 1914, triggered off World War I and within weeks Britain was involved. The assembly facilities at W.R.M. Motors Ltd were utilised to produce hand grenades, Stokes howitzer bomb cases and later, in 1916, mine sinkers in

Morris Oxford de luxe two-seater with electric lighting,
about 1915. (Jeffrey Freeman)

significant quantities. Car production continued in decreasing numbers as the conflict dragged on, the figure of 159 Oxfords made in 1915 dropped to thirteen in 1916 and only one in 1917; these were assembled from parts produced before the war.

The introduction of the Cowley is covered in detail elsewhere in this book so at this point in the narrative it is sufficient to record that the Cowley motor car, first shown in prototype form to the press in April 1915, was aimed at the untapped market of the family man not attracted by the small two-seater Oxford which was, presumably, to continue in production alongside the larger model. In the event, the supply of White & Poppe 60mm × 90mm, 8.92hp, engines dried-up when the White side of the partnership decided to sell the firm to Dennis Brothers Ltd of Guildford, in 1919. With the change of ownership went any plans Morris had of using a longstroke 60mm × 100mm White & Poppe engine in an envisaged post-war four-seater Oxford. Morris now turned to the Coventry branch of a French armament firm, Hotchkiss et Cie, who were prepared to supply an 11.9hp engine based on the American Continental Motors Company's 'Red Seal' design, which had been used in the 1,400 or so Cowleys produced by that time. With a supply of engines now assured, the post-war policy was formulated; the Cowley would be the basic model and the name Oxford revived for a more luxurious model.

Before leaving the White & Poppe engined Oxford, a few lines on the various 'specials' produced during the period will not be out of place. Although not

W.R.M. Motors Ltd 1915 catalogue illustration of the Oxford delivery van No 1 model.

specials in the accepted sense, a small number of delivery vans were made during 1914-15. The first (September 1913, chassis 321) was used by W.R.M Motors Ltd as their own parcels van until late 1916 when, reconditioned and given a new chassis (number 593), it was sold to a buyer in Sheffield. With this single exception, all commercials were on the de luxe chassis and records show that there were two canvas covered vehicles and eight other miscellaneous vans or van chassis. A specially ordered Oxford de luxe manufactured in the early months of 1914 was finished in white with black lining; two similar white vehicles were made about two months later; while a chassis with extra strong road springs was made in November 1914. Other vehicles had extras such as a child's seat, dash-mounted watch, extra tool box on the running board, petrol tin carrier, and 'one inch wood blocks in the pedals'.

The Earl of Macclesfield, president of W.R.M. Motors Ltd, had a de luxe Oxford with an extra long body made and finished in crimson lake — presumably to replace the standard Oxford purchased a year earlier. Elizabeth Maud Morris, obviously as interested in driving as her husband, took delivery of a white Oxford coupé in May 1914 — perhaps the one used by Mr Morris in the London-Edinburgh Trial. Two other named owners were Hugh Wordsworth Grey, the works manager at Cowley, who had a special blue painted body fitted by Raworth to chassis No 1828 in December 1914; while Lancelot W. Pratt (who was the Pratt of Hollick & Pratt, the Coventry bodybuilders) chose a Rover shade of biscuit for his cabriolet coupé body. Two incomplete assemblies went to White & Poppe Ltd in late 1915, both without engines and one without tyres or petrol tank.

In the mid-1919 W.R.M. Motors Ltd went into voluntary liquidation as a legal means of ending two onerous agency contracts which gave sole 'wholesale and shipping' rights outside London. It was under the name of the successor company 'Morris Motors Ltd' that an announcement appeared in *The Motor* of the 1919 programme indicating that the new Cowley and Oxford models would use common parts including frame, suspension, controls, engine, carburetter, gearbox, clutch, transmission, steering gear, axles, etc. Later that year the new series of Oxfords were listed, giving a choice of two-seater at £360 (£45 more than the basic Cowley), four-seater tourer at £390, or for an additional £60, the coupé version. All three models utilised the same chassis as the Cowley and

Rear-wheel-brakes-only model of the 1920 Oxford tourer.

shared the same Hotchkiss 69.5mm bore × 102mm stroke, 11.9hp, side-valve engine, and all were offered in a finish of sage green or elephant grey.

The Oxford specification for the next two years remained more or less as before. In November 1920 Hotchkiss et Cie announced that the new mounting position for the Lucas dynamotor was to be moved from above the bell housing, where it protruded through the floorboards, to a position adjacent to the gearbox and thus out of sight. The date would suggest that early 1921 models would continue to be fitted with the old design as listed in the early 1921 season catalogue. If the appearance had changed little, this was not reflected in the startling price changes that took place. With a works and supplier commitment geared to a production flow of something like sixty cars a week, the monthly sales figures in 1920 were dropping alarmingly. The 276 cars and chassis sold in September 1920 had fallen to a trickle of seventy-four four months later. The remedy to the dilemma was to cut the price and, in February 1921, price reductions of £25 on the Oxford two and four-seaters, and £80 on the Oxford coupé were advertised. Predictably the sales increased, reaching a peak in June, before dropping again at a rate not consistent with the normal seasonal fall before an annual Motor Show. The drastic price cuts then announced at Olympia surprised even the motor trade; the Oxford two-seater price fell from £510 to £415, and even larger was the reduction of £110 on the

Lines Brothers' interpretation of the Bullnose Oxford.

price of the four-seater. The idea of making these cuts, which put Morris Motors Ltd back into the black, was claimed by William Morris although Hugh Wordsworth Grey, then sales manager at Cowley, later said that the idea was his.

It was during this slump period that more valuable publicity came the way of the Oxford company when, under the sponsorship of Lord Northcliffe, John Prioleau, the *Daily Mail*'s motoring correspondent, set out to take a standard British car on an extensive tour abroad to test the reliability, or otherwise, of a British product on a variety of road conditions. The make of the car was kept from the readers of the *Daily Mail* until Prioleau's six month journey through France, Italy, Morocco, Algeria, Tunisia, and Spain had been completed. 'Imshi' as the car was named (Arabic for 'get a move on') left London on 19 December 1920 and at the end of 7,000 miles and a series of dispatches read by *Daily Mail* readers it was revealed that the car used was a two-seater Morris Oxford (registered FC2985). There was an 'Imshi II', a 1922 Morris Oxford four-seater, and even an 'Imshi III' (1925 Oxford four-seater), but as with all things it was never quite the same second time around.

One downright failure carrying the name 'Oxford' was the Oxford Silent-Six which first came into the public eye on stand 248 of the Olympia Motor Show in October 1922. The particular model on display was a china white cabriolet with Dunlop centre-lock wire wheels and powered by what was virtually a six-cylinder version of the Hotchkiss 11.9hp engine, known as the 'F' type. Experiments leading up to the announcement of the new Silent-Six had started with a prototype overhead-valve engine of 2,355cc fitted to a chassis 9in longer than that used on the 11.9hp cars, giving a 9ft 3in wheelbase. Subsequent development resulted in a side-valve engined prototype which still exists and has recently been restored by Kennings for the BL Heritage collection at Syon Park, Brentford. Registered BW5415, this coupé was once used extensively by William Morris.

Two factors appear to have been responsible for the failure of this promising entry by Morris into the six-cylinder market. The additional inches added to the chassis merely allowed for the longer engine in consequence the body size was restricted to that of the contemporary four-cylinder models. Hans Landstad, the chief designer at Cowley, had advised Morris against this restriction and suggested that prospective buyers wanted a bigger body. Morris, as usual, had his way. The second — and probably the more important factor — was the engine design itself, which had an appetite for crankshafts! Given that the dimensions of the crankpins and webs on the well tried Hotchkiss type 11.9hp engines were a borderline case (a fact appreciated some time later when, to compensate for the bored-out 13.9hp version, duralumin connecting rods and aluminium pistons were substituted for the heavier ferrous components), it was suicide to retain the same dimensions on a longer six-throw crankshaft less well supported.

The Oxford Silent-Six was listed for the 1923 and 1924 seasons in two-seater, four-seater, and coupé form. Production started in November 1922, but effectively ceased by July 1924. In all only about fifty cars and chassis were produced and the remainder of the components initially obtained for the first

*An excellent example of a restored 1922 model Morris
Oxford two-seater. Note the attention to detail including
the beaded-edge tyres.* (Chris Creevy)

500 were disposed of (in what became the traditional Morris manner) by using them as building foundations at the Osberton Radiator Company's premises in Bainton Road, Oxford.

The few changes made to the four-cylinder Oxford specification for the 1922 season were minor ones. The colour alternative of blue, in addition to the sage green or elephant grey available on all models, gave the purchaser a choice of three colours. Under the bonnet the Zenith carburetter had gone and in its place a bronze bodied SU 'sloper' carburetter with leather bellows. (A little known fact is that many of the leather bellows used in these carburetters were made on the family sewing machine by Mrs Skinner, wife of Carl Skinner, one of the founders of the Skinner's Union Carburetter Company.) Some months after the introduction of the 1922 model, the 11.9hp engines were being fitted with a four-port exhaust system in place of the earlier three-port arrangement.

For the 1923 season, the Oxford contrasted with the grey only Cowley offering. A cabriolet was introduced to swell the number of standard body types to four, and all offered the choice of green, grey, blue or claret. The SU carburetter of the previous season had, it was said, been found faulty and replaced on 1923 models with a Smiths five-jet carburetter working on the constant vacuum principle of later SU units. Another innovation was the provision of shock absorbers, a single-acting unit of American manufacture called the 'Gabriel Snubber', which consisted of a coiled belata belt designed to compress an auxiliary coil spring on rebound of the road spring. Standard

*A growing demand for closed cars resulted in the
introduction of a saloon model for 1924.*

fitments to the cars included a Smith's eight-day clock, two horns (electric and bulb), a running-board mounted 2-gallon petrol can painted red with SHELL MOTOR SPIRIT in black lettering, and a 'Gem' model Boyce 'Moto-Meter' to record the temperature in the radiator header tank. In January 1923 a bored-out version of the 11.9hp engine, rated at 13.9hp, was offered as an alternative for an additional £10. This larger version of the Hotchkiss type engine (although by now Morris had taken over the Coventry firm of Hotchkiss et Cie SA, and renamed the works Morris Engines Ltd, so they were, strictly speaking, Morris engines — but names die hard and enthusiasts always refer to this engine as the Hotchkiss type) was bored-out to 75mm diameter, thus increasing the cubic capacity by 255cc.

When they came to standardise the larger 13.9hp engine for the next season (1924), Morris Motors Ltd stated that this move had been decided upon because of the lack of demand for the smaller unit. Be that as it may, surviving post-January 1923 models show a majority of 11.9hp Oxfords. The road fund tax had been increased on 1 January 1921 to £1 per unit (or part unit) horse-power, which could account for the survival rate of the smaller horspower car, but not for the increased demand for the 13.9hp Oxford. Output of the larger engines by Morris Engines Ltd would have increased considerably in 1924 anyway, for during that year Morris Commercial Cars Ltd was formed for the production of a one-ton commercial vehicle utilising the 13.9hp Oxford power unit. During 1924 no fewer than 2,486 'T type Tonners' (as the new commerical was designated) left the works of E.G. Wrigley & Co Ltd at Foundry Lane, Soho, Birmingham.

Coincidentally, it was in 1924 that ex-Wrigley employee Cecil Kimber, general manager of Morris Garages in Oxford, started the successful MG sports car. Although Morris Garages had already produced a number of custom-built sporting specials on the Morris chassis, such as the Sports Cowley, it is generally accepted that the marque MG commenced with the MG 14/28 based on the 13.9hp Morris Oxford chassis. From 1924 to 1927 approximately 400 of these cars were built with two-seater, four-seater, or saloon bodies.

Of course many coachbuilding firms had supplied bodies for the Oxford chassis and in doing so had created interesting closed models, but so far Morris had not offered a standard production saloon. Not, that is, until the 1924 season when a two-door version of the Oxford saloon was announced. Upholstered in grey Bedford cord and finished in a choice of bronze green, grey, dark blue, or claret, the new model was fitted with mechanically operated front windows, sliding rear windows, luggage carrier, and internal fittings which included a luggage net and Wilmot 'smoker's companion'. One such model (registered DO4035) came to light in 1974, with a mere 9,661 miles on the clock, when the estate of a lady who had emigrated to America in 1932 was being sorted out. The car later fell under a Wisbech auctioneer's hammer for a (then) record £4,050. In common with the other four Oxford variations listed for the 1924 season, the Morris Oxford saloon had a larger radiator fitted to cater for the demands of the bigger engine.

*Oxford four/five seater of 1925. Note the warning triangle
above the rear number plate on this four-wheel-brake car.*
(Morris Motors Ltd)

Free insurance for one year with the General Accident, Fire and Life
Assurance Corporation, and a reduction in prices were two of the inducements
offered by Morris Motors Ltd when announcing their programme for 1925.
The 18th International Motor Show at Olympia during 17-25 October 1924
had displayed examples of the Morris range on stand 210. Initially, five models
were offered on the Oxford chassis; the two-seater and four/five-seater open
cars, a coupé, cabriolet, and a saloon — now with four doors. A considerable
number of changes had been made to the specification, not least of which was
the increase in frame length by 6in on all but the two-seater and coupé, to
accommodate larger coachwork. On all models, four-wheel braking had been
introduced (with the option of rear-wheel-braking-only versions at a reduced
price) with the front-end system made under Rubery patents (Alford & Alder).
Stresses imposed by the all-round braking system and a longer chassis dictated
the use of heavier gauge metal and a deeper section to the chassis frame, while
further strengthening was provided by the fitting of stronger front road springs
and a tie-bar between the front dumb-irons. Another major change was the
introduction of Dunlop ballon cord wired-on tyres (ie well-base rims) size 28 ×
4.95 on all but the saloon and cabriolet, which were shod with 30 × 5.25 covers.
Recalling driving a Morris with the newly introduced balloon tyres, Mr W.H.
Peterson of Tunbridge Wells remembers that he had three bursts in his first
year: 'The roads were not so well made as now and the extended walls did not
stand up to the roughness of the surface. I suppose it was a common fault and
later corrected by making the tyre walls stouter.'

Numerous detail changes were made, such as the alteration in the back axle
ratio (giving a 4.42:1 top gear), dash operated scuttle ventilator, all-weather
side curtains with improved sealing for the open models, and provision of a
Lucas vacuum operated windscreen wiper rendering the three-panel type windscreen unnecessary — replaced by a horizontally split two-panel arrangement
with water gutter. In place of the separate folding front seats, a full width
pattern on the 1925 models provided a mounting for the rigid rear screen on
the four/five-seater Oxford. The substitution of pleated upholstery for the
buttoned variety of previous years was obviously cheaper to produce and

*Landaulet body on the
1925 14/28hp Oxford
chassis.*
(Morris Motors Ltd)

*Luxurious interior of the
1925 14/28hp Oxford
landaulet. With front-wheel
brakes the model was listed
at £395.*
(Morris Motors Ltd)

facilitated the provision of removable panels on the open models.

Coming late in the 1925 season and aimed at a more sophisticated market was the Morris Oxford landaulet, sumptuously upholstered inside with Bedford cord covering the wide interior seat, augmented by a folding seat mounted under a sliding driver-division. Unfortunately for the chauffeur his compartment, despite its real leather upholstery, had semi-open drive and no side curtains. 'For town or country, it has no peer at anything approaching its price [£395] with a choice of four body colours', said *The Morris Owner* when

*Dashboard and controls arrangement on the 1925 Oxford.
Clearly visible is the 'smoker's companion' with a
container for matches, pipe rack and ash tray. The notice
on the windscreen warns the purchaser that a speed-
controlling washer is fitted to the carburetter and should
be left in position for the first 500 miles.*
(Morris Motors Ltd)

*The black spare petrol can with the gold Shell scallop on
this 1926 Oxford tourer was the last such fitment on
Morris cars; the following season's 'Flatnose' models had
a reserve tap on the petrol tank.* (Morris Motors Ltd)

announcing the new laudaulet in April 1925.

1926 was to prove the last season for the familiar 'D' or 'Bullnose' radiator
and while Hans Landstad and his chief draughtsman Stan Westby got plans
underway at Cowley for extensive changes to come, the attractions of the
enlarged range of 1926 Oxfords were published. Standard changes listed were
Barker dipping headlamps, Smiths 'Thermet' thermostatic water-flow control
for the cooling system, Dunlop reinforced balloon tyres on black enamelled
wheels, under-bonnet electric horn, Wilmot 'Calometer' radiator thermometer
(to use the original 1925 spelling; the 'r' was not added until later), moulded
ebonite battery box, improved slow-running control, parcel net and smoker's
companion on all closed cars, and detail improvement to the coachwork. One
other change which was perhaps insignificant at the time, but certainly of
interest to present day owners of vintage Morris cars, was the substitution of
the red spare petrol can for a black finished two-gallon container displaying the
Shell scallop in gold.

No less than eight Oxford variations were offered for the new season. As in
the previous model year the two-seater, four/five-seater, coupé, saloon, and
landaulet continued to be listed (all except the latter in a choice of blue, claret,
brown, or grey). Newcomers were a three-quarter coupé upholstered in
Bedford cord (leather upholstery £15 extra), 'to meet the demand for an
enclosed two-seater that has more than the usual refinements', and a special
saloon landaulet with four doors, in contrast to the landaulet now in its second
season, upholstered throughout in leather. Brown was the standard colour
scheme for the body but, to special order, it could be delivered in blue or claret
with upholstery to match. It should be noted that when Morris Motors quoted
'upholstery to match' it would not necessarily be the same colour as the body,
for until 1927 Connolly's range of leather colours were limited to brown, tan,
red, green and blue.

In many respects 1926 was a landmark in the history of the Morris marque. It had seen the company go public with a change in name for the second time, to Morris Motors (1926) Ltd; Osberton Radiators Ltd became Morris Radiators Branch; the old Hotchkiss works at Coventry (Morris Engines Ltd) became Morris Engines Branch; Hollick & Pratt Ltd changed its title to Morris Bodies Branch, and the SU Carburetter Co was purchased. On the commercial side, the War Department had placed the first of many orders for the 'D' type six-wheeler and Morris Commercial were working on the designs for another car to carry the Oxford name — the 'Empire Oxford'. But to Morris enthusiasts 1926 stands out as the year of the last Bullnose.

Bullnose Oxford. Engine Data	
White & Poppe engine	Four-cylinder water cooled, side-valve. 60mm bore × 90mm stroke. 1,018cc. 9hp. Three-bearing crankshaft. Cast iron pistons. Clutch of alternative steel and bronze plates running in oil. Magneto ignition. Three-speed gearbox (gate).
Hotchkiss type 11.9	Four-cylinder water cooled, side-valve. 69.2mm bore × 102mm stroke. 1,548cc. 11.9hp. Three-bearing crankshaft. Early engines to 66930 cast iron pistons, later engines aluminium pistons (with Cave-Baker gudgeon pins between engine 73101 & 120934). Three-port manifold to February 1922, later engines four-port manifold. Clutch cork insert running in oil. Magneto ignition. Three-speed gearbox (centre ball). Engine type 'CB'?
Hotchkiss type 13.9	As Hotchkiss type 13.9, but 75mm bore × 102mm stroke. 1,802cc. 13.9hp. Engine type 'CF'?

Bullnose Oxford. White & Poppe engine. Body Colours and Upholstery			
	1913	1914	1915
Two-seater	Pearl grey body. (Red bodywork and black bodywork also recorded.) Wheels either grey or blue.	De luxe models dark green with green leather upholstery. Standard and commercial models pearl grey with green leather upholstery.	De luxe models dark green with green leather upholstery. Popular models pearl grey or green with green leather upholstery.
Coupé		Colour to choice with Bedford cord upholstery.	Colour to choice with Bedford cord upholstery.
Sporting model		Dark green with green leather upholstery.	
Delivery van		Colour to choice with green leather upholstery. Mahogany panelled body.	Delivery van No 1, colour to choice. Mahogany panelled body. Delivery van No 2, dark green with truck type detachable canvas tilt. Express carrier No 3 was a two-seater type with goods box on rear.

Chassis Numbering		
Morris Oxford. White & Poppe engined vehicles.		Chassis Numbers
1913	Two-seater model production period 29 March 1913 - 22 December 1913	101-465, including 216A, 221A, 221B, 265A, 291A, 291B, 292A
1914	Two-seater de luxe from November 1913, sporting model, coupé de luxe	1001 onward
	Two-seater standard 3 January 1914 - 24 December 1914, two-seater commercial	400-533
	Delivery van 26 September 1914 - 26 October 1914	1742, 1786, 1767, 1768, 1801, 1825
1915	Two-seater de luxe, cabriolet coupé, limousine coupé	... to 2000
	Two-seater popular 2 January 1915 - 15 November 1915	534-591
	Delivery van No 1, delivery van No 2, express carrier No 3, 20 January 1915 - 1 June 1915	1864, 1865, 1910, 1990
Morris Oxford Silent-Six. Six-cylinder 17.9hp engine.		
1923 & 1924	Two-seater, four-seater, and cabriolet, 2 November 1922 - 2 October 1925 (Works experimental car chassis F126 finished 17 June 1926)	F101-F149

Chassis Numbering		
Morris Oxford. Hotchkiss type engine.		Chassis Numbers
1920 season	Two-seater model production period 28 October 1919 - 31 August 1920	5007-6372
	Four-seater 20 December 1919 - 31 August 1920	5033-6372
	Coupé 6 December 1919 - 31 August 1920	5031-6372
1921 season	1 September 1920 - 31 August 1921	7001-9288
1922 season	1 September 1921 - 31 August 1922	10001-14401
1923 season	1 September 1922 - 31 August 1923	15001-30000
1924 scason	1 September 1923 - 31 August 1924	30001-57100
1925 season	1 September 1924 - 31 August 1925 Landaulet April 1925	57101-105800
1926 season	1 September 1925 - December 1926	105801-156424

Note: Numbers quoted above are actually those given as 'Car No' on the identity plate of the vehicle. These numbers are quoted by Morris Motors in their spares list under the heading 'Chassis No'. The actual frame number stamped on the chassis dumb iron is always 3000 numerically less than the 'Car No' quoted on the identification plate.

Bullnose Oxford. Hotchkiss type engine. Body Colours and Upholstery

	1920	1921	1922	1923	1924	1925	1926
Two-seater	Sage green with green leather. Elephant grey with black leather.	Sage green or elephant grey bodywork.	Green, elephant grey, or blue bodywork.	Blue, claret, bottle green, or Morris grey. Latter with mottled grey leather.	Dark blue, claret, bronze green or grey bodywork.	Blue, claret, bronze green, or grey bodywork.	Blue, claret, brown, or grey bodywork.
Four-seater	Sage green with green leather. Elephant grey with black leather.	Sage green or elephant grey bodywork.	Green, elephant grey, or blue bodywork.	Blue, claret, green, or grey.	Dark blue, claret, bronze green or grey	(Four/five -seater) Blue, claret, bronze green, or grey body.	(Four/five -seater) Blue, claret, brown, or grey body.
Coupé	Sage green, or elephant grey bodywork.	Sage green, or elephant grey bodywork.	Green elephant grey, or blue bodywork.	Blue, claret, green, or grey.	Dark blue, claret, green, or grey. bodywork.	Blue, claret, bronze green or grey body. Hair-lined cloth upholstery.	Blue, claret, brown, or grey body. ¾-Coupé same colours, with Bedford cord.
Cabriolet				Blue, claret, green, or grey body.	Dark blue, claret, bronze green, or grey. Bedford cord.	Blue, claret, bronze green, or grey. Antique leather upholstery.	
Saloon					(Two-door) Dark blue, claret, bronze green or grey. Grey Bedford cord upholstery.	(Four-door) Blue, claret, bronze green, or grey. Hair-lined cloth upholstery.	(Four-door) blue, claret, brown, or grey. Bedford cord upholstery.
Laudaulet						Blue with Bedford cord upholstery. Driver's seat leather.	Special saloon laudaulet: brown with leather. Laudaulet: blue with Bedford cord upholstery. Driver's seat leather.

Bullnose Oxford. Magazine Bibliography

Year	Title	Magazine	Date	Page
1913	Road Test, Oxford Light Car	The Cyclecar	6/8/13	p278-9
	The Morris Oxford Light Car	The Motor	17/12/12	p978
	The Morris Oxford Light Car	The Autocar	19/4/13	p692-4
1914	The Morris Oxford Light Car for 1914	The Autocar	4/10/13	p635
	De luxe Chassis, Photograph	The Autocar	22/11/13	p1118
	Top Gear Running Extraordinary	The Motor	9/12/13	p981
1915-18	Light Car Talk by Runabout	The Autocar	20/11/15	p645
	Light Car Talk by Runabout	The Autocar	23/10/15	p505
	Light Car Talk by Runabout	The Autocar	29/1/16	p47
	Light Car Talk by Runabout	The Autocar	26/2/16	p269
	The Morris Oxford, Secondhand	The Light Car & Cyclecar	2/4/17	p390
	Lubrication Chart for Morris Oxford	The Light Car & Cyclecar	3/10/17	p289
	The Ideal Light Car - Morris Oxford	The Light Car & Cyclecar	10/12/17	p47-9
	A Sporting Morris Oxford	The Light Car & Cyclecar	22/7/18	p152
	An Improved Morris Oxford	The Light Car & Cyclecar	15/3/15	p406
1919	Correct Hand Cranking - Morris Oxford	The Light Car & Cyclecar	9/12/18	p57
	Morris Oxford & Cowley Cars for 1919	The Light Car & Cyclecar	31/5/19	p78&9
	Morris 1919 Light Cars	The Autocar	2/8/19	p169-72
1920	The 1920 Morris Oxford	The Motor	3/12/19	p869-71
	Impressions of the 1920 Morris Oxford	The Autocar	30/10/20	p753-5
	Important Hotchkiss Developments	The Motor	3/11/20	p673
1921	A Morris Six	The Motor	30/10/20	p778
	A Light Car for all weathers	The Motor	15/12/20	p1030
1922	Morris Oxford De luxe	The Autocar	24/6/22	p1068
1923	1923 Cars and New Models	The Motor	24/10/22	p531-2
	Morris Oxford Six	The Autocar	27/10/22	p797-9
	The Six Cylinder Morris Oxford Tested	The Motor	19/12/22	p1005-6
	The 13.9hp Morris Oxford Tested	The Motor	6/2/23	p16
	Bowden bodied all-steel saloon	The Motor	3/4/23	p365
	Care & Maintenance of Morris Cars	The Autocar	29/6/23	p1119-21
	How Morris Cowleys & Oxfords are made	The Motor	17/7/23	p1031-6

Year	Title	Journal	Date	Page
1924	Morris Cars in 1924	The Autocar	7/9/23	p428-9
	Accessories for Morris Cars	The Autocar	21/9/23	p501-3
	Small Cars of Yesterday	The Autocar	28/12/23	p1252-3
	Northwards on a Morris Six	The Autocar	8/2/24	p262-3
	Latest Morris Six, Photographs	The Autocar	22/2/24	p330
	Morris Cars	The AutoMotor Journal	12/6/24	p499-501
	Real Quality Production	The Motor	1/7/24	p940-2
	Cars on the Road - 13.9hp Oxford	The AutoMotor Journal	24/7/24	p613-5
1925	Morris Programme for 1925 Models	The Morris Owner	9/24	p641-4
	Road Test, 14/28hp Oxford	The Autocar	16/1/25	p89-91
	Road Test, Super Sports	The Autocar	1/5/25	p767-8
	Description of Oxford Landaulet 14/28	The Morris Owner	4/25	p172-3
	Morris Programme for 1925 Models	The Motor	2/10/24	p178-80
	Morris adopts Front Brakes	The Autocar	5/9/24	p428-9
	Testing Morris Front Wheel Brakes	The Motor	18/11/24	p829-30
	Cars on the Road - Oxford 14/28	The AutoMotor Journal	18/12/24	p1103-5
	A Car & Garage for under £200	The Motor	20/1/25	p1204
	Cars of Distinction - Morris Six Coupé	The Autocar	15/5/25	p869
	Motor Car Topics	The Motor	18/8/25	p95
1926	The New Morris Programme	The Morris Owner	9/25	p842-5
	The Morris Programme for 1926	The Motor	1/9/25	p179-81
	The New Morris Cars	The AutoMotor Journal	3/9/25	p737-8
	Morris Models for 1926	The Autocar	4/9/25	p395-7
	Morris Car Topics	The Motor	8/9/25	p224
	Morris Car Topics	The Motor	13/10/25	p636
	Morris Car Topics	The Motor	10/11/25	p783
	Morris Car Topics	The Motor	24/11/25	p889
	Morris Oxford at the Show	The Autocar	16/10/25	p777-8
	The 14/28 Morris Oxford Car	The AutoMotor Journal	10/12/25	p1089-92
	A Fabric Saloon Coupé	The Autocar	11/12/25	p1155
	A Test of the Morris Oxford	The Autocar	19/3/26	p473-4
	Testing Morris Engines	The Autocar	16/7/26	p89-91
General	The Bullnose Morris	Motor Sport	11/59	p846-57
	This Car Morris	Vintage	Vol 1, No 4	p4-10
			Vol 1, No 5	p25-31
	The Great Cars - Morris	On Four Wheels	Weekly part work. Published 1975. Vol 5	p1421-31

Morris Cowley
1915-26

Before the advent of the Morris car, Cowley was a little known southern suburb of the univeristy city of Oxford. Its main claim to fame was the use of its name by the Anglican Mission Order of the 'Cowley Fathers' in 1865. Hurst's Grammar School at Temple Cowley, later enlarged and transformed into a military training college, subsequently became the birth place of the first production Morris Cowley motor car — a model that can claim to be the only car ever to give its name to a railway station. In 1928 the Great Western Railway Company opened 'Morris-Cowley Station' close to the Morris Works, staffed by a stationmaster and a clerk to handle traffic. By 1935, with a complement of twelve clerks to handle goods traffic alone, Cowley Station could boast that more cars were handled there than at any other station in England.

In 1914, William Morris, intent on producing another model alongside his Oxford, approached various manufacturers with a view to obtaining components for his new four-seater Morris, to be called the 'Cowley', with which he hoped to attract the family motorist — a market that the two-seater Oxford could not satisfy. His approaches were met with scepticism by some firms when quantities were mentioned, while other firms admitted that the output he required was outside their capacity. American parts were the answer and following two visits to the United States, Morris placed an initial order with the Continental Motor Manufacturing Company of Detroit for 1,500 Continental 11.9hp four-cylinder engines. To other American manufacturers went contracts, in similar quantities, for front and rear axles, steering gear, transmission; gear boxes were made by the Detroit Gear & Machine Company.

A certain irony is to be found in the fact that every Trafford Park assembled American Ford 'T' was imported and Morris, it has been said, set out to show that an English manufacturer could compete with Ford at his own game. Yet here was Morris aiming to build a car with a high percentage of American components!

The effect of the assassination of Archduke Franz Ferdinand at Sarajevo and the subsequent world conflict was a factor William Morris had not reckoned with. Deliveries of the new 11.9hp Morris Cowley were due to commence in June 1915, but in fact war conditions were such that the first Continental engines were not delivered into Morris's hands until the following September, with the remainder spread over a period of four years — some of these finding a resting place in the Atlantic due to enemy U-boat action. A

limited production of the new Cowley commenced in the corner of the small
Cowley factory which, by 1915, had turned to the manufacture of hand
grenades and, later, bomb cases and mine sinkers.

The prototype Cowley, which as shown to the press in April 1915, was
greeted with enthusiasm. This new car was to be available in two-seater or
four-seater form complete with a five-lamp Lucas dynamo (type E20) lighting
system with fork mounted Lucas D50 headlamps and G230 side lamps
mounted on the windscreen supports. Detachable ten-spoke artillery wheels in
steel, one-man operated hood, windscreen, horn, and kit of tools formed part
of the specification. Prices were originally fixed at £168 18s for the two-seater
and £194 5s for the four-seater version. The imposition of the McKenna Duty
of $33\frac{1}{2}$ per cent early in the war (designed to prevent importation of goods not
essential for the war effort) was certainly partly responsible for the spiralling of
the listed price of the two-seater fixed at £163 16s in September to £194 5s the
following month, with similar increases for the four-seat tourer. By 1916 the
prices were £199 10s and £222 10s respectively — after the Armistice, catalogues were beginning to quote prices as 'approximate'.

As World War I came to a close, and with it the cessation of Government
contracts for munitions, a paradoxical situation evolved. Throughout the war
years Morris had continued in limited production, building something like
1,344 Cowleys with a trickle of components from America. Some of these had
been sold in chassis form while others, in addition to the open models, were
fitted with coupé or delivery van bodies (first listed in 1916). Now, with the
post-war boom and a demand for any kind of motor vehicle, Morris found
himself with a capacity for production and a dearth of components. The Continental Motor Manufacturing Company ceased production of the 11.9hp
engine and even if supplies had been available the Government restriction on
imports now made the use of this engine uneconomical. The answer to Morris's
dilemma came from an unexpected source. A French armament firm, with a
branch at Coventry, found that the abrupt termination of military orders left
them with a manufacturing potential and no orders. Hotchkiss et Cie offered to
produce a four-cylinder side-valve engine based on the Continental 'Red Seal'

Morris Cowley delivery van,
1916. This commercial would
have been fitted with the
Continental engine.
(W.R.M. Motors Ltd)

design for Morris, an arrangement particularly pleasing to Morris for not only did they undertake to get the unit price down to £50 but they required no deposit to help finance the change-over from guns to engines. Production of the new Hotchkiss engines commenced in September 1919, two months after delivery of the initial sample. An iron foundry established that year at Cowley by Morris to help monitor the costs of castings by outside suppliers of foundry-work proved useful in helping to cope with the Hotchkiss demand for cylinder block castings.

Continental engined Morris Cowley two-seater of 1917. (Geoff Ironmonger)

Although the Hotchkiss (69.5mm bore × 102mm stroke) engine was based on the Continental Red Seal design for which Morris had acquired the rights, it was by no means an exact copy. The dynamo was no longer a belt driven unit mounted on the cylinder head; the Coventry-built power unit utilised a Lucas Dynamotor driven by a Bamber or Reynold inverted-tooth chain from a toothed wheel fitted between the rear crankshaft bearing and flywheel or, in the case of those engines used on the earlier Cowleys, a dynamo combined as a unit with the magneto. The dynamotor, as the name suggests, had a dual role of starter motor and (once the engine was running) as it continued to rotate, provide the necessary charging current for the car battery. Mechanically it was a compromise, because the right gear ratio and size for its starting function resulted in the dynamo section being driven faster than desirable with the consequence of an unnecessarily high output. Lucas advised users to keep the dynamo charging at all times during the winter but to charge only during half their running time in the summer months. The answer was to provide a 'half-charge' switch and this was later incorporated into the equipment, but it still demanded discretion on the part of the driver to avoid overcharging the small battery mounted on the running-board, which more than occasionally occurred. The mounting position of the dynamotor was an unfortunate choice as it stuck out rearwards and projected through the floorboards in front of the handbrake and gear-change lever. Towards the end of 1920 a neater method of accommodating the dynamotor was adopted, repositioning it lower down onto an extension of the gearbox bellhousing casting. It is possible that a few of the early Hotchkiss engines (perhaps the early samples?) had a bore and stroke the same as the American engine; credence is given to this theory by the various references to '69 × 100' in the initial magazine reviews. Alternatively, the answer may be simply that, being an American engine, the bore and stroke of the Continental would have been listed in imperial units as $2^3/_4$in × 4in and the exact metric conversion of 69.85mm × 101.6mm rounded out.

An interesting and curious compromise resulted from the French firm's production of these engines. Hotchkiss et Cie's workshops were equipped throughout with French metric machine tools and the consequence was that threads on these Coventry-built engines were made to the French standard. British garages and motorists of the day would normally have Whitworth size spanners so a compromise was reached where all nuts and bolts — albeit with metric threads — were made with Whitworth size hexagons. Credit for this ingenious idea has been given to a young jig and tool draughtsman named Leonard Percy Lord, of whom much was to be heard of around Cowley in later years. The result of this was far reaching, as late as the Morris Eight engine certain threads remained metric.

With the introduction of the Hotchkiss engine there appears to have been something of a rethink as to the role the Oxford and Cowley should take. The Oxford was to be the de luxe offering while the Cowley name applied to a cheaper model. A preliminary announcement issued by W.R.M. Motors Ltd in May 1919 stated that both Oxford and Cowley models would share the same 8ft 6in chassis frame, suspension, engine, gearbox, carburetter, clutch, transmission, front and rear axles, brakes and so on, but with superior general finish and additional refinements on the more expensive Oxford models. The Cowley was to be in two-seater form only. The 1919 catalogue clarified this even further by listing the differences, which included narrower plain tyres in place of the Oxford's grooved Dunlop type, a three-lamp lighting set with the front lamps mounted on the mudguards and illumination and ignition being provided by a Lucas magdyno or BTH mag-generator (neither of which were really satisfactory) while the Oxford had comparatively large headlamps and the dynamotor and therefore a self-starter. The difference between the bodies was apparent in the sharp edges to the rear section of the Cowley, fewer louvres and the absence of a leather valance to fill the gap between body and running board. Leather gave way to 'Pegamoid' unholstery and the body was '. . . roughly painted in grey varnish colour only'. No deviation was to be made to the specification, neither would a self-starter be fitted. These differences enabled the Cowley two-seater to be cheaper by £50 compared with the Oxford two-seater's price of approximately £335 ex-works.

By 1921 Morris had relented a little and the season's Cowleys allowed for both two-seat and four-seat open models, the 700 × 80 tyres now had tread thanks to the fitting of Dunlop Magnum Cord beaded-edge (with a nobbly tread pattern best described as similar to today's motor cycle scramble tyres) as standard, the valance between body and running-board had returned, and a self-starter was available as an extra. The 6 volt three-lamp lighting system with lamps on the front mudguards remained, as did the no-choice single colour body finish (now a 'special shade of buff') and the Pegamoid upholstery.

By the time the Morris Motors Ltd catalogue for 1921 models had been printed it was fairly clear that the Cowley and Oxford had a four-plate cork-insert clutch running in oil. Whether this type was fitted to the first of the

Hotchkiss engines is not clear from the various contemporary magazine reports and reviews. The 1919 Morris catalogue was obviously drawn up with the assumption that the Continental engine would continue to be fitted and therefore refers to a four-plate Ferodo lined type running 'practically dry'. *The Autocar* review of the 1919 Morris cars, published in August of that year, indicates the use of the new Hotchkiss engine (albeit a 69mm bore × 100mm stroke unit) with a dry type of clutch. In December 1919 *The Motor* described the new engine as having a 69.5mm bore and 102mm stroke, but again the clutch was said to consist of Ferodo rings carried between steel plates. In their book *The Bullnose and Flatnose Morris* (published 1976) Lytton P. Jarman and Robin I. Barraclough suggest that the cork-insert clutch running in oil pre-dates the Hotchkiss engine and was the outcome of a failure of the rear main bearing oil seal arrangement on the Continental Red Seal engine. According to Jarman and Barraclough, oil passing the seal caused the Ferodo lined clutch to slip and William Morris himself thought up the solution, which was to fit a clutch which ran in oil intentionally instead of accidentally. But in copying the Continental engine Hotchkiss faithfully reproduced the complicated rear oil seal and the modification hole drilled underneath it for clutch lubrication! Owners of some of the early cars fitted with the wet clutch experienced trouble with oil getting onto the rear brake linings. The oil leaking into the gearbox then overflowed down the torque tube to thin out the heavy axle oil. A non-return valve between engine and clutch housing, and an oil thrower fitted on the front end of the gearbox shaft removed this nuisance.

Post-war inflation resulted in ever increasing prices. For example, the Cowley two-seater listed at £165 18s when first announced in 1915 became £199 10s in 1916, approximately £285 in May 1919, £315 in August 1919, and by late 1920 had reached £465. Some makers were selling hard-to-get cars at this time on the condition that private owners agreed to retain possession for at least six months. This was done to check profiteering by private buyers. Early in 1920 Morris Motors Ltd advised the motoring press that they had adopted this course of action. Then the bubble burst and things turned full circle. Although Morris maintained output, sales were down two-thirds and his embarrassingly large bank overdraft meant he could not pay his suppliers. In one of his rare broadcasts, Lord Nuffield described in 1955 how he overcame the crisis:

> I sent for my manager [Hugh Wordsworth Grey] and suggested to him that the price should be dropped by £100 per model, and he said: 'how can you do that with the profit you're making today?' And I said: 'you are making no profit at all because you're selling no cars! I therefore give you instructions to go and do exactly what I've suggested and also to double up the supplies.' He then went to the door and stood in the door and looked at me as much as to say 'He's gone at last!' With that he slammed the door and went out . . . . All my competitors called me cut-throat and everything they could possibly think of but, if I may say so, at this point it was the salvation of the motor trade, and at that time all manufacturers in this country were making a few cars at as high a price as they could possibly make. I reversed the battle order to make as many cars as possible at as low a price as possible.

In February 1921 the price of the Cowley two-seater was slashed by £90 and the four-seater version was reduced by £100. The Bean car management at Dudley had made a similar move four months previously and it had not helped them — why should it for Morris? Miraculously, it did work and the sales jumped from seventy-four cars and chassis in January to 400 by March 1921. The timing was right and the distributors accepted a reduction in their commission from 17 per cent to 15 per cent with the result that by the end of May the bank overdraft of £84,315 had been cleared. Surprised as his competitors were at these reductions, they could not have been prepared for the second big reduction Morris announced as the London Motor Show opened in 1921. The Cowley two-seater was reduced from £375 to £299 5s, while the four-seater showed a drop of £85 to £341. The Cowley, it appeared, was here to stay.

A third Cowley to come on the scene in March 1921 was the 'Morris Cowley Sports Model'. This stark two-seater had a polished aluminium body tapered to a wedge-shape point astern and thanks to the light body and the slightly higher rear axle ratio and 'specially tuned engine' fitted, the road tester for *The Motor* was able to refer to the 'liveliness' of the car and quote an estimated speed of over 60mph. The manufacturers claimed a petrol consumption of 40mpg at normal speeds. Equipment on the sports model was basically the same as the other Cowleys with a three-lamp system — the small headlamps perched incongruously on the forward ends of the long sporty (by standards of the period) mudguards which, together with the wheel discs, gave the vehicle a racy appearance. Doors were not deemed necessary, entry and egress being achieved by climbing or jumping! The model was also listed for 1922 and early 1923, but in fact any sold in the last year must have been left-over from the previous season as the last Sports Cowley was produced in October 1922. Initially priced at £398 10s in 1921, they were reduced to £315 after the end of production, by which time 107 had been manufactured.

To gain some promotion for the Sports Cowley, William Morris allowed Alfred Keen to tune a car specially and enter it in competitions. Keen had joined Morris in 1905 at the age of fifteen and was a natural mechanic, who had never had any formal engineering tuition. The modifications he carried out were substantial and in some areas drastic, examplified by the reduction in valve stem diameter to $\frac{3}{16}$in to reduce the weight of each valve by one ounce, and liberally drilled the con-rods until they weighed but $18\frac{1}{2}$oz each! The modified cylinder block had ports opened to 41mm diameter and the flywheel reduced in weight to 31lb, while the Ricardo slipper pistons he used were drilled until they turned the scales at a mere $6\frac{1}{2}$oz complete with gudgeon pin. As only one ring was used the engine was very noisy at low revs. By contrast he retained the standard valve timing (which was found by experiment to give the best results), normal lubrication arrangement, thermo syphon cooling system (with the fan removed), steel artillery wheels, gearbox and axles. Alterations to the chassis involved the replacement of the rear upper part of the $\frac{3}{4}$-elliptic spring with solid forgings to enable flattened $\frac{1}{2}$-elliptic springs to be used. The front leaf springs were also flattened, while the steering gear was raised and carried on a bracket to allow a centre location for the steering wheel. In this chassis frame, Keen fitted the engine 3in lower and 9in further back than normal — the torque tube being shortened to suit. With a slender single-seater body built onto the chassis, Keen entered the car (registered BW5748) in a local club speed trial, achieving a speed of over 90mph and beating a famous Bugatti into the bargain. In August 1923 the Eastern Centre ACU speed trials were held at Harling Heath, near Thetford, where the Keen Special averaged 63½mph in a standing start kilometre to put up the fastest time of the day of 35.52 seconds.

Some years later another racing Cowley, Wellsteed's famous 'Red Flash', made its mark at Brooklands between the years 1926 and 1928 driven by H.R. Wellsteed and Cyril Paul. This car (registered UH1736) is now part of the BL Heritage collection at Syon Park. One racing Cowley was built to satisfy a bet that a fairly standard Cowley could do 75mph. J. Crickmay used a 1926 Chummy model with 30,000 miles to its credit as the basis for his rebuild. Like Alfred Keen he modified the rear suspension to ½-elliptic springs, stiffened the chassis by introducing a bar between the front shackles, and lowered the steering; in his case he mounted the box upside down to bring the drop-arm inside instead of outside the frame, adding a bracket to hold the box to the side member as well as to the engine. Attention to the power unit included the drilling of the con-rods, aluminium drilled-skirt pistons, opened out oilways, widened and polished ports, double valve springs, and a flywheel turned down appreciably. Crickmay's special racing body was a long tailed two-seater made of ash covered in aluminium strips, with a streamlined undershield. Timed by six stop watches, his bet was safely won when the car did 93mph over half a mile at 4,800rpm. After some racing successes, a RAC observed six-hour run was made at Brooklands during which 387 miles were covered at 64.58mph, including three stops — one for a sump drain and refill. The lubrication system finally defeated Crickmay during a subsequent five-lap race when he pushed the engine to 5,000rpm.

An interesting review of the 1922 season's models appeared in *The Motor* in November 1921. The text refers to the Cowley range in terms of Cowley 'Standard' and Cowley 'De-Luxe' models which were practically up to the Oxford specification with electric starting, superior body finish, and in the case of the two-seater, complete with dickey seat, while the four-seater had separate armchair-type front seats in the manner of the previous season's Oxford. Could it be that the price cuts had created such a demand for the Cowley that this was a stop-gas measure using Oxford components to ease the situation? In the absence of corroborative evidence this can only be conjecture. The despatch figures for all 'Bullnose' cars jumped from 1,932 in 1920 to 3,077 during 1921, then more than doubled between January and October 1922 to 6,937. No mention was made of the 'De-Luxe' in the following season's (1923) catalogue, while in January 1922 it was announced that self-starters and dickey seats were being offered on the Cowley as optional extras.

The first Morris Cowley bodies had been built by Raworths of Oxford, and Hollick & Pratt of Coventry, but as early as 1919 it was evident to William Morris that the Coventry and Oxford bodybuilders were not going to be able to meet all future demands. To supplement the output, the first Morris body-building shop was erected on the site of garden allotments across the road from

the old military training college at Cowley. By mid-1922 Hollick & Pratt Ltd
were working solely on bodies to feed the Cowley production lines, when a fire
destroyed their main Coventry factory. Lancelot W. Pratt and William Morris
were good friends (Pratt had accompanied Morris to the United States as
adviser, and during one particularly tight period had made a loan of £10,000 to
Morris) and an agreement was reached whereby the complete firm was sold to
Morris for £100,000. Morris rebuilt the factory building in 1923 and ran both
Coventry and Cowley bodybuilding plants with Pratt in charge. The company
remained in Morris's personal ownership until three years later, when it was
purchased by Morris Motors Ltd to become the Morris Bodies Branch.

William Morris was keeping to his promise to make as many cars as possible
at as low a price as possible (Morris' output jumped from 6,956 in 1922 to
20,048 units in 1923, during which time the total British production figure
dropped) and the price of the Cowley and other Morris cars continued to be
reduced. So much so that the catalogue and advertising printing people had a
hard task in keeping up to date. The listed price of the 1923 season Cowley
two-seater, for example, started at £299 10s, was corrected to £278 5s, and
shortly the Morris advertisements were boasting; 'Two years ago the lowest
price Morris car sold for £465. Today we are marketing a BETTER car at £225
complete.' This referred to the basic Cowley two-seater which exactly matched
the price of Austin's new Seven.

As in previous years the chassis, engine, gearbox, clutch, steering gear, axles
and so on were identical to the more expensive Oxford; the savings were made
on the smaller section tyres, a less expensive design of body upholstered in grey
leathercloth, single-panel windscreen with a frameless top to the glass (unlike

*Dr Blair at the wheel of his 1923 four-seater Cowley
outside the Kilsyth Motor & Cycle Depot building, now a
Chinese restaurant. (Bill Gardener)*

*Compton & Hermon Ltd, Walton-on-Thames, were
responsible for this two-seater sports body with a single
dickey seat. The chassis is a rear-wheel-brake-only
Cowley of about 1923.* (George Compton)

the Oxford's framed adjustable split screen), only one colour (grey), lighting equipment which consisted of front lamps on the cheaper D-shaped mud-guards and a single rear light. Although a hood and hood bag were included in the price, side screens and carpets were not. The four-seater had a two-door body with both doors on the near-side; for the two-seater Cowley a single door on the near-side was considered sufficient. Instrumentation on the basic cars certainly did not compare with the contemporary Austin Sevens, but for an additional £30 on the two-seater and £25 on top of the basic price of £255 for the four-seater, a starter motor and dashboard equipment were fitted. On the two-seater, these extras included a single dickey seat. Dashboard equipment consisted of a speedometer, eight-day clock, petrol gauge, oil gauge, etc, while the provision of the dynamotor and instruments altered the electrics from 6 volts on the basic model to a 12 volt system. Another extra available that season were Gabriel snubbers (single-acting shock absorbers) fitted all round for an additional £6 10s. Despite all the cost cutting a feature about even the cheaper motor cars of the period was the comprehensive tool kit supplied. The Cowley, for example came equipped with no less than thirty items housed in a running board mounted tool box on the four-seater car and in a tool bag with the two-seater.

The fortunes of Joseph Lucas Ltd were tied to the success of the Morris motor car following Morris's contract for Lucas dynamos and lamps for the first Cowley. The order for fifty sets a week, at £10 10s a set, was considered enormous, and some people even thought it beyond the capacity of the Lucas Great King Street factory! By 1923 half the Lucas output of starting and lighting components was destined for the assembly lines at Cowley. Oliver Lucas had acquired a Morris Cowley for his personal use and was therefore conversant with its shortcomings, so when during a chat with William Morris in 1923 he suggested that the Cowley could be improved if the Bowden cable ignition control was changed to a system of rods running alongside the steering column, full five-lamp lighting sets fitted, and a scuttle ventilator provided. Morris was enthusiastic about the ideas, resulting in the standardisation of these features on the 1924 season Cowleys.

The 1924 model year also saw other detail changes to the Cowley programme, not least of which was a new version called the 'Occasional Four'. It gained the nickname 'Chummy' (a word borrowed from America) for the good reason that the occasional rear passengers, seated on 14in-wide folding seats which faced inwards, would be in close proximity to chums in the two independent front seats. This was due to the short body dictated by the inward sloping back, not unlike the Morris Eight two-seater of a decade later. 'Market research' (to use a modern term) had already been done on such a vehicle for, prior to the 1924 season, Morris Garages had been selling a similar vehicle based on the Cowley chassis with bodies by Carbodies of Coventry. Both Chummy and four-seater for 1924 had the Auster framed horizontally-split windscreen previously only fitted to the Oxford chassis, but the two-seater still retained a single-panel frameless top design. Other new features were applicable to all three Cowley models: side-screens were now supplied as were the dash instruments, Boyce Motometer temperature gauge, steering column stay, half-gallon tin of Shell lubrication oil, and a red painted spare two-gallon petrol can marked 'Shell Motor Spirit' with a carrier on the off-side running board. The American-made Gabriel snubbers were still extra and the body on all three models remained a one-colour, grey only, finish. Surprisingly, the superior specification was not reflected in a price increase; quite the contrary, with a decrease of approximately £30 on the lowest 1923 prices, and further reductions were yet to come!

'There is no doubt that the announcement of the new Morris programme, as effective from September 1st 1924, is the most arresting that has ever been made in the annals of British automobilism.' So said the Morris publicity people when giving details of the 1925 models which, in Cowley terms, consisted of the two-seater, four-seater and occasional four. All three versions were now priced below £200 with the added bonus of the free comprehensive insurance for a year, mentioned earlier. The first obvious change was the redesign of the two-seater car which had lost the curve behind the bench seat in favour of a flat sided design tapering to the rear dickey seat compartment, with its higher back and detachable cushion, now a standard fitting. That a change in body design had been made was accentuated by the addition of Dunlop Cord balloon tyres and the adoption of the double-panel adjustable front screen on the two-seater. All the 1925 Cowley models had (in addition to the balloon tyres) Gabriel rebound snubbers (replaced on late season chassis by Smith shock absorbers), a driving mirror and a hand-operated windscreen wiper. A novel feature provided on the four-seater was the ability (due to an extra light) to join the forward sections of the rear sidescreen at the centre to form a 'V'-shaped windscreen for rear passengers when the car was used in open form. In 1925 there was a change from button to removable pleated type upholstery on the Cowley models, with the exception of the Chummy which retained button upholstery. At this time the side-mounted spare wheel on the

Chummy was moved to the sloping rear of the car body.

So far, as we have seen, the Morris Cowley as bodied by Morris had, with one early exception, remained an open car in its various forms. However, in March 1925 a fixed-head coupé was added to the range, followed some four

The frameless top to the windscreen identifies this two-seater Cowley as a 1924 model, photographed in 1926.
(Ken Searles)

weeks later by a saloon version. The Morris Cowley coupé (£210) with a single door on the near-side, and windows on both sides operated by a handle from inside, deviated from the usual Cowley grey colour and came finished in blue with black top. This, no doubt, had some bearing on the decision to offer the option of blue on the other Cowley cars about the same time. Internally, the coupé body was trimmed with pleated blue leathercloth upholstery, with an arm rest for the driver, above which a small oval cubby hole had been provided for small items. At the rear a wide dickey for two persons was '. . . provided to accommodate those extra passengers who make calls on the generosity of owners of all two-seater cars'! The other closed model, the four-light two-door saloon priced at £250, following the example of the other Cowley models for 1925, was fitted with the new Dunlop Cord balloon tyres on well-base rims. The blue/black body (the darker shade on the superstructure) was well appointed having a carpeted floor to set off the blue leathercloth trimming and upholstery and such luxuries as an electric roof lamp and adjustable ventilators in both roof and scuttle, the latter controlled by a knob on the dash. Another standard fitting on the saloon was a windscreen wiper which Lucas had just brought out, operated from the manifold vacuum.

A popular motor car provides a lucrative market for extra equipment, and as

Rear-wheel-brakes-only on this 1925 season Morris Cowley. (R. Hunt)

today with the Mini, so in the 'twenties it was the Cowley. One interesting piece of equipment for the two-seater Cowley in 1925 was a purpose-made hood for the dickey seat, marketed by W. Jarvis of Oxford. The hood was held in place by two slots added to the existing rear hood stick and a series of press studs. Celluloid side screens ensured a dry ride for the occupants. There were, of course, many other contrivances to protect the rear passengers from the elements, notably the Auster screen for touring cars which were standard fitments on some Morris models. Similar attachments were made by Easting Windscreen Ltd, Joseph Gibson & Co, etc. Some accessories were more practical than others. Maltbys Ltd of Kent would convert the Cowley gear and handbrake levers to operate on the right-hand side, while J. McDowell or Wefco Motor Accessories could reduce the annoyance of brake squeak with mufflers, which were steel bands fitted tightly around the drums (these too became standard on the Cowley from mid-1926 models). Delco-Remy & Hyatt Ltd desired to eliminate magneto problems by supplying a coil ignition replacement unit, and from other sources there were horn rings for the steering wheel, headlight dimmer switches, windscreen wipers driven by cable from a friction wheel on the fan pulley, and radiator shutters operated by the driver via a Bowden cable. Devices to fit on the gear lever appear to have been popular, some simply providing a catch to prevent accidental selection of reverse ('Gearguide'). Others combined this function with a thief-proof locking arrangement for all gears ('Smith'), yet others like the 'Morrilock' secured both handbrake and gear lever with a conventional padlock! One combination lock obtainable from F.H. Harding of Cheshire ensured that only the Cowley owner was able to operate the starter. By 1925 'bumpers' had found their way across the Atlantic and were generally of wide flat spring steel, but all kinds of variations were available, ranging from heavy tubular rubber to sophisticated hinged devices with recoil springs. Luggage carriers too came in all shapes and sizes and the manufacturers of motoring accessories were not slow in realising that as well as providing a location for the tool box and spare petrol can, the wide running boards on the Cowley could be utilised as luggage space with the aid of folding luggage grids. Step mats for the running board were a regular line, usually rubber, sorbo, or hair matting, set in aluminium frames.

*The first standard saloon body on the Cowley chassis was
introduced for the 1925 range. This model with front-
wheel brakes and it's only two doors on the near-side dates
it to the 1926 season.* (Morris Motors Ltd)

By 1926 the Morris Commercial, now in its second year of production at
Soho, was becoming well known. For the lighter commercial market Morris
Motors Ltd had introduced in 1923 a light 8cwt van on the Cowley chassis,
graced with a flatter version of the familiar bullnose radiator, which, in more
recent times, has been nicknamed the 'squashed bullnose'. In standard form
the van had the usual period draughty cab with a low door on the nearside
only. A better quality body with curved panels, roof rails for additional light
packages, a sliding door between the driver's compartment and the rear goods
area, and a short door on each side of the cab were provided on the de luxe
model. For commercial travellers a Morris 'traveller's car' was listed which, in
essence, was a two-seater Cowley fitted with a 'box' in place of the dickey seat.
Unlike the contemporary Cowley two-seater it was equipped with the earlier
type of frameless-top single-panel windscreen.

A curious feature of the 1926 Morris Cowley is that one of the Smith's
carburetter fixings has a hole drilled through the bolt head. When the car left
the assembly line a wire passing through this hole fixed a label with a lead seal
informing the new owner that the engine was fitted with a 'speed controlling
washer', designed to restrict the speed to 27-30mph on the level. The object
(said Morris Motors Ltd) was to 'ensure that irresponsible people did not, by
their wanton brutality when driving the car away from the works, transform a

Standard version of the 1926 8cwt van.

*Sole survivor of the Cowley commercial traveller's car is
this 1925 model owned by Grant Brothers of Croydon
from new.*

Dickey seat hood marketed in 1925 by Jarvis of Oxford.

'Hotchkiss' type 11.9hp engine fitted to a surviving Cowley model. The cross-flow design can be clearly illustrated with the Smiths carburetter on this side of the block while the exhaust manifold is hidden from sight on the far side. (John Lowrie)

beautiful piece of mechanism into a noisy, rough and unsatisfactory, although reliable, power unit'! New owners were instructed to remove the washer after 500 miles. These speed restricting washers were obviously fitted to new motors for several years for in mid-1928 a convict, making his escape from Dartmoor, discovered too late that the Morris saloon he had stolen was so fitted and according to *The Western Morning News* he was arrested while in the act of removing the device!

Front-wheel brakes were introduced to the Cowley for the 1926 season as standard although the two-seater, four-seater, occasional-four open models, and the commercial traveller's car were also available in rear-wheel-brakes-only form. Front-wheel brakes were standard on the remainder of the Cowley variations, that is the ¾ fixed-head coupé and the two-door saloon. The latter,

*Miss May H. Cameron of Kilmarnock with Lord Nuffield
and her 1925 model Morris Cowley tourer which, in 1949,
she had owned from new. Four years later she took
delivery of a new Morris Minor and presented the old car
to the Nuffield Organisation, retaining the registration
number SD8751 for use on her new car.*

*Two surviving Cowley four-seater tourers, 1925 and 1926
models.*

unlike the previous year, now had both doors on the near-side of the body. A
subtle change to the two-seater body brought the top of the door in line with
the rear body panel, and on the coupé driving visibility was considerably
improved by the provision of windows in the rear quarters.

A number of detailed changes make easier recognition of the 1926 Cowley.
For the first time the Wilmot Breeden 'Calometer' was fitted (and later in the
year became subject of a court action, which is another story), the original
wooden battery box gave way to one of ebonite construction, the front end of
the chassis had a tie bar between the dumb irons, while doors were fitted with

outside handles. A new design of hand throttle control was fitted, and the familiar red spare petrol can was now black with a gold colour Shell scallop. All Cowley models were fitted with Dunlop reinforced balloon tyres of four-ply instead of the two-ply rating previously normal equipment, and on the closed models a parcel net and smoker's companion were added.

The most striking change to the Morris car image came with the Motor Show of 1926. The shape of the radiator was looked upon as a distinctive emblem of a car manufacturer and jealously guarded, for a recognisable radiator had considerable advertising value. Now, for the 1927 season, that familiar radiator was a thing of the past and in its place was a new flat-fronted radiator with an increased cooling capacity of over 60 per cent.

Morris Cowley 'Red Flash' originally raced by the late H.R. Wellstead still exists in the BL Collection at Syon Park.

Morris Cowley. 1915-26. Chassis Numbering.	
Model	Chassis Numbers
1915-19 Continental engine	3001-4485
1920 Hotchkiss engine	5001-6372
1921 Hotchkiss engine	7001-9288
1922 Hotchkiss engine	10001-14401
1923 Hotchkiss engine	15001-30000
1924 Hotchkiss engine	30001-57000 (except 55095-55664 inc.)
1925 Hotchkiss engine	57101-105800 (and including 55095-55664 inc.)
1926 Hotchkiss engine	105801-156424

Note: Numbers quoted above are actually those given as 'Car No' on the identity plate of the vehicle. These numbers are quoted by Morris Motors Ltd in their spares lists under the heading 'Chassis No'. The actual chassis frame number stamped on the chassis dumb-iron is always 3000 numerically less than the 'Car No' given on the identification plate.

All sports Cowleys, manufactured between March 1921 and October 1922, had numbers between 7398 and 15006.

Morris Cowley. 1915-26. Specifications.

Continental Engine Models

'Red Seal' engine, Type U. Makers: Continental Motors Company of Muskegon and Detroit, USA. Side-valve, four-cylinder, 69mm bore × 100mm stroke*, 11.9hp, 1,495cc. Cast iron pistons. Zenith carburetter. Magneto ignition by American Bosch, type NU4, Dixie type 40A, or Thomson-Bennett type AD4C. Cooling by thermo-syphon and fan. Three-speed Detroit Gear Company gearbox, central ball change. Clutch by same manufacturer, comprising two Ferodo-lined plates running dry. Transmission by enclosed torque tube, bronze universal joint, spiral bevel crown wheel and pinion (53/12 or 57/12). Rear-wheel brakes only, with hand brake and foot brake operating separate shoes in 11in diameter rear drums. Wheels three-stud steel artillery. Dunlop beaded edge tyres 700mm × 80mm (85mm or 90mm section extra). Lucas electric 6 volt five-lamp lighting system with Lucas type E20 dynamo on cylinder head, belt driven. Wheelbase 8ft 6in, track 4ft 0in.

Hotchkiss Engine Models

Original engines made by Hotchkiss et Cie of Coventry, until May 1923 when the company ownership changed hands and became Morris Engines Ltd. Side-valve, four-cylinder, 69.5mm bore × 102mm stroke, 11.9hp, 1,548cc. Cast iron pistons to engine 66930. Later pistons aluminium alloy. Carburetter 1919-21, Zenith. 1922, SU Sloper. 1923-6, Smiths. Ignition by Lucas or BTH magdyno or, with self-starter fitted, magneto. Early engines had three-port exhaust manifold, this was superseded February 1922 by four-port exhaust manifold. Cooling by thermo-syphon and fan. Three-speed gearbox with central ball change. Clutch, two driven plates with cork inserts running in oil. Transmission by enclosed torque tube, spiral bevel crown wheel and pinion (57/12, except on Sports Cowley which was 53/12). Rear-wheel-brakes-only models, where applicable, had hand brake and foot brake operating separate shoes in 9in diameter rear drums. Four-wheel-brake models, where applicable, had reversed Elliott type front axle with 9in diameter drums with two shoes. Steering, worm and wheel. Wheels three-stud steel artillery. Tyres 1919-22 two-seater 700mm × 80mm Dunlop Magnum. 1921-2 four-seater 700mm × 80mm Dunlop Magnum Cord. 1923 two-seater 700mm × 80mm Dunlop Cord. 1923 four-seater 28in × 3½in Dunlop Clipper Cord. 1924 all models, 28in × 3½in Dunlop Clipper Cord. 1925 all models 19in × 3½in wheels with 27 × 4.40 Dunlop Cord balloon tyres. 1926 all models 19in × 3½in wheels with 27 × 4.40 Dunlop reinforced balloon tyres. Wheelbase 8ft 6in, track 4ft 0in.

* See p31-2

Morris Cowley. Hotchkiss Engine Type. Body Colours and Upholstery

	1920	1921	1922	1923
Two-seater	Probably grey body with Pegamoid leathercloth upholstery.	Buff colour body with Pegamoid leathercloth upholstery.	Standard model: grey body with Pegamoid leathercloth. De luxe model: brown body with Pegamoid leathercloth.	Grey body with grey Rexine antique leathercloth upholstery.
Four-seater		Buff colour body with Pegamoid leathercloth upholstery.	Standard model: grey body with Pegamoid leathercloth. De luxe model: brown body with Pegamoid leathercloth.	Grey body with grey Rexine antique leathercloth upholstery.
Sports		Aluminium body.	Aluminium body.	Aluminium body.

Morris Cowley. Hotchkiss Engine Type. Body Colours and Upholstery

	1924	1925	1926
Two-seater Four-seater, Occasional four	Grey body with grey leathercloth upholstery.	Grey body with grey leathercloth upholstery. Blue body colour added March 1925.	Grey body with grey leathercloth upholstery. Blue body with blue leathercloth upholstery.
Traveller's car		Shop undercoat. Lead colour.	Shop undercoat. Lead colour.
Saloon, Coupé		Blue bodywork with black top. Blue leathercloth upholstery.	Blue bodywork with blue leathercloth upholstery.

Colour descriptions of the Continental-engined Cowleys; ie two-seater, four-seater and cabriolet coupé, differ. The following have been noted:
1915 Light chocolate/grey body with similar coloured leather upholstery. Black mudguards.
1916 Buff body and mudguards with upholstery to match.
1917 Greenish shade of khaki.

Morris Cowley, 1915-26. Magazine Bibliography

Year	Title	Publication	Date	Page
1915	The 11.3hp Morris Cowley Car	*The Autocar*	24/4/15	p454-5
1915	Morris Cowley Light Car	*The Motor*	27/4/15	p346-7
1915	On the Road	*The Autocar*	14/10/16	p383-5
1916	The Anglo-American Car	*The Light Car & Cyclecar*	13/9/15	p352-3
1916	About the Morris Cowley	*The Light Car*	13/10/15	p337
1916	Perfect Lighting of a Light Car	*The Light Car & Cyclecar*	24/1/16	p210-11
1916	A Winter drive in a Morris Cowley	*The Motor News*	8/4/16	p538, 540 & 542
1916	The 11.9 Morris Cowley Car	*Motoring in South Africa*	1/4/16	p21-2
1916	A Car of Superlative Merit	*The Light Car & Cyclecar*	18/9/16	p385-8
1916	Impressions of a Morris Cowley	*The Light Car*	27/9/16	p288-90
1917	The Morris Cowley Second-Hand Value £300	*The Light Car & Cyclecar*	2/4/17	p391
1918	Lubrication chart of the Cowley	*The Light Car*	30/1/18	p92
1919	A cruiser type of body	*The Motor*	8/10/18	p307
1919	A Morris Cowley & cruiser type body	*The Light Car & Cyclecar*	7/10/18	p366-8
1919	Morris Oxford & Cowley Cars in 1919	*The Autocar*	17/5/19	p . . .
1919	Morris 1919 Light Cars	*The Autocar*	2/8/19	p169-72
1919	Programme of Morris Motors	*The Light Car & Cyclecar*	31/5/19	p7-8
1920	A striking two-seater	*The Autocar*	20/3/20	p513-14
1920	Review of 1920 Cars	*The Motor*	3/12/19	p869-71
1921	A short trial of the Sporting Morris Cowley	*The Motor*	23/3/21	p381
1921	A new Sporting Model	*The Autocar*	12/3/21	p466
1921	After 3,000 miles in a Morris Cowley	*The Motor*	31/8/21	p194-96
1922	The Morris Cowley at 285 Guineas	*The Motor*	2/11/21	p593
1923	1923 Cars & New Models	*The Motor*	24/10/22	p531-2
1923	Care & Maintenance of Morris Cars	*The Autocar*	29/6/23	p1119-21
1923	How Morris Cowley & Morris Oxford Cars are made	*The Motor*	17/7/23	p1013-16
1924	Morris Cars in 1924	*The Autocar*	7/9/23	p428-9
1924	The 11.9hp Morris Cowley Chassis	*The Automobile Engineer*	3/24	p64-72
1924	Morris Cars	*The AutoMotor Journal*	12/6/24	p499-501
1924	Cowley, Christmas Road Test	*The Motor*	24/12/58	p794
1924	Real Quality Production	*The Motor*	1/7/24	p940-2
1925	The New Morris Programme	*The Morris Owner*	9/24	p651-4
1925	Servicing the Morris Cowley	*Motor Commerce*	22/8/25	p127-30
1925	Morris Programme for 1925	*The Motor*	2/9/24	p178-80
1925	Morris Cowley Coupé	*The Morris Owner*	3/25	p28-9
1925	Morris Cowley Saloon	*The Morris Owner*	4/25	p172-3
1925	Morris Cowley Saloon	*The Motor*	28/4/25	p566
1926	The New Morris Programme	*The Morris Owner*	9/25	p842-5
1926	The Morris 1926 Programme	*The Motor*	1/9/25	p179-81
1926	The New Morris Cars	*The AutoMotor Journal*	3/9/25	p737-8
1926	Morris 1926 Models	*The Autocar*	4/9/25	p393-7
1926	After 30,000 miles	*The Autocar*	30/4/26	p746-7
	Servicing the Morris Cowley	*Motor Commerce*	22/8/25	p127-30

Empire Oxford 1927~9

The acquisition of the Soho, Birmingham, works of E.G. Wrigley & Co Ltd in 1923, after the firm had gone into liquidation, provided William Morris with the development department he badly needed to develop new models and provided the necessary modern equipment and plant adapted to the production of motor components. In February 1924, Morris Commerical Cars Ltd was formed with William Cannell, transferred from Cowley, as managing director.

Morris had produced small commercial vans from the very start, using the Oxford car chassis, but with the facilities now available to him at Soho, Morris was able to move into the commercial section of the motor vehicle market. The first Morris one-tonner, powered by the same engine as the Oxford car, was completed in the amazingly short period of four months, and the next twelve months' production changed a loss of £22,000 into a profit of £39,000.

William Cannell and the sales manager at Morris Commercial Cars Ltd, C.F. Lawrence-King, pressed William Morris to be allowed to develop a more powerful engine for commercial vehicles. Morris, not convinced of the need, at first refused to sanction the project but eventually relented and the first Morris commercial engine of 15.9hp was designed. This new side-valve engine, with

One of the many magnificent technical illustrations by the late Alec F. Houlberg shows the 15.9hp Oxford chassis.
(Morris Motors Ltd)

similarities to the existing 13.9hp and 11.9hp 'Hotchkiss' type, nevertheless differed in many respects, not least of which was the dry-plate clutch consisting of a single metal plate connected to the transmission shaft and gripped between linings on the flywheel face and the face of the pressure plate. To overcome the considerable spring pressure a system of levers reduced the force required on the clutch pedal. However, this dry-plate clutch arrangement was only used with the earliest of engines on the new 15.9hp Oxford. Modifications were introduced fitting the conventional Morris cork-insert clutch designed to run in oil. A consequence of this change-over to the tried and tested clutch arrangement was the addition of a non-return valve fitted in the partition between the engine sump and the clutchcase to ensure retention of oil when the vehicle was on an inclined surface. These later engines were identified by a white star painted on the off-side of the cylinder block near the front end. In its original form lubrication for the new engine was by a camshaft-driven plunger pump feeding the three main bearings and camshaft bearings under pressure and continually replenishing the big-end troughs. Later engines (as used on the 15.9hp Oxford with cork-insert clutch) had a drilled crankshaft allowing the big-end bearings to be pressure fed. Engine cooling was facilitated by a large four-bladed fan and water impeller sharing a common belt driven pulley.

Instead of a cross shaft being utilised to drive the magneto, as in the case of the 'Hotchkiss' type engine, the helical timing gear was extended sideways on the near-side of the engine to drive the Lucas Type C45L dynamo and vernier coupled magneto mounted in tandem. Sharing this drive, forward of the engine, was a Smith's Maxfield patent type inflator which could be brought into operation by means of a gear clutch. Unlike the contemporary Morris car engine with its chain driven dynamotor, the new 15.9hp unit had a separate starter motor mounted on the opposite side of the flywheel housing where it engaged via a Bendix drive with teeth cut on the flywheel.

The new engine was first used in production for the Z type Morris Commercial 25/30cwt truck announced around May 1926, but prior to this, in March 1926, two models of the curious 'Morris-Martel' tankette powered by the 15.9hp engine had been delivered to the War Office for trials. Subsequent vehicles to use the same engine included the TX and R type commercials.

There is a certain irony to be found in that the first Morris Cowley was constructed mainly from American components, and when Morris could see the production advantages of an all-steel car body it was American technology that brought about the establishment of the Pressed Steel Company at Cowley. Indeed, American manufactured Gabriel snubber shock absorbers were still being fitted to the Oxford and Cowley in 1925. Still later, in the 'twenties, when all Morris cars were finished in 'Bripal' cellulose, the British Paint & Lacquer Co Ltd at Cowley was founded by the trinity of Morris Motors, Pressed Steel, and the Cleveland Varnish Company of Ohio. Yet it was the competition from American cars that frustrated Morris in his attempts to get a larger share in what was then called the Dominion market. Half a century later the gibing references to American cars in the pages of *The Morris Owner* magazine emanating from Cowley make curious reading.

To the Morris Commercial management, with their development facilities, fell the task of creating a Morris car specially suitable for conditions likely to be encountered in overseas markets, particularly Australia in which William Morris was so interested. The outcome was the Empire Oxford powered by the new Morris commercial engine. This was displayed in chassis form at the 1926 Olympia Motor Show where W.R. Morris in the company of the then Prime Minister of New Zealand, Mr J.G. Coates, and his opposite number Mr S.M. Bruce, Prime Minister of Australia (conveniently in the UK at the time) was able to demonstrate the features of the new model on the Morris stand.

The new 15.9 h.p. Morris-Oxford Five-Seater.

15.9hp Oxford five-seater as originally presented in 1926.
(Morris Motors Ltd)

The '15.9hp Morris Oxford' was the name generally given by the motoring press when describing the car at the time of its introduction. Perhaps the appellation 'Empire Oxford' was confined to those vehicles destined for export as the nickel plated hub caps on the earlier models proclaimed MORRIS EMPIRE OXFORD for all to see. Be that as it may, in its original form the car was offered as a four/five-seater tourer with deliveries to commence January 1927, before which date the Scottish Motor Exhibition had taken place and a four-door saloon version displayed.

The specification for the saloon body included a full width, high squab, front seat which was adjustable by means of side straps (a feature to be incorporated in the following season's Cowley), blinds and opening mechanism on all windows, locks on all doors, a roof ventilator, robe-rail, roof parcel net, pile carpets, and polished mahogany panelling. Upholstery was in soft furniture hide. The design also incorporated features new to the Morris driver of the period such as the four-speed gearbox (albeit gate change), a rear mounted ten-gallon petrol tank with gauge, steering column adjustable for rake (although there had been a means of adjustment on the Bullnose cars), and right-hand accelerator position. The instruments, illuminated by indirect lighting, comprised a clock, speedometer, ammeter and oil gauge within an oval facia

flanked by cubby holes. The 12 volt Lucas RB55 headlamps were physically
dipped by the lever operated Barker double-dipping system and the solid nickel
radiator shell had a round Calormeter in the radiator cap. All of which was
reflected in the list price of £375 for the saloon model compared with the £265
asked for the 14/28hp Oxford saloon of the same season.

On paper at least the Morris Commercial designers appear to have done
their homework well. The high ground clearance they felt necessary on a
vehicle destined for exacting overseas conditions was achieved by positioning
the Perrot brake operating rods above the axle instead of below as on other
Morris models, an overhead worm-drive rear axle, and large 21in diameter
five-stud artillery wheels carrying 31 × 5.25 Dunlop reinforced balloon tyres.
The resulting clearance, the lowest point being 10¼in at the centre of the axle
casing, was demonstrated by driving the vehicle over a standard building brick
stood on end! Four-wheel brakes of Rubery patent type were fitted with the
pedal simultaneously actuating 12in drums at the front and 14in drums at the
rear. The hand-brake lever operated independent side-by-side shoes in the rear
brake drums. According to the contemporary Morris catalogue: '. . . the
actuating mechanism is so arranged that the total distribution of braking
power is 50% front and 50% rear. This has been found to give maximum
stopping power.' The track of 4ft 8in, an increase of 8in on the contemporary
Cowley and Oxford, was chosen to approximate with George Stephenson's
standard railway gauge thus allowing for the substitution of flanged wheels
should the necessity arise. The shock treatment the car would receive on
extremely rough terrain was catered for by the deep section channel-steel side
members of the robust chassis, the Smith's shock absorbers mounted front and
rear and the large amount of spring deflection allowed for in the design.

All in all it was a well designed vehicle and one which should have sold well
overseas. But this did not prove to be the case. According to Jarman and
Barraclough in *The Bullnose and Flatnose Morris,* there were many shipped to
Australia: 'Unfortunately of the Empire Oxfords exported to Australia at least
205 of them, which was the vast majority, were returned to Cowley and
dismantled. The cars would not sell.'

During the winter of 1926/7 in an attempt to gain worthwhile publicity for
the 15.9hp Oxford, a trio consisting of Hans Landstad (who had joined Morris
as an employee in December 1914 and by 1926 was the works manager at
Cowley), Miles Thomas (then editor of *The Morris Owner* magazine), and one
other, set out from London with Landstad at the wheel of the (presumably)
prototype tourer heading for Scotland. The first day was fairly uneventful,
covering the 278 miles to Newcastle. Perth was the second day's destination,
then on the third day they found the poor conditions they were looking for to
prove the vehicle. They headed north-east to Pitlochry with wheel spin on hard
ice, up the Pass of Glenshee with Parsons chains fitted where, despite these
aids, the car got stuck in the snow drifts. Ultimately Braemar was reached.
Next day — the fourth — Cockbridge Ladder, one of the steepest hills in
Scotland was tackled. By the morning of the sixth day they were in Glasgow
and the 382 miles from there to Oxford was completed by late evening,
reporting back a round trip of 1,224 miles using one gallon of oil (which would

seem fairly high at 153 miles to a pint), and a petrol consumption of 20mpg.

Early in 1927 a 15.9hp Oxford was driven by B.W. Scappel in the Annual Swedish Reliability Trial from Stockholm to Gothenburg, over a course which involved the negotiation of all kinds of terrain and seasonal weather including deep snow and ice encrusted hills. This was good publicity and it gained the driver first prize and the additional prize of the Swedish Royal Automobile Club Cup for the best performance by a British car. Several months later 'Mirande', an Oxford registered 15.9hp tourer sponsored by Morris Motors Ltd, was driven by the Earl of Cardigan from Cowley to Constantinople.

When the 1928 season catalogue was published there was no change in price, model type, or specification for the 15.9hp Oxford (although as the model year progressed changes were introduced to the shock-absorbers and the spare wheel carrier). As in the previous programme the four-door tourer was priced at £315 with a choice of four colour alternatives with leather upholstery to match. Unlike the saloon, the front seats on the open model were separately adjustable and could easily be removed from the body. The saloon was offered in the same colour choices with furniture hide upholstery.

On both models a weatherproof toolbox was fitted at the forward end of the off-side running board (the spare wheel being located at the rear) containing a collection of standard tools which a modern do-it-yourself mechanic would envy, ranging from a jack with universal handle to an oilcan, a cold chisel, a 6in steel punch and hammer. Despite the provision of an engine-driven Maxfield

'Empire Oxford' four-door tourer, 1928.
(Morris Motors Ltd)

1928 season 15.9hp Morris Oxford saloon.
(Morris Motors Ltd)

tyre inflator, the tool kit also boasted a hand pump.

During the early months of 1928, Morris advertisements for the 15.9hp Oxford four/five-seater tourer made much of the large order for a fleet of these cars to be supplied to the Royal Air Force for staff duties. 'In their choice of these Morris cars' said the advertisement 'the Air Force, whose technical knowledge of motoring matters is undoubtedly greater than that of any other arm of the Services, were guided by the well proven efficiency, economy and reliability of this model.' Another special order came from the Anglo-Persian Oil Company who demanded non-standard gilled-tube radiators.

One, and almost the only, coachbuilder to use the 15.9hp Oxford chassis was the Hoyal Body Corporation Ltd, who as early as February 1927 were offering a 'D' back saloon with partition 'suitable for an owner-driver tour or for town motoring with a chauffeur at the wheel'. Special features were listed as winding windows in locking doors, step mats, kick plates, spring blinds, carpets, roof lights, rope pulls and ventilators. The body, which was supplied finished in grey, blue, maroon, or two-tone Hoyal brown with leather upholstery to match, cost £465, making it by far the most expensive 15.9hp Oxford; £120 more than the standard Morris saloon version. Also available from Hoyal was a saloon of the 'D' back design with conventional bucket seats at the front.

For the third and final year (1929) the 15.9hp Oxford (or the '16/40 Oxford' as it was now described) had been changed in minor detail and equipment.

Special body on the 1928 15.9hp Oxford. This was probably the work of A.P. Compton when designing for Jarvis of Wimbledon.
(George Compton)

Immediately apparent was the introduction of double-bar bumpers front and rear, luggage grid modification and a curve added to the rear of the front mudguards. With new legislation in the offing, Triplex glass was an optional extra. In addition to the tourer (now called a four-seater) and the saloon, there was a fabric saloon model which only appears to have been listed for the first few months of the 1929 season, also a four-door, four-light, body with imitation hood irons at the rear quarters and sliding front windows.

The last 15.9hp Oxford left the production line in July 1929. Only 1,742 had been produced and William Morris's idea of a special Empire car had died. Sir Miles Thomas (later Lord Thomas of Remenham), in his autobiography *Out on a Wing,* put its failure in Australia and other markets down to an engine which was rough and a body with square-cut box-like lines in competition with the then freely exported American cars. Other writers subscribe to the idea that they were under powered for colonial use. But was it just the car? In a report by an Australian correspondent in the *Motor Trader* in March 1929, an example is given of an Australian agent with all his capital invested in a British car franchise, waiting three months for British chassis because the manufacturer's first concern was for his home market. In the end the agent was compelled to take on an American car agency to keep in business. The report quoted the names of nine Sydney car distributors who had just abandoned British agencies and the writer added that British manufacturers had reached the stage when it was dangerous to wave the Union Jack in Australia unless they had goods to deliver!

However, all the development work that had gone before was not wasted. The 13.9hp Oxford was the basis of the new Morris Commercial G type International taxicab which took over where the Empire Oxford left off.

The Phoenix that arose out of the Empire Oxford, the 'G' type International Taxicab. (Dave Illsley)

15.9hp Morris Oxford. Engine Data

Four-cyclinder, side-valve. 80mm bore, 125mm stroke, 2,513cc, 15.9hp. Three-bearing crankshaft. Duralumin connecting rods. Aluminium pistons. Crankshaft and big-end bearings, white metal in bronze shells.

SU type HV carburetter. Magneto ignition. Single dry-plate Ferodo clutch on early models, later a cork-insert clutch running in oil was fitted. Engine three-point mounting.

Four-speed gearbox with gate change. Ratios: top 4.5:1, third 7.65:1, second 10.8:1, first 15.75:1, reverse 19.025:1.

Rear mounted petrol tank with Autovac feed to carburetter. Artillery wheels, 5 stud, 31 × 5.25 Dunlop reinforced balloon tyres.

Chassis Numbers. Body Colours and Upholstery

Chassis Numbers	1927 Models EO101-EO1075 15.9hp Oxford	1928 Models EO1076-EO1565 15.9hp Oxford	1929 Models EO1566-EO1841 16/40hp Oxford
Tourer 4-door	(Four/five-seater) Blue, grey, maroon, or brown. Upholstered in furniture hide.	(Four/five-seater) Blue, grey, maroon, or brown. Upholstered in 'leather to match'.	(Four-seater) Wine maroon duotone body with claret antique leather. Blue/black duotone body with blue antique leather upholstery. Deep maroon/bronze duotone with brown antique leather.
Saloon 4-door	Blue, grey, maroon, or brown. Upholstered in furniture hide.	Blue, grey, maroon, or brown. Upholstered in brown furniture hide.	Wine/maroon duotone body with brown furniture hide. Stone/brown duotone body with brown furniture hide. Deep maroon/bronze body with brown furniture hide.
Saloon			(Not listed after about May 1929) Brown fabric body with beige leather upholstery.

Note: Apart from the fabric saloon the bodies were all finished with cellulose paint.

15.9hp Morris Oxford. Magazine Bibliography

1927 model, technical description	The Motor	19/10/26	p547
1927 model, technical description, five-seater	The Morris Owner	11/26	p1215-18
1927 model, road test, saloon	The Autocar	11/3/27	p385-86
1927 model, road test, saloon	The Motor	19/4/27	p559-60

Morris Oxford Four-cylinder 1927~9

The familiar radiator, referred to as 'D fronted' in contemporary Morris literature and generally known as the 'Bullnose' was replaced by the flat-fronted radiator towards the end of 1926 for the 1927 season. The customary dealers' hanging sign depicting a Bullnose radiator surmounting the words 'Morris Service' lived on for a short period, before being replaced in early 1928 by a large replica of the round badge carrying the ox. When and who coined the word 'Bullnose' is open to conjecture, the earliest reference the writer can find appears in *The Autocar* of late 1926 where, recording the new type of radiator, the reviewer wrote: 'Though lots of enthusiasts will mourn the death of the bull-nose Morris . . . .'

Four-cylinder 14/28 Oxfords for the 1927 season, like the Cowleys, were presented on a completely redesigned chassis frame which was wider and upswept at the rear, to allow for long (47in) underslung and gaitered semi-elliptic road springs in place of the long established three-quarter elliptic rear springs. Onto this much more substantial frame a rigid superstructure formed an additional cross brace as well as carrying the seven-gallon scuttle mounted petrol tank, a pressed-steel facia board with cubby holes, and an anchorage for the steering column which, by using alternative mounting holes, could be adjusted for rake. The steering box was mounted onto the engine block with an interleaving gasket of rubber; unfortunately when this perished with age it made the steering misbehave.

'To define the difference between the Oxford and Cowley in the briefest way', wrote the Morris scribe of the time, 'one may say that if the Cowley model represents comfort, the Oxford model represents luxury. Both are of the same fundamental design, but the Oxford model is rather larger than the Cowley; the engine is larger, so is the frame, so are the tyres. In consequence a larger body is fitted to the Oxford chassis, so that what on the Cowley is a four-seater body is on the Oxford a five-seater body. The Oxford has been made a car of luxury in every respect. It is not only larger than the Morris Cowley, it is heavier and faster, in details in which the Morris Cowley is good and servicable it is elaborated and reasonably ornate.'

One of the minor features to differentiate the Cowley from the Oxford 'Flatnose' models was the number of studs used to attach the pressed-steel artillery wheels. On the Cowley the wheels, following the precedent of earlier models, had plain pressed-steel hubcaps and were held in place by three-stud fixings. On the Oxford a more decorative hub cap, having a nickel-plated dome, was standard with five-stud wheels. A simple change to be sure but one which cost an employee at Cowley his job. William Morris, driving one of the new Oxfords, had the misfortune to puncture a tyre (not an uncommon occurance in those days) and he discovered, when attempting to change the wheel, that someone at Cowley had provided the car with a useless three-stud Cowley wheel as a spare.

The use of a well tested engine, gearbox, and back axle, obviously cut down the redevelopment costs involved in the considerable up-dating of the 'old fashioned' Bullnose. Some attention was given to the braking system which was simplified so that a tubular shaft, spanning the chassis, provided a fulcrum for rods running to the brake drums. In the case of the 12in diameter rear drums there were two sets of shoes, with the foot pedal operating one pair while the second pair were provided for handbraking. To facilitate adjustment, the footbrake pedal, and that for the clutch, could be altered for length while the out-of-sight turnbuckles had been replaced with accessible wing nuts at the end of the rods.

External tidying up of the Oxford design changed the mudguard configuration and to complement the new radiator a splash guard was added at the bottom between the dumb irons. In view of the provision of a two-way normal/reserve petrol tap under the gravity-feed petrol tank, the need for a spare two-gallon can and Pennant carrier on the running board had gone. Additional tidying of the running board was achieved by accommodating the twelve-volt battery within the chassis frame in a carrier beneath the driver's seat, although the tool box remained in its accustomed position. Generally, the tool boxes fitted on the 1927 cars were made of solid mahogany by Frederick Restall Ltd of Birmingham and cellulosed black. The luggage grid, standard fitting on all 1927 Oxfords, was a folding device with the number plate, and single tail lamp, attached so that it took up its correct visible position whether the grid was folded or opened for use.

Body styles for the season (1927) comprised three open models, the two-seater, four/five seat tourer and ¾ coupé. Closed models were the saloon, cabriolet, and saloon landaulet. The high sided rear end of the two-seater, by comparison with its Bullnose predecessor, commenced with a downward curve from the base of the fabric hood; there was a dickey seat and an aesthetically pleasing styling line ran the length of the car from the tall radiator, through the bonnet side, to the rear end. This styling line was a feature of the four-cylinder Oxfords (not found on the Cowleys), providing an easy identification reference. Both two-seater and four/five seater had larger doors than hitherto and windscreens of the horizontally split two-piece type with the upper portion adjustable. A Lucas nickel-plated vacuum windscreen wiper deriving its power from the manifold via rubber tubing, and hood/sidescreen equipment (manufactured by the Coventry Motor & Sundries Co of Coventry) made driving the open models in inclement weather at least tolerable. A dickey seat at the rear and a folding hood formed part of the specification of the ¾ coupé, providing the owner with ample open-air motoring; in fact the only model in the 1927 listing without some form of folding head was the saloon. On the saloon landaulet, as the name suggests, the folding portion was above the rear seats.

The specification varied depending upon the body type, heavier springs for example were used on those chassis destined to carry weightier closed models and while 28 × 4.95 tyres (19in × 3½in wheels) sufficed for the two-seater, tourer and ¾ coupé, the other models had larger 20in × 4in wheels shod with 29 × 4.95 tyres; the spare wheel was carried on the near-side on all body styles.

Internal features, which represented Oxford 'luxury' as opposed to Cowley 'comfort', included an oval instrument panel illuminated by hidden rim lamps, Wilmot-Breeden smoker's companion on closed models, Barker dipping system on the Lucas RB55 headlamps, door pockets, leather upholstery, and carpets in both open and closed cars.

Immediately beneath the large four-spoked steering wheel on all 1927 Morris Oxford cars was to be found a neat bracket carrying short levers to control magneto and carburetter settings. The rod running the length of the steering column was in fact a rod running within a rod and was a development of a similar system fitted to earlier season's Oxfords. The story behind this fitting began when Oliver Lucas was driving a Morris Cowley he had acquired in the 'twenties. He thought of ways in which it might be improved and later, during a talk with William Morris, Lucas suggested the desirability of ignition advance/retard and throttle controls by rods and levers alongside the steering column. Morris liked the idea and so Jack Orme of Lucas and a draughtsman named Slater worked out the details, resulting in the neat design. Another addition to the Morris car suggested by Oliver Lucas was a scuttle ventilator, and Orme and Slater were seconded to Cowley where they worked with Morris and his chief engineer.

'Buy British and be proud of it' was the slogan used by Morris Motors Ltd from about 1924, but it was not until 1927, with the fitting of single-acting shock absorbers manufactured by the Yorkshire firm of Frank Smith & Co Ltd, that the Morris car finally became entirely British. The previous season's Gabriel snubbers were the product of a Cleveland, Ohio, factory, marketed in Great Britain by Brown Brothers Ltd.

Various add-on extras could be bought by the Oxford owner. Some were given the blessing of Morris Motors Ltd, while others attracted a downright condemnation and nullifaction of guarantee rights if fitted. One such device in the latter category was a special cylinder head with overhead valve gear for the conversion of the Hotchkiss type engine. The popularity of some components influenced the design of later Morris cars. Several firms such as The Central Motor Institute Ltd and Wray Park Garages of Reigate (now part of the Wadham group) offered kits to convert the accelerator pedal from the as-supplied centre position to a 'more restful' right-hand location. Around January 1927 all Morris Oxfords and Cowleys (chassis 173053 onward) were fitted with a right-hand accelerator as standard. Surprisingly, many subsequent Morris models such as the Minor, Ten, etc continued to be fitted with a centre pedal, while the Series II 8/10cwt van had a central accelerator pedal until 1939. Maltbys Ltd of Folkestone offered a conversion set approved by Morris Motors Ltd which put both gear change and handbrake levers on to the right-hand side; this was not adopted by the car manufacturers at Cowley. Bumpers of all shapes were available and the carburetter makers such as Brown &

Barlow, Zenith, and SU were not slow to advertise the alleged advantages of increased power, economy, speed and acceleration. The 1927 Oxford used the Smiths five-jet carburetter, but later models changed to the SU.

MG 14/40 Mark IV based on the Morris Oxford 14/28 chassis. (Wadham Stringer)

Approximately 700 MG cars of the 14/40 Mark IV type were made using the 14/28 Oxford chassis as a basis, although Cecil Kimber made considerable modifications to the standard Morris chassis. The light aluminium body fitted to this sporting model allowed the springs to be flattened and the braking system was completely modified to include a Dewandre vacuum servo, but the complexity of rods was subsequently simplified to a purely mechanical, and more successful, arrangement. The Oxford 13.9hp engines were dismantled by Morris Garages to have the crankshaft balanced, valve ports and other parts polished, and stronger valve springs added. Other modifications included a cast aluminium dash, changes in the electrical system, Smiths windscreen wiper acuated by the speedometer drive, altered steering rake, Rene Thomas flat sporting steering wheel, screwdown bonnet fasteners, and wire wheels. The use of a straight-through silencer may have caused some owners problems if the police decided that the car was not complying with the regulations, not necessarily because they were actually noisy but because the 'scientific' test employed by some forces was to insert a stick or similar object up the exhaust pipe and the absence of baffle plates was sufficient 'proof' that an offence had been committed!

Some notable personalities of the day were Morris Oxford owners and the

consequential publicity was obviously welcome. One such person was Harry
Tate, the comedian of the 'twenties who was noted for his comedy with a
motoring slant. It has been said that Tate was one of the first 'autonumer-
ologists'; his cars carried the personalised HT8 Bristol registration: a number
which, presumably, was used on the Oxford coupé he bought in mid-1927.
When the Prince of Wales (later the uncrowned King Edward VIII) visited the
Cowley Works in May 1927, William Morris drove His Royal Highness back to
London in a Morris motor car and, subsequently, the prince took delivery of a
Morris Oxford coupé specially upholstered in grey velvet calf hide.

Oxford saloon of 1928. An
early example of all-steel
construction. (Ken Martin)

For the 1928 season, the 14/28hp Oxford continued to be listed in the form
of two-seater, four/five seat tourer, $^3/_4$-coupé with folding-head, four-door
saloon, and saloon landaulet. The cabriolet was discontinued. A new variation
which put the Oxford into a superior commercial field was the Morris Oxford
traveller's brougham, its van-like body with side windows in the manner of a
modern brake, but with the added dignity of curtains on both the side windows
and the large rear door window. The driving cab, unlike the usual commercial,
had a specification equal to that of the Oxford saloon, while an illustration in
the period catalogue shows a chauffeur-driven brougham awaiting return of a
representative who is demonstrating the company products. Morris Motors
Ltd themselves made use of one of these as a mobile film display unit,
transporting films and projection equipment to dealers throughout the
country. Was it a coincidence that earlier, in 1927, Messrs Lyne, Frank &
Wagstaff of Crouch End were making a similar design of body for mounting
on both the Oxford 14/28 and 15.9hp 'Empire Oxford' chassis?
    Generally, the tourer and saloon bodies were of the newly developed all-steel
construction, a body mounting and finishing plant being installed at Cowley
around this time to deal with the steel fabrication. The use of this body manufacturing technique revolutionised some of the traditional trimming methods,
for example, upholstery was held in place by drive screws and steel clips instead
of being tacked onto a wooden framework. The rear seat back was simply held
in place by two clips on the rear sill. The roof was made as a separate unit and
completely trimmed and upholstered before being attached to the steel body by
means of screws through the peak panel, door roof header, rear quarter header

*Open mode demonstrated on
a surviving example of the
1928 Oxford ³/₄-coupe.*
(K.R. Westwood)

*The status commercial, a 1928 Morris Oxford traveller's
brougham.* (Morris Motors Ltd)

and the use of panel pins around the back of the rear quarter and across the top
of the rear panel; all fixings were readily accessible by removing the headers
from inside the body and drip moulding from around the rear. On the four/five
seater tourer an ingenious secondary use was made of the rear side screens.
These were arranged so that it was possible to fold them inwards and fit into
special brackets thus forming a 'V' windshield for the rear passengers. Further
protection was afforded by an apron attached along the bottom edges with
press studs. As a result of the all-steel body construction, the saloon windows
had large radii in the corners and lost their square appearance, while the
windscreen became a top-hinged single-panel screen.

By March 1928 the 14/28hp Oxford range was augmented by a saloon de
luxe (chassis 266960 onward). Described by the catalogue as 'luxurious almost
to a point of being ornate', the immediately apparent difference from the
cheaper saloon was the three-piece vee-fronted windscreen, a wider body
covering more of the rear mudguard surface, and additional refinements such

14/28hp Oxford saloon landaulet. Ex-works price in 1928
was £285. (Morris Motors Ltd)

as remote control for the rear blind, blinds over the rear quarter-windows, and glove pockets and garnish fillets in mahogany. Following the 1927 criterion, the closed models had larger diameter artillery wheels and wider tyres. The body of this de luxe Oxford was heavier than the standard model and in consequence the manufacturers attempted to compensate for the loss of performance by installing a 5.27:1 rear axle.

One of the curiosities of the 1928 season, and for that season only, was the Oxford offered in 11.9hp tourer and saloon form at a price falling between the Cowley and the 13.9hp Oxford. By a coincidence, the then Chancellor of the Exchequer, Winston Churchill, was shortly to put a duty of 4d a gallon on petrol in the 1928 Budget. With the addition of minor items such as an electric horn, luggage grid and (in the case of the saloon) a smoker's companion, the specification was generally that of the Cowley including the smaller radiator. However, the larger Oxford chassis was utilised and in deference to the smaller engine — 7bhp less at maximum revs — a change was made to the rear axle ratio from the normal 4.75:1 to 5:1. Wheels were of the smaller size (19in diameter) as fitted to closed Cowley models and touring versions of the 13.9hp engined Oxford.

Mechanically the specification for the 1928 14/28 Oxford differed little from the previous season, and these changes were fairly minor such as the re-arrangement of the braking system to include a short length of cycle-type chain to provide compensation (actually from chassis 219077) and a return to an SU carburetter (at chassis 223513) after the unsuccessful use of the SU 'sloper' type in 1922.

The last four-cylinder Oxford was that of the 1929 season; henceforth this name would only be applied to six-cylinder cars, for this was also the last season of the Oxford 16/40 or 'Empire Oxford'. New-style rounded wings linked by a lamp mounting bar and the incorporation of twin-blade bumpers front and rear (Wilmot Breeden were given large contracts for the latter) gave the Oxford a face lift. One body style, the saloon landaulet, disappeared and a

*A 1929 Morris Oxford coupé parked outside The Castle
pub in Newbury.* (Eric Jackson)

new one, the fabric saloon, was added to the range. Detail changes to a generally unaltered mechanical arrangement were made, but these were not always introduced immediately on the new season's cars. A considerable number of 1929 Oxfords left the production line before a new facia plate and Lucas instrument panel with locking ignition switch made its appearance at chassis 276925. There was now an earth return system for the electrical circuits, where the chassis and other metalwork provided the return conductor. The Lucas R50BDS headlights dispensed with the mechanical lever-controlled Barker dipping system in favour of a pneumatic method evolved by Lucas in 1927, where the centre hinged reflector of the near-side headlamp was operated via rubber tubing from a pump-like control cylinder mounted on the steering column. Like the electric solenoid-operated arrangement to be used in later years, this dipping action of the reflector simultaneously switched out the off-side lamp.

Prior to the 1929 season, Oxford cars had a duplication of audible warning devices; in addition to a Lucas 'No 7' electric type which operated on the principle of a vibrating armature, bulb horns were mounted through a hole in the base of the windscreen on the off-side. These varied in detail depending on the body, the touring model, for example, required a special cranked extension.

Of the specialist coachbuilders to utilise the Morris Oxford 14/28hp chassis Hoyal Body Corporation Ltd appear to have been the most prolific. These special bodied Oxfords were generally, but not exclusively, channelled through Stewart & Ardern outlets and often sold as Stewart & Ardern special models. For the 1928 season there was a 'Semi-Sports Fabric Saloon' of four-light design consisting of a light frame with plywood panelwork and metal con-toured corner panels covered in 'high quality' fabric in brown, blue or maroon, with soft cloth upholstery to match. Furniture hide was an optional extra with a slight increase on the basic price of £272 10s. A six-light version also appears to have been available as a black fabric covered saloon body with dark blue mudguards and bonnet was displayed at the 1927 Motor Show on the Hoyal stand. Another 1928 Stewart & Ardern offering was a four-door 'D' or 'dome-back' coachbuilt saloon in cellulose finish with leather upholstery. Morris Motors Ltd listed a similar model for the following season, and not for the first time a special bodied Morris appeared in standard Morris form in the next season's list (the Oxford Traveller's Brougham has already been mentioned). The same Morris 1929 programme included a closed coupé, but Stewart & Ardern obviously saw the need for a coupé with folding-head which, unlike its Cowley works counterport, was based on the heavier Oxford closed-body chassis and as a consequence had the wider 28 × 5.25 tyres and wheels; the spare wheel was mounted at the rear together with a luggage carrier. The choice of colours was left to the purchaser although brown, blue, black or claret were suggested with leather interior and dickey seat to match, and it is interesting that, at a time when cellulose was firmly established at Cowley, Stewart & Ardern offered these colours in either cellulose or paint and varnish. Also available from London's main Morris agent in 1929 was a six-light fabric saloon de luxe, even though the standard and cheaper Morris version (also six-light) catered for this market. Both the Morris and Stewart & Ardern versions had the sliding-roof facility (optional on the S&A special) and the major differences, such as winding windows all round and doors shutting onto the centre pillar on the special body, do not appear to justify the duplication, especially as the special was listed £45 above the Cowley ex-works price. To judge by the used car advertisements of the early 'thirties and reinforced by the relatively large survival rate, considerable numbers of Stewart & Ardern/Hoyal special bodied Oxfords were made.

Various 'one-offs' made use of the Oxford as a basis. The most striking must have been the ornamental 'horseless coach' body supplied by Stewart & Ardern Ltd in 1928 for Dorothy Perkins, the hosiery and underwear specialists of Oxford Street. A 'one-off' high performance special was sponsored by the Huddersfield main dealers, C.H. Mitchell of East Parade. This stark hotted-up Oxford was capable of 80mph and proved it by putting up the fastest time of the day and course record when driven by Miss M. Mitchell in the Bradford and District Motor Club's summer event in 1928. The opening event on that occasion was a 1 in $1\frac{1}{2}$ gradient at Hepolite Scar when, in the hands of another member of the family, A.G. Mitchell, the special defeated the hill.

Other coachbuilders also found a market for their own particular variations. Salmons & Sons of Newport Pagnell fitted Tickford bodies to the Oxford using an upholstery material called 'Fabrikoid', which was said by the manufacturers (Welin-Higgins & Co Ltd) to wear, feel, and look like real hide. It consisted of a closely woven cloth base with leather-like covering built up by the application of twenty-seven coats of 'Pyroxylin'. At least one Parkes Utility vehicle on the 14/28hp chassis is known to have survived to the present day. These utilities, named after the coachbuilding firm of Parkes of Sandling, taken over by The County Garage of Maidstone Ltd, consisted of a coachbuilt structure extending to the waist line of ash framing with pine panels. The folding roof of khaki twill, customary side screens, and rear seats were easily removed to adapt the vehicle to a light lorry, tender or shooting brake. Morgan's contribution from their Leighton Buzzard coachworks in 1928/9 was a fabric covered, four-door four-light, body in coupé style with dummy hood irons (the same body suitably modified was used for the 12hp Austin, 16-45 Silent Six Wolseley and the 16hp six-cylinder Sunbeam, to mention a few). In addition for the 1929 season was a six-light Weymann saloon, built under licence, and a sportsman's coupé whose lines anticipated the Morris special coupé bodies of some three years later. The normally sombre black finished artillery wheels and brake drums were brightened by the application of a contrasting light colour, and bumpers were offered as an optional extra. Moore's Presto Motor Works Ltd of Croydon claimed an astonishing choice of fifty colour schemes for their fabric bodied version of the Oxford in 1928.

Using the standard 1929 Oxford domehead coupé as a basis, a Mr Gregory invented a curious conversion which was constructed and distributed by the Westminster Carriage Co Ltd of London. The conversion involved the fitting of a rear portion made up as a separate unit of light steel framing covered in fabric. This extra piece was arranged to run in telescopic fashion into the front area of the roof. The 'enclosed' sliding head was carried on two roller-bearing carriages running in concealed tracks. As a saloon, the unit was pulled back uncovering the rear quarter lights and allowed the rear seat to be unfolded. As a coupé, with Mr Gregory's device locked in the forward position, a detachable section of decking covered the area of the rear seat and the whole locked into position with various catches. The complete operation, it was claimed, took a mere two minutes.

Before leaving the four-cylinder Oxfords, brief mention should be made of the export-only wide-track (56in) 14/28 Oxfords made for the 1927 and 1928 seasons. These were available with English four-seat tourer or saloon bodies, but a considerable number went as chassis to Australia where the heavy import duties severely restricted sales of complete vehicles. A total of 768 chassis, two tourers, and twenty-nine saloons were exported to Australia. The story of the wide-track variations is covered in the chapter on the Morris Cowley and Twelve-Four 1927-35.

Morris Oxford, Four-cylinder, 1927-9. Engine Data	
11.9hp	Four-cylinder, side-valve. 69.5mm bore, 102mm stroke, 1,550cc. Steel connecting rods. Aluminium pistons. Three-bearing crankshaft. White-metal bearings in bronze shells. Clutch, multiplate cork insert running in oil. Inlet and exhaust valves 33mm diameter. SU 'HV' carburetter, type 2M. Gearbox 3-speed and reverse. GA4 Lucas magneto ignition. Four-point suspension. 26bhp at 2,800rpm (approx).
13.9hp (14/28)	Four-cylinder, side-valve. 75mm bore, 102mm stroke, 1,802cc. Duralumin connecting rods. Aluminium pistons. Three-bearing crankshaft. White-metal bearings in bronze shells. Clutch, multiplate cork insert running in oil. Inlet and exhaust valves 33mm diameter. Smiths straight-through 5-jet carburetter to chassis 223512, SU 'HV' carburetter type 2M on chassis 223513 onwards. Gearbox, 3-speed and reverse. Speedometer connection to universal joint from engine 309080 onwards. GA4 Lucas magneto ignition. Four-point suspension. 33bhp at 3,000rpm (approx).

Morris Oxford, Four-cylinder. 1927-9. Magazine Bibliography.			
Morris programme for 1927	*The Autocar*	3/9/26	p371-3
Morris programme for 1927	*The Morris Owner*	9/26	p906-10
Morris programme for 1928	*The Autocar*	2/9/27	p437-40
Morris programme for 1928	*The Morris Owner*	9/27	p908-11
Morris programme for 1929	*The Autocar*	7/9/28	p499-500
Morris programme for 1929	*The Morris Owner*	9/28	p . . .

Specification

Oxford 1927 models. Chassis 156501-215000.

13.9hp 'Hotchkiss' type engine. Mechanical four-wheel brakes, foot pedal to all wheels, hand brake to independent shoes at rear. 12in diameter brake drums. Centre accelerator pedal to chassis 173052, right-hand accelerator pedal from chassis 173053 onwards. Half-elliptic springs front and rear. Heavier springs on chassis intended for closed bodies. Leather spring gaiters. Worm and wheel steering box attached to engine with rubber packing. Totally enclosed torque tube transmission. Smiths shock absorbers, single acting friction. Scuttle mounted petrol tank with two-way tap, 7 gallons with 1 gallon reserve. Petrol gauge in tank. Pressed steel artillery wheels. 19in × 3½in 5-stud fixings, 28 × 4.95 Dunlop tyres on the two-seater, four-seater, and ¾ coupé. 20in × 4in wheels on the saloon landaulet, saloon and cabriolet, with Dunlop 29 × 4.95 tyres. Hub caps black and nickel plate. Oval panel with illuminated instruments. Calormeter temperature gauge and wings on 'flatnose' radiator. Barker dipping, headlights, Lucas type RB50 headlamps. Vacuum windscreen wiper, nickel plated. 12-volt battery under driver's seat. Lucas A900R Dynamotor. Lucas electric horn. Bulb horn Lucas type 24T on two-seater, 24T with cranked extension on tourer, 24T long on coupé and saloon. 64B type on all-metal saloon. Folding luggage grid standard. Wheelbase 8ft 10½in, track 4ft. Two-wire system for electric wiring.

Oxford 1928 models. Chassis 215001-268465. (Saloon de luxe starts chassis 266960.)

13.9hp and 11.9hp 'Hotchkiss' type engines. Mechanical four-wheel brakes, foot pedal to all four wheels, hand brake to independent shoes at rear. 12in diameter brake drums. Right-hand accelerator pedal. Half-elliptic springs front and rear. Leather spring gaiters. Worm and wheel steering box attached to engine with rubber packing. Totally enclosed torque tube transmission. Smiths shock absorbers, single acting friction. Scuttle mounted petrol tank with two-way tap, 7 gallons with 1 gallon reserve. Petrol gauge in tank. Pressed steel artillery wheels. 19in × 3½in 5-stud fixings, 28 × 4.95 Dunlop tyres on two-seater, four-seater, ¾ coupé, and the 11.9hp four-seater and saloon. 20in × 4in with 29 × 4.95 or 30 × 5 Dunlop tyres on the saloon landaulet, saloon, and saloon de luxe. Hub caps black and nickel plate. Oval plate with illuminated instruments. Calormeter temperature gauge and wings on 'flatnose' radiator. Barker dipping headlights, Lucas type RB50 on 14/28 models, Lucas type R45 on 11.9hp models. Smaller Cowley size radiator on the 11.9hp models. Rear axle on saloon de luxe, 5.27:1. Other Oxford 14/28 cars, 4.74:1. 11.9hp Oxfords, 4:1. Vacuum windscreen wiper, nickel plated. 12-volt battery under driver's seat. Lucas A900R dynamotor. Lucas electric horn. Bulb horn type 24T on two-seater, 24T with cranked extension on tourer, 60B on coupé and saloon, 24B on saloon landaulet. Folded luggage grid standard. Wheelbase 8ft 10½in, track 4ft. Two-wire system for electric wiring.

Oxford 1929 models. Chassis 268466-313896.

13.9hp 'Hotchkiss' type engines. Mechanical four-wheel brakes, foot pedal to all four wheels, hand brake to independent shoes as rear. 12in diameter brake drums. Right-hand accelerator pedal. Half-elliptic springs front and rear, rear 11 leaves except on saloon de luxe which had 8 leaves. Leather spring gaiters. Worm and wheel steering box attached to engine with rubber packing. Totally enclosed torque tube transmission. Smiths single acting friction shock absorbers with paddle. Scuttle mounted petrol tank with two-way tap, 7 gallons with 1 gallon reserve. Petrol gauge in tank. Pressed steel 18 × 4 artillery wheels, 5-stud fixing. Hub caps black and nickel plate. Oval panel with illuminated instruments, change in design with centre ammeter, Lucas key ignition switch and inspection lamp sockets from chassis 276925. Front wheel drive for speedometer finishes at chassis 299091. Pressed rounded pattern mudguards. Calormeter temperature gauge and wings on 'flatnose' radiator. Inverted 'U' section lamp bar spanning front wings. Pneumatic dipping system for headlights, 'cut and dip' arrangement, Lucas R50 BDS headlamps. Lucas Sparton electric horn. 12-volt battery under driver's seat. Lucas A900R dynamotor. Twin-blade bumpers front and rear quarters. Triplex glass extra. Wheelbase 8ft 10½in, track 4ft. Earth return system for electric wiring.

Note. Morris Motors Ltd always quoted the number given on the identification plate as 'Car No' when referring to 'chassis numbers' in their parts lists. The actual chassis frame number stamped on a dumb iron (usually with a prefix letter) was always 3,000 less numerically than the Car Number. Above numbers are those that would have appeared in the log book.

Morris Oxford. Four Cylinder Models. Body Colours and Upholstery.

	1927	1928	1929	
Two-seater	Blue, grey, maroon or brown cellulose with leather to match.	Blue cellulose with blue leather. Maroon cellulose with leather. Brown cellulose with brown leather. Beige cellulose with leather.	Stone-brown*, wine-maroon*, deep maroon-bronze*, blue-black* duotone cellulose with antique leather. * Superstructure	
Four/five seater tourer	Blue, grey, maroon or brown cellulose with leather to match.	*11.9hp model tourer* Blue or maroon cellulose with Karhyde to match. *13.9hp model* Blue, maroon, brown or beige cellulose with matching antique leather.	Stone-brown*, wine-maroon*, deep maroon-bronze* or blue-black* duotone cellulose. with antique leather. * Superstructure	
Coupé	*¾ coupé model* Blue, grey, brown or maroon cellulose with self-toned moquette. (Leather upholstery available as an extra.)	*¾ coupé model* Blue, maroon, brown or beige cellulose with matching leather.	*Dome-back coupé model* Stone-brown*, wine-maroon*, deep maroon-bronze*, blue-black* duotone cellulose with Monochrome cloth. * Superstructure (Natural grain leather to match the above colours available as an extra.)	
Cabriolet	Blue, grey, brown or maroon cellulose with leather.			

Saloon	Blue, grey, brown or maroon cellulose with self-patterned moquette to match. (Leather upholstery available as an extra.)	*11.9hp model* Blue or maroon cellulose with self-patterned moquette or Karhyde *13.9hp model* Blue, maroon, brown or beige cellulose with leather or self-patterned moquette to choice. *13.9hp saloon de luxe* Blue cellulose with brown furniture hide.	*Fabric saloon* Blue-grey or red fabric body with leather upholstery to match. *Saloon dome-back* Stone-brown*, wine-maroon*, deep maroon-bronze* or blue-black* duotone cellulose with Monochrome cloth upholstery. (Natural grain leather upholstery to match body colour available as extra.) *Saloon de luxe* Blue-black* duotone cellulose with blue leather upholstery. Wine-maroon* duotone cellulose with red leather upholstery. * Superstructure
Saloon landaulet	Brown, blue, grey or maroon cellulose with furniture hide to match.	Brown cellulose with brown furniture hide.	
Traveller's Brougham	Blue, maroon or brown coachpainted with upholstery to match.	Blue, maroon or brown coachpainted with upholstery to match.	Blue, maroon or brown coachpainted with upholstery to match.

Morris Six
1928-9

Up to the mid-1920s the traditional body building techniques involving the use of an ash frame and steel or aluminium panels remained the principle system used by the larger motor manufacturers in Britain. Although other methods of fabrication had been tried, none were suitable for mass-production. Various alternatives included the 'Zephyr' principle of light-gauge steel tubes, braced by wire, replacing the conventional timber frame, and the all-steel bodywork introduced on a BSA car as early as 1912. ('All-steel' was not strictly true as the two- and four-seater bodies built under German patent had a frame of T-section members with considerable reliance on cast aluminium corner brackets and panels.) The Weymann fabric-covered silent bodies evolved by Charles T. Weymann (where the interface of individual timbers was held apart by metal straps) were successful for a few years following their introduction on the French Talbot in 1922; this resulted in a spate of fabric-covered versions on the ordinary ash frame, to avoid dues payable for special licences required when building on the patent Weymann system. But the short life of the fabric used and deterioration caused by damp resulted in a fall off of demand for this type of bodywork, although it had helped to popularise the closed car.

In America the Hupp Motor Car Corporation began marketing the steel-bodied Hupmobile in 1913, followed by the Dodge brothers, who broke their association with Fords and began to manufacture a Dodge car with all-steel bodywork in 1914. In France, the Citroën Company were the first European manufacturers to adopt steel construction soon after World War I, and it was not long before the Citroën Company's works at Slough were assembling component parts brought over from France. All of which did not go unnoticed by William Morris who, realising the importance of the new technical developments, visited the United States in 1925 to study the American methods and in particular those of the Edward G. Budd Manufacturing Company of Philadelphia who owned valuable patents to the system; and from whom André Citroën had bought rights to manufacture all-steel bodies for his B12 saloons and tourers.

Edward Gowan Budd, who was born in Delaware in 1870, had an early career with the American Pulley Company who built up a thriving business merchanting pulleys stamped out of sheet steel. Then followed a spell with the Hale & Kilburn Manufacturing Company where he and another engineer, Joseph Ledwinka, were employed on the design and planning of steel body

Side and front views of the 1928 Morris-Six saloon showing the Barker headlamp dipping mechanism. (Eric Miller)

pressings. With this background and the foresight to see the expanding use of steel body pressings in the motor industry, it is not surprising that in 1912 he and Ledwinka set up their own company employing twelve men and a capital of £15,000. The Dodge brothers, John and Horace, who approached Budd for pressed steel body panels were persuaded that the all-steel body was a practical proposition.

As a result of Morris's visit to the United States, The Pressed Steel Co of Great Britain was established in 1926 under the auspices of the Budd organisation, J. Henry Schroeder & Co (merchant bankers), and Morris Motors Ltd (Morris withdrew their interest in 1930 as it inhibited the sale of bodies and pressings to other motor manufacturers). The new company was located at Cowley, adjoining the Morris works, where some of the largest presses in the country (including a 1,600-ton American Hamilton, weighing 249 tons) were installed, making it, subsequently, the largest body plant in Europe. Meanwhile, in Germany, Budd was also setting up the Ambi-Budd Presswerke in Berlin. In the initial stages, in order to get an early start, Morris purchased dies and jigs for two types of steel body from Budds for £120,000, together with over four thousand drawings which had to be converted for English manufacture. Teething troubles were inevitable, especially in view of the unfamiliar work for the local labour force, and it is recorded that the Morris management looked in horror at the first examples of the new all-steel body. In recent years, Lytton Jarman (co-author of *The Bullnose and Flatnose Morris*) has had a sample of the original metal used analysed and concluded that an entirely unsuitable material, lacking the ductility required for press work, had been used. Welding was another area fraught with problems, but eventually the completed bodies were rolling off the line.

The all-steel body was truly named, as even the mouldings and facia could be made, if required, of steel and grained to resemble wood. The floor was also of sheet steel. The first of the new bodies were employed on the Morris Light-Six which was introduced at the Motor Show, Olympia, in the late months of 1927, ostensibly, for the following season. A six-cylinder car was a new departure for Morris Motors, if the abortive 'Silent-Six' of 1923-4 is ignored, and this one almost suffered the same fate. The engine used was a new 17.7hp overhead camshaft unit designed by Frank George Woollard and Arthur George Pendrell at Morris Engines Branch in Gosford Street, Coventry. Cecil Kimber, at that time general manager of Morris Garages, who had worked with Woollard at E.G. Wrigley & Co (the original suppliers of axles and other components to Morris, and later taken over by Morris Motors Ltd) persuaded Woollard to embark on a six-cylinder engine project. Woollard went ahead with the development without William Morris's prior knowledge and the result was the 'JA' 2,468cc power unit developing something like 50bhp at 3,200rpm. This was the engine destined to be used in the Light-Six and, subsequently, the Isis models as well as being the basis of the early Morris marine engine, the 'Commodore'. Unfortunately the first Light-Six cars were built with the new engine in what was basically an elongated Oxford chasis retaining the original 48in track, which resulted in a car with sparkling performance, but impossible to control. Eleven such models were built, eight of which were subsequently dismantled. Kimber used one of the remaining three as a basis of the prototype MG 18/80; not surprisingly, the 750 cars later built by MG in Mark I and II from between 1928 and 1932, had their own chassis design. The revised Light-Six was based on a chassis with an 8in wider track and a longer 9ft 9in wheelbase.

As the first of these were not dispatched from Cowley until 13 March 1928, it is safe to assume that the show vehicles on stand 98 at Olympia for the 1927 Motor Show were narrow-track versions, and the date would also account for the late (May 1928) Road Test carried out by *The Autocar,* in which the tester's veiled remarks suggest that he was not specially pleased with the handling at low speeds. Morris sources claimed a speed of over 70mph in top gear and a petrol consumption of 22mpg.

In production form the Light-Six — or Morris-Six as it was now called — was available as a four-door saloon and a coupé with dickey seat. The original listing of a tourer body, in publicity material depicting the narrow-track car, was not repeated. In addition to increasing the track by 8in and the wheelbase by 3in, thus precluding the utilisation of Oxford car bodies, the new design had larger section Dunlop Balloon 30 × 5.25 tyres (in place of the 29 × 4.95) on the same 20in × 4in steel artillery wheels. The back axle was also modified from

*Hollick & Pratt copper and aluminium body orginally
fitted to a Silent-Six now survives mounted on a 1928
Morris-Six chassis.*

4.42:1 to a lower geared 4.77:1 ratio and coil ignition had been substituted for the magneto. Costs were also reflected in a price increase of £45 on the saloon, now £395, and the coupé which despite the loss of the drop-head facility was up £55 to £385.

The fixed-head coupé model, while primarily designed to carry two persons could, at a pinch, accommodate six, by use of the occasional seats provided behind the front bucket seats inside and the space for two passengers in the dickey seat; the spare wheel was side mounted. In the roomy body of the saloon, which had the same overall length of 13ft 5in, the seating arrangements were described as amply suitable for three abreast in the rear and two in the separately adjustable front seats. The use of furniture hide for the upholstery and polished mahogany panelling helped to substantiate the maker's claim that the car compared favourably with 'models costing very much larger sums'. Certainly the equipment was not skimped and the Morris-Six owner had a fairly comprehensive instrumentation; a speedometer with trip, clock, oil pressure gauge, ammeter, dash ventilator control, and lighting/charging switches on an instrument panel with concealed illumination. Elsewhere the equipment included dipping headlights, a Wilmot Breeden Calormeter to monitor the radiator temperature, road springs protected by 'Wefco' leather gaiters (named after the inventor in 1918, a Mr W.F. Cattrell), winding windows, Lucas vacuum-type windscreen wiper for the single-panel windscreen, and a luggage grid on the saloon. The 1928 season saw a trend towards rear mounted fuel tanks, in place of the then customary location on the dash. On both the 15.9hp Oxford and the Morris-Six the petrol feed to the SU carburetter was by means of an Autovac from the rear tank (of 11 gallons capacity on the latter model) which incorporated a large filler cap, a petrol gauge, and reserve tap.

One of the facilities offered by Morris Motors Ltd at this time was a hire-purchase plan which the company had arranged with the United Dominions Trust Ltd. Common enough these days, but for the prospective purchaser in the late 'twenties of (say) a new Morris-Six for a deposit of just under £99 and payments of £14 per month compared very favourably with the £150 or so being asked for a two-year old Oxford on the 1928 second-hand market.

There is a story of a customer requiring a Morris-Six coupé from Wilson's showrooms in Chapel Allerton, Leeds, in 1928 and insisting on delivery the same day. The proprietor, Arnold G. Wilson, had not got such a model in stock but a telephone call to Cowley and a drive to the local aero club was followed by a flight in a Blackburn Bluebird biplane from Leeds to Port Meadow, just outside Oxford. By early that evening the appropriate car had been driven from the Morris works to the customer, who was presumably happy to pay the sixpence (2½p) a mile delivery charge. About the same time a customer for another Morris-Six coupé in Luton was taking delivery of his car from Dickinson & Adams Ltd. Hearing that although the Bedfordshire 'TM' registrations had reached over 4,000 in the series, no one would accept the registration number 'TM13' he specially requested this mark and, defying superstition, took delivery of the car on a Friday.

Was a fan really needed for cooling on the six-cylinder engine or was the water pump on its own sufficient? Parts were made available in November 1928 to allow for the fitment of a four-bladed fan, driven by a Whittle belt, by the addition of a pulley attached to the forward end of the horizontal auxiliary drive shaft. Ostensibly this was to cater for those owners who felt the need of additional cooling when holidaying in warmer foreign climes. However, it is noticeable that when the 17.7hp engine was later used in the Isis-Six it had a fan fitted as standard (initially four-bladed but later twin-bladed), although the MG 18/80 had no fan.

The Motor Show at Olympia was held during 12-20 October 1928 and the

Morris-Six coupé of the 1929 season. The visor was a standard fitting on these models. (Len Barr)

1929 season's versions of the Morris-Six were to be seen on stand 125. Both saloon and coupé examples on display revealed the addition of bumpers (a feature of all 1929 Morris cars) and a larger steering wheel. The basic prices had been reduced, the coupé from £385 to £365, and that of the saloon by a similar amount. However, Triplex safety glass was extra, as were the wire wheels which could be fitted in place of the standard five-stud pressed steel artillery type to both models for an additional £10. A wider choice of five duotone colour schemes was originally listed at the time of the Motor Show, including the brown and beige of the previous season but, curiously, this one appears to have been dropped from the list by early 1929. If the contemporary catalogue photographs are to be taken as a guide to standard equipment (not a wise thing to do as often the previous year's photographs were re-used in a touched-up form) then the new models had the addition of an external glass sun visor and a black leather (or leathercloth) cover on the side-mounted spare wheel — although these items are not mentioned in the specification and may well have been optional extras. Michael Sedgwick, writing in the *Veteran & Vintage* magazine, wondered if owners would have enthused quite so much over the latter gimmick had they known that in America it was the hallmark of a successful mortician! Was there a change of mind at Cowley as to the most suitable side to mount the spare wheel? The saloon used for the Road Test in *The Motor* in March 1929 was a late 1928 registered car (UD 2329), carrying the spare on the off-side. Elsewhere, published photographs (including the catalogue printed for September 1928) depict the wheel on the near-side of the body, which, if surviving examples are any guide, is where it came to rest.

An additional model for the 1929 season was the Gordon England Morris-Six club coupé which at £399, because the small-centre wire wheels and safety glass was included as standard, was actually cheaper than the saloon with these options. Unlike the Austin Motor Company who often included 'custom' coachwork within their catalogue pages (usually by Gordon & Co of Sparkbrook, Birmingham — not to be confused with Gordon England Ltd of Wembley) the listing of the club coupé was something of a precedent. A maximum speed of well over a mile-a-minute was claimed by Morris Motors Ltd and a petrol consumption of 20mpg under normal touring conditions. In common with most Gordon England bodywork of the period the four available two-tone colour schemes were arranged to give a light colour to the roof, bonnet top and wheels, complementing the darker colour of the remainder of the body while the mudguards and aprons followed the Morris all-black custom. Additional equipment included pneumatic inners (by the Stockport firm of Moseley) to the Vaumol leather upholstered seats, pile carpets, scuttle and roof ventilators, smoker's companion, lady's companion, cigar lighters, and twin windscreen wipers. The 'companions' were usually an ash receptacle with pipe rack and match box attachments or, in the case of the 'lady's companion', a container fitted with scent bottles, mirror, note book, etc. Headlamps, combined with the Barker dipping mechanism (fitted to all Morris-Sixes), described as 'extra large', were the Lucas 9in diameter type RB67 lamps as fitted to the standard Morris versions.

Of course, there were other body builders who found the Morris-Six chassis an ideal medium for their designs, but these never made the Morris catalogue. Indeed, Gordon England marketed a four-door close-coupled saloon with sliding-head, in 1929. Morgan & Co Ltd of Leighton Buzzard produced a Weymann body six-light four-door saloon with the complete roof folding to give a landaulette-like silhouette, and a similar fabric bodied fixed-head saloon also made under Weymann license. From the Hoyal works at Weybridge came a number of body styles on the Six chassis such as the Hoyal $\frac{3}{4}$-coupé with folding-head of leathercloth and metal panelled body in black, red, blue or brown; the 'Bournemouth de luxe Fabric Saloon', 'Sportsmans Coupé', and the 'Family Fabric Saloon' — all over £400, with wire wheels extra. Stewart & Ardern were offering an 'S&A Morris-Six Fabric Saloon' in November 1928, but as the body was by Hoyal it was probably one of the two Hoyal saloons already mentioned.

Of the surviving Morris-Sixes known to the writer, the copper-aluminium bodied two-seater is the most interesting, as it provides a link between Morris's first abortive venture into six-cylinder cars with the 'F' type of 1923/4 (see page 17) and the second, more successful, Morris-Six. The body, which is constructed with the upper panels of polished aluminium, side panelling of copper, and the extensive use of mahogany, was originally built on the third Silent-Six F-type chassis (registered BW6477) by the coachbuilders Hollick & Pratt for Lancelot W. Pratt in 1923. He was one of the few close friends claimed by William Morris and was, at that time, deputy governing managing director at Cowley. Sadly, he did not live long enough to enjoy using this unique car for in 1924, at the age of 44 years, he died of cancer. The vehicle then appears to have gone to William Morris. What is known is that in 1928 the body was transferred to a new Morris-Six chassis and registered UD2133. After World War II the car got into private hands, and eventually, in the early 'sixties, was sold to a Kidderminster enthusiast who paid £75 for the car and estimated to have spent £2,000 over a period of three years restoring it. For some time the vehicle was on loan to the Birmingham Museum of Science & Industry, by which time a 1903 Middlesex registration number H84 had replaced the Oxford number. In July 1977 the hybrid was auctioned at the Measham Motor Auction and realised what must be a record price for a Morris motor car, £17,000.

In the two seasons that the 'JA' or Morris-Six was available, 3,650 cars or chassis left the lines at Cowley, almost two-thirds of these in the second year. Production ceased in the summer of 1929 to make way for the new 17.7hp Six, the Morris Isis.

Morris-Six. 1928-9 Models. Chassis Numbers, Body Colours and Upholstery

	1927 Light-Six, All-Steel Chassis F201-F211	1928 Morris-Six Chassis JA101-JA1576	1929 Morris-Six Chassis JA1577-JA3750	
Saloon, four-door Coupé, two-door	Blue & grey or brown & beige duotone cellulose with leather.	Blue & grey or brown & beige duotone cellulose with furniture hide.	Grey & blue duotone cellulose with blue natural grain leather. Wine & maroon duotone cellulose with red natural grain leather. Deep maroon & bronze duotone cellulose with brown furniture hide. Blue & black duotone cellulose with blue natural grain leather. (Beige & brown duotone cellulose with brown furniture hide was orginally listed in September 1928 but not quoted in early 1929.)	
Club coupé, two-door			Black with green or cream superstructure. Dark blue with light blue superstructure. Burgundy with grey superstructure. All with Vaumol leather.	
Tourer, four-door	Blue & grey or brown & beige duotone cellulose with leather.			

Morris-Six Specification
Engine: 'JA' six-cylinder, overhead camshaft, inclined valves. 69mm bore, 110mm stroke, 2,468cc 17.7hp. Aluminium pistons with three rings. Steel connecting rods. Four-bearing crankshaft. Water impeller. Coil ignition. SU type M3 Carburetter. Three-speed gearbox. Clutch, multiplate with cork inserts running in oil. Axle ratio 4.77:1. Rear-mounted petrol tank in conjunction with Autovac. Smiths single-acting friction shock absorbers. Foot brake by rod to four wheels, hand brake by rod to independent shoes at rear. 12in drums. Divided bronze-ring type universal joint. Barker dipping mechanism on headlamps. Bumpers standard on 1929 models. Dunlop Balloon tyres 30 × 5.25 on 20in × 4in wheels. Artillery 5-stud standard on saloon and coupé. Wire wheels standard on club coupé for 1929 and optional extra on other 1929 models. Side-mounted spare. Wheelbase 117in, track 56in.

Morris-Six 1928-9. Magazine Bibliography				
Model Year	Subject	Source	Date	Pages
1928	Morris Light-Six Specification	*The Morris Owner*	11/27	p1226-9
1928	To Spain in a Morris-Six	*The Morris Owner*	6/28	p434-7
1928	Morris Light-Six Specification	*The Autocar*	14/10/27	p739-44
1928	Morris-Six Road Test	*The Autocar*	25/5/28	p1065-6
1929	Morris Programme for 1929	*The Morris Owner*	9/28	p . . .
1929	Morris-Six Club Coupé	*The Autocar*	1/2/29	p230
1929	Morris-Six Road Test	*The Motor*	12/3/29	p261-3

Morris Cowley Twelve-four 1927-35

When, in September 1926, publicity was given to the completely re-designed models with the new flat radiator, this revolutionary change brought with it some discontent from recent purchasers of Morris cars. Morris Motors went to great lengths to pacify those buyers who had bought old-type cars and who now felt they were saddled with a car perhaps only a week old and yet obsolete. In addition to showing with detailed figures that a depreciation in second-hand value was not necessarily a loss to the owner, because the additional amount he had to find in replacing the car with a new model was the criterion, the company stated: 'To produce cheaply a maker must keep his factory running twelve months of the year. Therefore he cannot announce his intentions beforehand, for then he would be unable to produce and sell anything at all between the announcement and the production of the new models.'

For a car which had basically changed little in appearance since its inception in 1915, this was no modification, but a complete re-design of the suspension, chassis and body, with the added benefit of a price reduction which brought (for example) the cheapest Cowley two-seater down to a record low of £148 10s. The engine, gearbox and rear axle were about the only components to remain unchanged. Starting with the frame, the designers had made this more rigid by introducing a deeper channel section, helped by the running board irons running transversely on the underside. The rear of the chassis was up-swept to accommodate the rear axle, which was fitted with long under-slung semi-elliptic springs in place of the Bullnose's three-quarter elliptic rear springs. A more direct system of operation for the 9in diameter drum brakes was instituted, each brake having individual adjusting wing nuts at the ends of the brake rods. An all-steel dash which, together with its cubby holes flanking the oval instrument panel, formed part of the rigid superstructure carrying the seven-gallon gravity-feed petrol tank. A two-way petrol tap on this tank isolated a reserve gallon which did away with the need for a petrol can and carrier on the running board.

The immediately noticeable change was, of course, the flat radiator which allowed an easier match to the bonnet and square lines of the new bodies. William Morris, who disliked the idea of changing the shape of his 'Bullnose' radiator anyway, was aghast when he saw the first 'Flat' prototype, likened it to a gravestone and insisted that it be made 2in narrower. The toolroom, with little time to spare before the 1926 Motor Show, set about modifying the press tools, working continuously for two days and two nights and stopping only for meal breaks. The modified pressings were more to Morris's liking and he gave the men involved a £5 bonus. These early radiator shells had very sharp edges making mass production difficult, so the press tools were subsequently altered once more (at a more leisurely pace) to produce a shell with larger radii on the bends.

Mudguards, although of a similar pattern to previous Cowleys, changed from a radius to a sloping section from about the 11 o'clock position eventually to join the front of the running board, while the adoption of a splash guard below the radiator further altered the frontal appearance.

Both clutch and brake pedals were adjustable for reach and by means of alternative fixing holes in the steering column support bracket it was possible to change the steering column rake. The Lucas electrical equipment, still an insulated return circuit, had been simplified for servicing by the use of an accessible separate cut-out/fuse box. Not quite so accessible was the position of the battery, located in a cradle inside the frame forward of the driver's seat.

Morris Cowley in 1932 two-seater form. (Eric Miller)

One result of the re-designed Cowley chassis frame was an increased wheelbase of 3in. Wheels were still of the pressed steel artillery type with three-stud fixing (the 1927 model Oxfords being increased to five-stud fixing) with 27 × 4.40 Dunlop reinforced balloon tyres; combined with the 'all-British' Smiths adjustable shock absorbers and long springs, the suspension was described as providing 'ideally comfortable riding and road holding qualities'. The spare wheel was carried on the near-side running board.

The 1927 Cowleys were listed in two groups. The 'popular' models finished in grey in two-seater and four-seater form, had rear-wheel brakes and simplified equipment. The second group comprised a two-seater and four-seater in

Morris Cowley in 1927 two-seater form.

grey or blue, and a fixed-head coupé and two-door saloon in blue finish only. Four-wheel brakes and full equipment were standard fittings to vehicles in this second group. Commercial variants were the commercial traveller's car with two doors, and the standard 8cwt van (still retaining its curious squashed bullnose radiator). Both commercials had the option of front-wheel-brakes and full Cowley equipment.

'Simplified equipment' referred to the Lucas R515 combined double-filament head/side lamps mounted on the wings and the absence of clock, speedometer, spring gaiters etc; otherwise the upholstery, trimming, etc was common to both front-wheel-brake and rear-wheel-brake-only versions. On the two-door four-seater the front seats were of the semi-bucket style, adjustable for position and hinged on the passenger's side to allow access to the rear seat. A similar arrangement was to be found on the closed models, which had doors on both sides. By January 1927, Morris Motors (1926) Ltd listed a four-door version of the Cowley four-seater, in time, as it happened, for such a tourer to become the 200,000th Morris car to be made.

Naturally some coachbuilders installed special bodies on the Cowley chassis. As early as 1916 Stewart & Ardern had sold a Continental engined Cowley with a saloon body. Doors, in the reverse order to those on the later Morris bodied saloon models of the Cowley, were mounted one on the off-side front and one on the near-side rear. This car was restored about 1965. Stewart & Ardern were to continue supplying special bodies on the Morris chassis, subcontracting the actual bodybuilding to firms such as Hoyle Body Co Ltd (successors to Chalmer & Hoyer) of Weybridge. Stewart & Ardern Cowley specials included a two-seater boat-tail and four-seater sports models of 1924-5, available in either aluminium or mahogany panelling. Many Stewart & Ardern designs appear to have inspired Morris to produce similar standard models later. A case in point is the S&A fabric bodied three-door Cowley traveller's saloon of 1929 with a large door at the rear and removable rear seat to enable goods to be carried. This type of saloon was subsequently added to the Morris programme of 1930 with a coachbuilt body. Willys Overland Crossley Ltd of Stockport also liked the idea for by 1931 they were offering a 'Willys Commerce Saloon' with rear opening door on the 15.7hp six-cylinder (£198) and the 15/40hp four-cylinder chassis.

*Evolution of the 8-10cwt
1928 model light van.*

*8-10cwt light van, 1929
model.*

*8-10cwt light van, 1931
model. Note the introduction
of the flat radiator.*

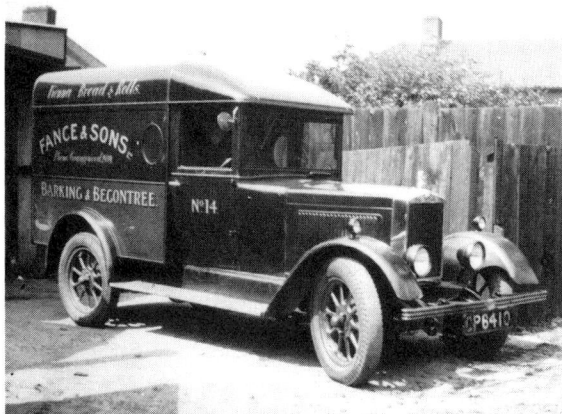

8-10cwt light van, 1932 model. (Pat Weale)

8-10cwt light van, 1933 model with larger capacity body redesigned.

8-10cwt light van. 1935 model with redesigned chasis and body of 90cu ft capacity.

Carnival decorated special bodied light van chassis in the shape of a Daren bread loaf. These advertising vehicles were supplied by John C. Beadle Ltd, Dartford Morris Dealers, in 1928.
(Pat Weale)

Several firms offered special conversions for the Cowley in the late 'twenties, consisting of either two separate bodies (tourer and van) bolted into position as required, or as a structure which fitted on top of the existing open body to create a van. Of these firms, Messrs Holtby White & Co of Bridlington, Wilde & Bennett of Hadfield, near Manchester, Ellison & Smith of Gatley (also near Manchester), and Whitehead & Furness of Dukinfield are noted, suggesting something of a demand for these conversions in the North of England. It appears that to comply with the law of the time, a car used alternatively as a commercial and private car had to be licenced as a commercial vehicle only.

A Morris Cowley figured prominently in the trial of Browne and Kennedy for the murder of a policeman in September 1927, resulting in their execution in May 1928. They had stolen a blue four-seater from a Dr Lovell's house in Billericay, Essex, and on the journey home to London they killed Constable Gutteridge with a revolver at point-blank range when he stopped the car on the Romford-Ongar road in the early hours of the morning. The Cowley (TW6120) was later found abandoned outside a house in Foxley Road, Brixton. Another Cowley, this one a Bullnose, to make the news headlines in the 'twenties was an open model belonging to the mystery writer Agatha Christie. She had left her home in her Morris Cowley on the night of 3 December 1926 and disappeared. Twenty miles from her home in Surrey the police found the car hanging over the edge of a 120ft deep chalk pit with the handbrake off and the gearbox in neutral. Hundreds of police with a number of dogs searched the area without success. Eleven days later she was discovered at a hotel in Harrogate, registered under the name of Teresa Neele. Right up to her death several years ago, Agatha Christie stuck to her story of suffering from amnesia, thus leaving behind a real life mystery.

The Morris Cowley range for the 1928 season offered a four-door four-seater saloon in addition to the two-door model. Prices were £185 and £177 10s respectively. A feature of note on this new saloon was the single-panel windscreen (but not on the two-door saloon which continued to have the split windscreen) hinged at the top and adjusted by means of quadrant fasteners at the sides. Also provided were a Lucas vacuum windscreen wiper, carpets, roof lamp, a roof net above the front seats, doors with expanding pockets and winding windows in the doors, although the rear windows were fixed. The

1927 Morris Cowley. This four-seater was the actual car stolen and used by the murderers of PC Gutteridge. Note the sharp edges of the early flat radiator surround. (Essex Constabulary)

Morris Cowley two-seater receiving assistance on the 1 in 4 Park Rash hillclimb near Kettlewell, Yorkshire. The original price of this 1928 model was £152 10s.

choice of wheel and tyre size is interesting, for although all other Cowleys had 27 × 4.40 tyres, the new four-door saloon had larger 28 × 4.95 tyres. The two-seater was the only 1928 Cowley offered in 'simplified' rear-wheel-brake-only form, or as a standard version with brakes all round. Apart from the van (now reclassified as 10cwt and fitted with full doors to the cab, with winding windows) and the commercial traveller's car, the other variants on the Cowley chassis were the four-seater now obtainable in blue or beige, and the 'Three-quarter Coupé' with rear quarter lights. For the saloon, four-seater and coupé there was an ingenious but simple means of adjusting the reach, tilt and angle of the front bench seat involving the use of straps fitted either side to support the back squab.

The 1928 programme introduced an 11.9hp version of the Oxford, which combined the 'Oxford roominess with the Cowley running costs'.

For the following season (1929) the bumper was a standard fixture on Morris cars. Although the range was restricted (commercials excluded) to a two-seater, four-seater, saloon and coupé, the colour range was considerably extended with a choice of two-tone cellulose finishes of stone/brown, blue/black, or stone/maroon on the closed models. The first colour was used on the super-structure; wheels, mudguards and aprons, as always were stove enamelled black. The side-strapped front bench seats of the previous season did not last long, being replaced by adjustable bucket seats on runners with pockets on the backs. Other improvements were dipping headlights, optional safety glass, improved lines of the metal-framed khaki hood and, in addition, the all-weather equipment added two quarter lights on the tourer. The inclusion of a Lucas Sparton electric horn suggests that the bulb horn had been superseded and, indeed, spares lists (for chassis after 268431) confirm this. However, many period photographs of 1929 season cars show the bulb horn fitted. As well as the new bumpers, a major change in appearance was the redesigned one-piece domed mudguards reinforced by the use of an inverted 'U' section bar spanning them and providing a useful mounting for the Lucas R40 headlamps.

Bodies on the 1929 Morris cars had lost some of the sharp corners and taken on a more rounded look, accentuated by the domed mudguards. Morris Motors described the saloon as a 'dome-back' and this rounding-off of the body shape was particularly noticeable on the coupé model '. . . particularly suited to the lady driver's requirements'. The commercial variants of the Cowley included the commercial traveller's car, up dated with the equipment to be found on the private cars, and the 'Morris Light Van' (as it was now described) still retaining the 'squashed Bullnose' radiator shape, but with a re-designed rear which brought the bodywork down to the level of the forward section of the body thus giving an additional $3\frac{3}{4}$cu ft capacity. Some of the new Cowley features were not incorporated in the light van, for example, no bumpers were fitted and the headlamps and mudguards followed the fashion of the previous season, so that the former were still mounted on individual stalks. However, a single-panel adjustable windscreen was fitted.

Although quaint and seeming crude, but in practice quite efficient, the dipping system fitted to the Cowley headlamps never fails to intrigue the modern motorist. This Lucas device, first introduced by them in 1927, was rather like a

The 11.9hp Morris Cowley traveller's car of 1929. Original price was £167 10s. (Morris Motors Ltd)

pump fitted to the steering column and by means of a pumping or sucking action via a rubber tube the driver was able to move the pivoted reflector to the desired full or dipped position.

Centric Superchargers Ltd of Ribble Bank Mills, Preston, used a 1929 Morris Cowley to test the new Centric supercharger which they introduced in 1934. The four-year-old Cowley was standard in every respect save for special road springs and shock absorbers. During a run with a checked speedometer the maximum speeds were shown to be 45mph in second and nearly 70mph in top gear. Acceleration figures were given as 20-40mph in 5.6 seconds in second, and 20-40mph in 34.6 seconds in top gear. Using all gears 50mph was reached from a standing start in 17.6 seconds. On Saturday 24 May 1939 a race was organised at Brooklands for standard Morris Cowley cars. A 1926 'Chummy' model driven by H.R. Webster won the race at an average speed of 48.83mph. His car, with 50,000 miles on the clock, and in normal circumstances a means

Display chassis of the 1929 Morris Cowley which was presented to the Science Museum, London, by Morris Motors Ltd. (Science Museum)

Two-seater Cowley in use in Cairo. The car is a 1929 model. (Barry W. Watson)

of transport for a traveller, covered the flying lap at 51.34mph.

Of special interest are the wide-track models made specially for the export market, notably Australia, in the late 'twenties and early 'thirties. These Cowleys, with a 56in track (the normal models wer 48in) were made in reasonably large numbers for the 1928-32 seasons; a fact substantiated by the figures from the records of G.M. Holdens Ltd, which show that for the years 1927-30 totals of 2,737 tourers, 919 two-seaters (roadsters), and 91 saloons on the Cowley wide-track chassis, were bodied at their South Australian Woodville plant. They differed from the home product in having the same axle ratios as the Empire Oxford, but with the front-brake cross-shafts above, rather than below, the axle, for greater ground clearance. Many of these wide-track Cowleys (but not all) appear to have been fitted, for climatic reasons, with a taller and wider radiator mounted on the chassis cross member, which had a pressed-in

Cowley fixed-head saloon of 1929. The visor was a standard fitting on the fixed-head saloon. (John Hagger)

depression to avoid the higher bonnet line. Australian tourer bodywork built on the Cowley and Oxford flatnose chassis were very American in appearance, with heavier and more serviceable looking side screens; the hood overlapped the top of the windscreen with a pronounced peak. The rear window for the hood was of framed glass rather than the usual celluloid, no doubt in such a sunny climate the ultra-violet would reduce the life of the celluloid. Despite the wide track the bodies tended to be the same width as the home models, the additional width being absorbed by wider mudguards and running boards. Another feature to be found on these wider chassis was the arrangement of the rear road springs, which did not run beneath the frame side members but ran alongside, with the rear shackles on outriders at the end of a bar running across the chassis rear end.

The first wide-track flatnose cars were made for the 1927 season on the 14/28 Oxford chassis and it was not until the following model year that the Cowley received similar treatment, the latter having the smaller 11.9hp engine. From the 1929 season onward the larger 13.9hp engine from the Oxford was combined with the Cowley chassis to become the Cowley 14/28, 56in track model, and the Oxford was dropped.

Many of the wide-track Cowleys were made as complete vehicles for export outlets such as South Africa, New Zealand, India, and Argentina, but few complete vehicles were imported into Australia due to the heavy import duties imposed. Apart from four tourers and one saloon in 1929, all the wide Cowleys imported into Australia were bodied by the vigorous Australian motor body industry. A total of 1,210 11.9hp and 3,139 14/28 Cowley wide-track chassis went to Australia between late 1927 and 1932. Initially, for the 1928 season, the export 11.9hp 56in track, front-wheel-braked Cowley, was offered with the same choice of bodywork as the United Kingdom market; that is, two-seater, tourer, coupé or saloon. With the introduction of the 14/28 wide-track Cowley for the 1929 season, the choice was limited to tourer and saloon.

Introducing the programme for 1930, Morris Motors announced that: '. . . for 1930 two important and far reaching improvements have been introduced to the entire Morris range of cars. They comprise chromium plating on all external bright parts and Triplex glass on all models.' Cynics may point out that chromium plating on brass was cheaper than the nickel based metal previously used (William Morris is alleged to have said that the radiator could be polished until you reached the water), and anyway, new regulations were to make safety glass compulsory on all new cars. In practice, some of the bright parts, such as the door handles, being fitted to 1930 models were still nickel plated, no doubt as a means of using existing stocks. The new chromium-plated models for the 1930 season on the Cowley chassis included a commercial traveller's saloon to replace the two-seater traveller's car. This was in essence the season's fixed-head Cowley saloon with the addition of a side-hinged wide door in the rear panel of the vehicle to allow loading once the detachable rear seat had been removed, the extra door accounting for the additional £15 on the list price of the fixed-head saloon. A folding-head version of the saloon was new for 1930, identified by the absence of the external metal sun-visor over the windscreen, and this same 'Kopalapso' folding roof (manufactured by Donne

*Seven years' restoration work by Harry Diamond, of
County Antrim, brought this 1930 Cowley fixed-head
saloon back to its original condition.* (H. Diamond)

& Willans Ltd) working on the 'lazytongs' principle was now also fitted to the
Morris Cowley coupé. There was little to distinguish between the 1929 and
1930 open models, the absence of any superstructure dictating the continued
use of the horizontally split windscreen. For the commercial market the light
van continued unchanged in design or price (£165), although the pedantic may
point out that the decorative wings for the Calometer had been dropped from
the specification.

As with the 'Bullnose' so with the 'Flatnose' radiator, both terms of affection
coined in later years. The end of 1931 was to see the demise of the flat radiator
on the cars in favour of the more rounded radiator of the 'thirties, heralded, it
could be said, by the curves given to the previous year's Isis. Frankly, with all
other models now sporting wire wheels as standard, the Cowley was beginning
to look decidedly old fashioned on its artillery wheels, differing little in
appearance to those fitted to the first Cowley some sixteen years before. The
choice of Cowley models for the 1931 programme was the same, but some
attempt to up-date the design had been made. Externally, this manifested itself
in the taller radiator (with a new winged badge), but to retain a horizontal line
for the bonnet yet not increase the overall height of the vehicle, the windscreen

was made shallower, giving the vehicle an illusion of sitting lower on the road. Mechanically, the annual minor changes had been made involving a modified cylinder head, provision of group lubrication nipples on a panel above the differential casing, and a single wing nut compensation for the foot brake gear. As a further concession to fashion wire wheels were offered as an optional extra, orginally on all models, but by July 1931 this type of wheel was made standard equipment on the Cowley folding-head saloon.

The Morris light van (variously rated in Morris publications as 8cwt, 10cwt, and 8/10cwt), also up-dated with its less severe roof line was, nevertheless, following the apparent Morris policy of lagging behind the private car as an outlet for the obsolete components. The 1931 season's models had at last lost its 'squashed Bullnose' radiator and received the short flat radiator with a round badge, but retained the old style mudguards and the stalk mounted headlamps. The major change, however, was that a modifed version of the larger 13.9hp Oxford engine had been substituted for the smaller 11.9hp unit. The van body was constructed of ash framing assembled in master jigs and panelled on the flat surfaces with 4mm three-ply board faced with lead coated sheet steel, called Plymax. Curved surfaces were pressed out of 22-gauge sheet steel. The interior of the van body was painted a light stone colour and the outside coated in a shop grey primer in readiness for the purchaser's own colour and livery. One subtle change to the 1931 body was the inclusion of ventilating louvres in the rear doors. Wire wheels were not optional on the light van, but coachbuilders and others could still purchase the van chassis with the lower horsepower engine.

Considerable use was made of the Morris Cowley open models by the Metropolitan Police in the early 'thirties. For example, the Motor Patrol fleet in 1931 consisted of 119 vehicles made up of forty-five Cowley two- and four-seaters, eight BSA three-wheelers, and the remainder solo and combination motorcycles.

Two coupé variations were available in the 1932 Morris programme, a 'Sports Coupé' and the four-light model shown here. (Morris Motors Ltd)

Just prior to the Motor Show at Olympia in October 1931 the Morris programme for the 1932 season announced the Cowley in its new modern image. A
completely re-designed body and chassis incorporated longer springs, Lockheed hydraulic brakes, 'organ' type accelerator pedal, a new type radiator,
Magna wire wheels, and, where applicable, a Pytchley sliding-roof. Although
the 'Hotchkiss' type engine and three-speed gearbox were retained, even here
modifications had been made. The cylinder head now carried the combined air-
cleaner, pre-heater and fume-consumer linked to the SU carburetter; and in
order that coil ignition might be incorporated in an engine designed for
magneto working, an ingenious device was built to fit the magneto mounting
and linkage, the unit (Lucas A-12-O) containing the coil and necessary contact
breaker apparatus. A reversal of the Oxford arrangement in 1928 now applied
for the new Cowley engine capacity was 13.9hp or 11.9hp to choice.

The slump and high unemployment figures had their effect in the early
'thirties on the total British production of cars and this dip in the curve was
reflected in the sales of Morris cars. In terms of units produced and pre-tax
profits, 1931 was to prove a disastrous year for Morris Motors Ltd. A car
suitable for the export market was not necessarily ideal for the home motorist,
particularly in view of taxation based on horsepower, and a consequence of
attempting to satisfy the differing demands was to be seen at the end of 1932,
when Morris were listing no less than nine basic types with twenty-two
different styles ranging from the 8hp Morris Minor to the 17.7hp Isis. Another
factor was the general trend away from the open car, so it is not surprising to
find that for the 1932 season the four-seater had been dropped from the Cowley range, and now comprised the two-seater (£165), saloon (with or without
sliding-head at £179 10s and £185 respectively), commercial traveller's saloon
(£195), coupé (£190), and a new 'Sports Coupé' model (£215). Prices remained
the same regardless of engine capacity and as a further incentive the guarantee
period was increased from twelve months to two years.

With the 'hansome new radiator', carrying a shield shaped badge and a
Wilmot Breeden 'Regent' Calormeter with matching 'President' wings, came
many design innovations. The closed models had what was termed an 'eddyfree' front, referring to the absence of any protrusions such as peaks or visors
above the windscreen which would materially increase the wind resistance.
Another important advance was the re-location of the 7-gallon petrol tank at
the rear of the chassis, with an SU Petrolift under the bonnet feeding the SU
carburetter. Also under the bonnet was housed a comprehensive tool kit in
either a metal box or, in the case of the larger items such as the wheel brace,
pump, and jack, secured with special clips or spring bands. With the
introduction of five-stud Magna type wire wheels the Dunlop tyre size, 29 ×
500 (5.00-19), was standardised for all models; the spare was carried in a
pressed well in the near-side front mudguard. Changes in the four-wheel foot
braking from the previous season's mechanical arrangement to a Lockheed
hydraulic system dictated an increase in diameter of the brake drums from 9in
to 10in, while the linkage from the handbrake lever to the rear wheels was by
cable. Contemporary tests of braking efficiency of the hydraulic system on the
1932 Cowley compared with the 1931 model showed that while the earlier car

required 120ft to stop from 40mph, the hydraulic braked vehicle stopped within 75ft. The overall height of the closed cars had been reduced by 2in which, when added to the visually prominent twin-blade bumpers fitted fore and aft, suggested a much longer and lower car, despite only 1in increase in the wheelbase. Internally, the trim was of Karhyde, and (except on the sports coupé) pleated on the squab and seats of the adjustable front arm-chair type seats, rear seats, and door pockets. The entirely new steering wheel had three spokes with the controls for ignition and the dipping of headlights in the centre. The Lucas LB140 headlamps had an electrically operated dipping reflector on the near-side combined with contacts which, on dipping, extinguished the off-side headlight. Special to the close-coupled Cowley sports coupé was the Celstra leather upholstery in a choice of blue or brown (all the other Cowley models for 1932 were restricted to brown), rope pulls, and remote door-handle control. Inbuilt luggage containers were fitted at the rear of both the four-light coupé and the two-light sports coupé.

Restored example of the 1933 Cowley, complete with Wilcot Robot direction indicators. (Chris Creevy)

1933 Cowley side-valve engine. 11.9hp with air cleaner/fume consumer head. (John Lowrie)

None of this applied to the 1932 light van, which was the mixture as before with artillery wheels and the original rod-operated brakes. True, the electric horn button had moved from its 1931 position on the driver's door to a more convenient spot on the steering column, together with a dimming switch, and the specification quoted oval windows in the rear doors with no mention being made of similar side windows although, if surviving examples are any guide, they were fitted, for the last time, on these 1932 vans.

Once the 1933 season was well under way it was evident that the once popular Cowley was becoming less popular than the new Morris Ten. If the chassis numbers can be used as a guide then the sales of the Cowley were halved compared with the 1932 figures. The cost of tools and jigs for the 1932 re-design had to be recouped so it is not surprising to find that the 1933 Cowley models were unchanged in style or model types, although there were the expected annual modifications. Foremost was a further alteration to the engine (now only listed as an 11.9hp 'CL' unit for the home market, although export cars continued to have the optional 13.9hp 'CJ' engine) where a distributor, driven off the chain drive camshaft, allowed a conventional coil ignition system to be used. A 'twin-top' four-speed gearbox was now fitted, but the light van still had a three-speed box. As with all the larger 1933 season Morris cars, the Cowley front wings had taken on sideshields with internal gutters to carry away water and prevent splashing, and outrigged Wilcot indicators (later superseded by Lucas trafficators) were standard fitments.

The extremely strong and serviceable pressed-steel artillery wheels continued as standard for the Morris light van (perhaps the words 'old-fashioned' should be avoided as this type of wheel was still in use by Nuffield sixteen years later on the Oxford Taxicab!). Much work had gone into the 1933 van body, now well rounded in appearance — a compromise between the flat radiator and the rest of the body, and combining the previous season's domed wings, lamp bar with centre-mounted horn, and 'eddyfree' radius above the single-panel adjustable windscreen. Underneath, of course, it remained the old Cowley chassis with rod-operated mechanical brakes. Bumpers were not fitted on the light van, nor for that matter were they standard fittings on the cheaper fixed-head or traveller's version of the saloon.

Morris had decided on a thorough reorganisation of the factory in 1933, which he estimated would take about a year to complete. The variations of the models in the 1934 programme suggests that the new facilities were soon being utilised. For the first time the name Cowley was applied to a six-cylinder car giving the intending Morris purchaser of 1934 a choice of the Ten, Cowley, Oxford, Isis, and Twenty-Five with six cylinders. The four-cylinder Cowley continued, but in saloon form only and although similar in style, the Cowley-Six had a wider track and a wheelbase 4in longer, the same dimensions as the Morris Major which it replaced.

A completely new engine was designed for the 1934 Cowley four-cylinder model which came in saloon form only, with the option of fixed or sliding-head. The bore of 69.5mm and stroke of 102mm of the new 'TA' engine followed the dimensions of its 11.9hp predecessors, but a slight increase in maximum revs and other characteristics increased the power output from 27.7

to 34bhp. In a way the 1934 Cowley was a test bed for the new engine which would find its home, with various minor modifications, in subsequent Morris vehicles such as the Twelve-Four of 1935 as the 'TB' and 'TE' (and 'TG' for the left-hand drive cars), the 10cwt van version of the same year ('TF'), the three-speed version of the Series II Twelve-Four 'TJ', and in 'TK' form in the Series II and Series Y 10cwt vans: the latter well into the post-war years. The new 'TA' engine had at last discarded the Lucas dynamotor (the Cowley also lost its characteristic sound), replaced by a separate Lucas $4\frac{1}{8}$in diameter starter motor and a Lucas $4\frac{1}{2}$in diameter dynamo. Other changes included four-ring aluminium pistons in closer bores, a larger sump capacity, distributor (with automatic advance/retard) driven off the oil pump shaft, and higher valve lift. The smooth action of the cork-insert clutch running in oil remained, but it was now reduced to a single plate mated with a four-speed synchromesh gearbox.

The four-cylinder car, following the trend of the time, had its radiator further forward, almost hiding the dumb irons. This tended to nullify any reduction in the spacious body suggested by the shorter (by 3in) wheelbase of the new reinforced chassis frame. Bumpers were of the single bar pattern and the Lucas trafficators were concealed within the door pillars. The Minor apart, almost all the Morris cars for the 1934 model year had the headlamps mounted on a round bar spanning the front wings and providing a central location for the Lucas Altette horn. The Cowley was no exception, although the bar formed a higher arc than, for example, the Tens. Internally, real leather had at last replaced the Karhyde for upholstery and other new-for-the-year fitments to the Cowley included the use of a SU electric petrol pump and a new design of handbrake lever with thumb release.

Considerable changes were made to the appearance of the light van — or to use the correct 1934 season nomenclature, the '8/10 Cwt Van' — which, because of the use of the previous season's chassis design, had a longer (105in) wheelbase than the car. Although devoid of bumpers the frontal aspect followed that of the 1933 Cowley with the lamp support tube horizontally across the front of the radiator, and centre-mounted horn. The most out-standing contribution to the new image was the use of 19in diameter, five-stud, Magna wire wheels. The spare was mounted on the rear of the near-side front wing. There was now Lockheed hydraulic brakes all round and cables for the hand-brake linkage to the common rear brake shoes, Armstrong shock absorbers, and a rear mounted petrol tank in conjunction with a SU Petrolift (there is reason to believe that late in the season an electric SU petrol pump was substituted). For £3 10s on top of the ex-works price of £160, purchasers could obtain the van in a choice of cellulose finishes as well as the basic shop grey undercoat. The 13.9hp power unit on rubber mountings for this season was a modification of the side-valve engine ('CS') from the 1933 Cowley, having a chain-driven camshaft and coil ignition, mated to a three-speed gearbox. The use of this 'Hotchkiss' based engine and an earlier chassis probably accounts for the continuation of the earlier Cowley series of chassis numbers (382398 onwards) at a time when the contemporary car version had started a new series carrying the year prefix at 34/C501.

By this time most of the Morris models were being produced in both right-hand-drive and left-hand-drive form. In the case of the 8/10cwt van this export version had certain additions such as an engine undershield to protect the sump and the provision of an impeller in the cooling system to assist the fan.

1934 was the last year for the 'Cowley' name until its post-war resurrection for the 10cwt van in the 1950s, followed in 1954 by the 1,200cc saloon. In the mid-1930s both Morris and Wolseley moved away from model names in favour of numbers denoting the horsepower rating and number of cylinders, such as the 'Ten-Four', 'Fifteen-Six', etc. (MG on the other hand, had a preference for letter designations, after starting in the 'twenties with numbers.) The abandoning of the RAC horsepower formula and the substitution of a cubic capacity tax in 1946, turned manufacturers back to model names, as engine capacities of large denominations are rather a mouthful for colloquial use. Model names considered by the Cowley management in those immediate post-war years were 'Iffley', 'Carfax', 'Marston', 'Littlemore', 'Mosquito' (the name originally allocated to the Morris Minor designed by Alec Issigonis, but never used), 'Cygnet', and 'Gazelle' (later used by Singer in the 'sixties). In the end they chose a revival of the old names of Minor, Oxford, and Cowley, when the first post-war designs were produced. Later, in 1954, the name Isis was reused.

But to return to 1935, there was the new season's Morris Twelve-Four which, despite its name was the Cowley replacement in what was to prove a short model year before the 'Series' models came on the scene. The Twelve-Four was, in fact, the previous season's Cowley with minor gimmicks such as the ingenious window winding mechanism which, if operated beyond the point where the window had reached the top of its travel, would slide the glass

Successor to the Cowley was the 1935 Morris Twelve-Four, available as a saloon only. (W.E. Spapé)

backwards an inch or so opening a ventilation gap between the glass and the forward pillar. Internal changes appear to have been confined to a change in the shape of the rear seat arm rests, a driver's 'Rollsvisor' sun blind, and a right-hand accelerator pedal. The obvious features of identification were the repositioning of the round headlamp mounting bar (which passed through the radiator shell but forward of the radiator proper) and the rear bumper with a centre section which retracted when opening out the luggage carrier. Electrical equipment improvements made the original 'summer/winter' charging switch obsolete, as the compensated voltage-control dynamo automatically regulated the output to suit the demands made upon the battery; the latter could now be isolated by means of a safety master switch.

At last the 8/10cwt van had caught up with the car specification having almost all the major components interchangeable with the 1935 Twelve-Four. Custom or necessity dictated some variations such as the absence of bumpers, the location of the side-mount spare wheel, and a flatter radiator shell; although even here the lamp bar followed the car design by passing behind the radiator shell. The air filter/fume consumer was not deemed necessary on the commercial model, but the engine was basically the same four-cylinder side-valve of 69.5mm × 102mm stroke, thus bringing the van engine back into the 11.9hp category.

Morris Cowley and Twelve-Four. 1927-35. Chassis numbering	
Model year	Chassis numbers
1927 cars and vans	156501-215000 *
1928 cars and vans	215001-268430 *
1929 cars and vans	268431-313896 *
1930 cars and vans	313897-341406
1931 cars and vans	341407-358192
1932 cars and vans	358193-373499
1933 cars and vans	373500-382397 (vans start at 373525)
1934 cars	34/C501-34/C7120
1934 vans	382398- . . .
1935 cars	35/TW7721-35/TW15970 **
1935 vans	35/TWV7721-35/TWV15970 **

*   Numbers shared with Oxford four-cylinder models.
** Numbers shared with Fifteen-Six models.
Note: Numbers quoted above are actually those given as 'Car No' on the identification plate of the vehicle. These numbers are quoted by Morris Motors Ltd in their spares lists under the heading 'Chassis No'. The actual chassis (frame) number, on cars and vans up to and including 1933 models, as stamped on the dumb iron, is always 3,000 numerically less than the 'Car No' given on the identification plate. This difference probably also applies in the case of 1934 vans.

Morris Cowley and Twelve-Four. 1927-35. Engine Data		
All engines: Four-cylinder. Water cooled. Three-bearing crankshaft.		
Model year	Engine type	Specification
1927-31 cars 1927-30 vans		11.9hp, 69.5mm bore × 102mm stroke, 1,548cc. Aluminium alloy pistons. Spiral-gear driven camshaft. Compression ratio 5:1. 28bhp at 3,000rmp. 1927 models, Smiths 5-jet carburetter. 1928-31 models, SU carburetter. Lucas dynamotor. Magneto ignition. Clutch comprising two cork insert plates running in oil. Three-speed gerbox.
1931 vans		13.9hp, 75mm bore × 102mm stroke, 1,802cc. Spiral-gear driven camshaft. SU carburetter. Lucas dynamotor. Magneto ignition. Clutch comprising two cork insert plates running in oil. Three-speed gearbox.
1932 cars	CH	11.9hp, 69.5mm bore × 102mm stroke, 1,548cc. Aluminium alloy pistons. Spiral-gear driven camshaft. Compression ratio 5:1. 27.75bhp at peak 3,400rpm. SU carburetter. Lucas dynamotor. Coil ignition unit DC4, spiral-gear driven off camshaft gear. Clutch comprising two cork insert plates running in oil. Three-speed gearbox.
1932 cars	CJ	13.9hp, 75mm bore × 102mm stroke, 1,802cc. Aluminium alloy pistons. Spiral-gear driven camshaft. Compression ratio 5:1. 34.75bhp at peak 3,150rpm. SU carburetter. Lucas dynamotor. Coil ignition unit DC4, spiral-gear driven off camshaft gear. Clutch comprising two cork insert plates running in oil. Three-speed gearbox.
1932 vans	CG	13.9hp 75mm bore × 102mm stroke. 1,802cc. Spiral-gear driven camshaft. SU carburetter. Lucas dynamotor. Coil ignition unit DC4, spiral-gear driven off camshaft gear. Clutch comprising two cork insert plates running in oil. Three-speed gearbox.
1933 cars	CL	11.9hp, 69.5mm bore × 102mm stroke, 1,548cc. Aluminium alloy three-ring pistons. Chain driven camshaft. Compression ratio 5:1. 31bhp at peak 3,400rpm. SU carburetter 1⅛in diameter. Air cleaner and fume consumer. Lucas dynamotor. Distributor off camshaft, coil ignition. Water impeller on export models. Clutch comprising two cork insert plates running in oil. Four-speed 'twin-top' gearbox.

1933 cars (Export)	CM	13.9hp, 75mm bore × 102mm stroke, 1,802cc. Aluminium alloy three-ring pistons. Chain driven camshaft. Compression ratio 5:1. 36bhp at peak 3,300rpm. SU carburetter $1\frac{1}{8}$in diameter. Lucas dynamotor. Coil ignition with distributor driven off camshaft. Clutch comprising two cork insert plates running in oil. Four-speed 'twin-top' gearbox. Air cleaner and fume consumer.
1933 vans	CN	13.9hp, 75mm bore × 102mm stroke, 1,802cc. Three-ring cast iron pistons. Chain driven camshaft. Compression ratio 5:1. 36.3bhp at 3,200rpm. SU carburetter $1\frac{1}{8}$in diameter. Lucas dynamotor. Coil ignition with distributor driven off camshaft. Clutch comprising two cork insert plates running in oil. Three-speed gearbox.
1934 cars	TA	11.9hp, 69.5mm bore × 102mm stroke, 1,548cc. Aluminium alloy four-ring pistons. Chain driven camshaft. Compression ratio 5.65:1. 34bhp at 3,800rpm. SU carburetter $1\frac{1}{8}$in diameter. Lucas belt driven dynamo and Lucas starter motor. Coil ignition with distributor driven off oil pump shaft. Automatic advance/retard. Clutch comprising single cork insert plate running in oil. Four-speed gearbox with synchromesh in top and third. Rubber mounted engine. Air cleaner and fume consumer.
1934 vans	CS	13.9hp, 75mm bore × 102mm stroke. 1,802cc. Three-ring cast iron or aluminium pistons. Chain driven camshaft. Compression ratio 5:1. 36bhp at 3,200rpm. SU carburetter $1\frac{1}{8}$in diameter. Lucas dynamotor. Coil ignition with distributor driven off camshaft. Clutch comprises two cork insert plates running in oil. Three-speed gearbox. Water impeller on export version.
1935 cars 1935 cars, LHD 1935 vans	TB & TE TG TF	11.9hp, 69.5mm bore × 102mm stroke, 1,548cc. Aluminium alloy four-ring pistons. Chain driven camshaft. Compression ratio 5.65:1. 33bhp at 3,800rpm. SU carburetter $1\frac{1}{8}$in diameter. Lucas belt driven dynamo and Lucas starter motor. Coil ignition with distributor driven off oil pump shaft. Automatic advance/retard. Clutch comprising single cork insert plate running in oil. Four-speed gearbox with synchromesh in top and third. Rubber mounted engine.

Morris Cowley and Twelve-Four. 1927-35. Body Colours and Upholstery

	1927	1928	1929
Two-seater	*Simplified Model* Grey cellulose with leathercloth to match. *Four-wheel brake model* Grey or blue cellulose with leathercloth to match.	*Simplified model* Blue cellulose with blue leathercloth. *Four-wheel brake model* Blue or beige cellulose with leathercloth to match.	Stone-brown or stone-maroon cellulose with brown Karhyde. Blue-black cellulose with blue Karhyde.
Four-seater tourer	*Simplified model* Grey cellulose with leathercloth to match. *Four-wheel brake model (2 & 4 door)* Grey or blue cellulose with leathercloth to match.	Blue or beige with leathercloth to match.	Stone-brown or stone-maroon cellulose, all with brown Karhyde and khaki hood.
Coupé	Blue cellulose with blue leathercloth.	Blue or beige cellulose with leathercloth to match.	Stone-brown or stone-maroon cellulose with brown Karhyde. Blue-black cellulose with blue Karhyde.
Saloon	Blue cellulose with blue leathercloth.	*Two-door model* Blue cellulose with leathercloth. *Four-door model* Blue or beige cellulose with leathercloth.	Stone-brown or stone-maroon cellulose with brown Karhyde. Blue-black cellulose with blue Karhyde.
Commercial traveller's car	Shop grey.	Shop grey.	Shop grey.
Van	*8 Cwt* Shop grey.	*10 Cwt* Shop grey.	*Light Van* Shop grey.

	1930	1931	1932
Two-seater	Morris brown cellulose with brown Karhyde. Niagara blue cellulose with blue Karhyde.	Blue cellulose with blue Karhyde. Morris maroon cellulose with red Karhyde.	Blue, brown or black with brown Karhyde.
Four-seater tourer	Morris brown cellulose with brown Karhyde. Niagara blue cellulose with blue Karhyde.	Blue cellulose with blue Karhyde. Morris maroon cellulose with red Karhyde.	
Coupé	Morris brown cellulose with brown Karhyde. Niagara blue cellulose with blue Karhyde.	Morris maroon cellulose with red Karhyde. Blue cellulose with blue Karhyde.	*Coupé* Blue, brown or black cellulose with brown Karhyde. *Sports coupé* Blue-black, black, or beige-brown cellulose with brown Celstra leather. Grey-dove with blue Celstra leather.
Saloon	*Fixed-head* Morris brown cellulose with brown Karhyde. Niagara blue cellulose with blue Karhyde. *Folding-head* Morris maroon cellulose with Karhyde.	*Fixed-head* Blue cellulose with blue Karhyde. Morris maroon cellulose with red Karhyde *Folding-head* Dark maroon cellulose with brown Karhyde.	*Fixed-head and sliding-head* Blue, brown or black cellulose with brown Karhyde.
Commercial traveller's saloon	Niagara blue cellulose with matching Karhyde.	Blue cellulose with blue Karhyde.	Brown cellulose with brown Karhyde.
Van	*Light Van* Shop grey.	*Light Van* Shop grey.	*Light Van* Shop grey.

	1933	1934	1935 Twelve-Four
Two-seater	Blue or brown cellulose with brown Karhyde. Black cellulose with green Karhyde.		
Coupé	*Coupé* Blue or brown cellulose with brown Karhyde. Black cellulose with green Karhyde. *Special coupé* Green duotone or red duotone cellulose with matching leather. Grey duotone cellulose with blue leather. Black duotone cellulose with brown leather.		
Saloon	*Fixed-head and sliding-head* Blue or brown cellulose with brown Karhyde. Black cellulose with green Karhyde.	*Fixed-head and sliding-head* Blue or green with matching leather. Black cellulose with brown leather.	*Fixed-head and sliding-head* Green-black cellulose with green leather. Red-black cellulose with red leather. Blue cellulose with blue leather. Black cellulose with green leather.
Commercial traveller's saloon	Brown cellulose with brown Karhyde.		
Van	*Light Van* Shop grey.	*8-10 Cwt* Shop grey. Finish in blue, black, brown or green at extra cost.	*8-10 Cwt* Shop grey. Finish in blue, green, red or yellow at extra cost.

Morris Cowley and Twelve-Four. 1927-35. Magazine Bibliography

Model Year	Article	Magazine	Date	Page
1927	Morris Programme for 1927	The Morris Owner	9/26	p906-11
1927	Morris Programme for 1927	The Autocar	3/9/26	p371-3
1927	Cowley Saloon Road Test	The Autocar	20/5/27	p851-3
1927	Cowley Saloon Road Test	The Motor	21/12/26	p1021-2
1927	Cowley Four-door Tourer	The Autocar	4/2/27	p195
1928	Morris Programme for 1928	The Morris Owner	9/27	p908-11
1928	Morris Programme for 1928	The Autocar	2/9/27	p437-9
1929	Morris Programme for 1929	The Morris Owner	9/28	Supplement (5 pages)
1929	Cowley Saloon Road Test	The Autocar	1/3/29	p423-4
1929	Morris Programme for 1929	The Autocar	7/9/28	p499-500
1930	Morris Programme for 1930	The Morris Owner	9/29	p840-5
1930	Cowley Traveller's Saloon	The Morris Owner	2/30	p1514
1930	Morris Light Van	The Morris Owner	7/30	p616
1930	Morris Programme for 1930	The Autocar	30/8/29	p401-9
1931	Morris Programme for 1931	The Morris Owner	9/30	p847-51
1931	Morris Light Van	The Commercial Motor	2/9/30	p75
1931	Morris Programme for 1931	The Autocar	5/9/30	p445-8
1932	Morris Programme for 1932	The Morris Owner	9/31	p787-92
1932	Morris Programme for 1932	The Motor	1/9/31	p199-206
1932	Morris Programme for 1932	The Autocar	4/9/31	p391-5
1932	Cowley Road Test	The Motor	9/2/32	p46-52
1932	Cowley Saloon Road Test	The Autocar	29/1/32	p167-8
1933	Morris Programme for 1933	The Morris Owner	9/32	p709-16
1933	Morris Programme for 1933	The Autocar	2/9/32	p411-14
1933	Morris Programme for 1933	The Motor	6/9/32	p214-20
1934	Morris Programme for 1934	The Morris Owner	9/33	p819-26
1935	Morris Programme for 1935	The Morris Owner	9/34	p655-62
1935	Twelve-Four Road Test	The Practical Motorist	13/4/35	p787-9

Morris Minor 1929~34

There can be no doubt that the success of Herbert Austin's 'Seven' was responsible for William Morris's diminutive Morris Minor being developed to capture part of the new market; this was at a time when the trend at Cowley was to produce larger six-cylinder models to complement the established Oxford and Cowley to compete with American cars in the 'dominion' markets.

It is also clear that the writer of a news item in *The Motor* in May 1928 had the Austin in mind when he referred to the 'New 7hp Morris', mention being made to a prototype which had already been made at the Morris Commercial Cars works, at Soho, with a wheelbase of 78in and a track of 42in, smaller than the subsequent production version. Morris's aim was '. . . to produce a car that can be housed in the average tool-shed or motor-cycle shelter at the side of a suburban villa'.

Advance details of the new Morris Minor appeared in *The Morris Owner* the following month. A jingoistic caption to a photograph, showing William Morris beside a saloon version of the car, made play of the popular song of the period 'Yes Sir, That's my Baby'. The brief specification boasted an overhead camshaft engine developing 20bhp at 3,000rpm, coil ignition, wire wheels, three-speed gearbox, and full electrical equipment including two large headlights.

Wolseley, at Adderley Park, were responsible for the new Minor engine. Its design owed much to the techniques employed on earlier Wolseley engines, which, in turn, followed the general layout of the overhead camshaft arrangement originated in the big Hispano-Suiza aero engine made by Wolseley during World War I. A six-cylinder version of the overhead camshaft engine was also developed for the Wolseley Hornet which was introduced in 1930. The body followed the lines of the Morris Minor with a longer bonnet to cater for the extended block of the six-cylinder 1,271.3cc engine. The basic chassis was an elongated Minor type. Mr K.E. Davis was in the drawing office of Wolseley at the time and he recalls (in a letter to the *Wolseley Register Newsletter,* Spring 1979) the first Hornet with the new six-cylinder engine fitted into a Morris Minor chassis: 'We had to cut an arch in the dash to take the end of the cylinder block. When you accelerated the whole contraption almost became airborne.'

Luckily for the designers, fabric bodies were in vogue in the late 'twenties. The considerable expense involved in tooling-up for a steel body was avoided on the first models by evolving a light fabric covered saloon but this was not without its later problems, for in seeking to avoid sharp angularity on the corners a lot of felt and wool padding was employed, which rural owners of the cars were later to discover was ideal for rooks and other birds building their nests.

When the Morris Minor appeared at the Motor Show at Olympia in late 1928, the fabric saloon was joined by a four-seater tourer which was £10 cheaper than the saloon price of £135. Anticipating regulations already in the pipe-line, both models were offered with the option of Triplex safety glass at extra cost. Four-wheel cable-operated brakes were fitted and an interesting feature was the hand brake which operated external contracting shoes on a drum behind the gearbox. In fitting a dry plate clutch, Morris had deviated from their normal practice and found it necessary to warn owners that oil must not be poured into the clutch case.

No effort had been spared to make the best use of the available body space on the 78in wheelbase chassis. Entrance to the rear seats was made easy by arranging the back of the front bucket seats to fold down and then the whole to tip forward. Additional leg room for the passengers was achieved by use of foot wells either side of the transmission shaft. Even the space beneath the leather-cloth covered seats was not wasted as the battery rested in a container under the driver; while a box beneath the front passenger seat provided room for the very comprehensive tool kit supplied. On the inside of both doors large elastic topped pockets were fitted. Body apart, the tourer model differed only in that it had a horizontally split two-piece windscreen.

The dynamo served a dual electrical and mechanical function. The overhead camshaft drive was transmitted through the dynamo armature via bevel gears and the close proximity of the unit to the engine block necessitated a small diameter Lucas DEL dynamo with its consequent low output, resulting in discharged batteries in winter. A larger DDS type dynamo was subsequently fitted giving a 10-amp output at full speed, but this was only achieved by the introduction of a concavity in the engine block casting. Another problem was that of oil getting past the camshaft bevel gear drive and into the dynamo. Various modifications were made by Morris during the currency of the engine, but none of these (which included protective cowls, additional oil drain holes, modified gaskets, oil grooves, and felt washers) appear to have solved the problem completely.

1929 fabric bodies ohv Minor saloon. (Morris Register)

The potential of the lively little engine and the Minor chassis was immediately appreciated by Cecil Kimber, managing director of Morris Garages. The subsequently successful MG 'M' type Midget which made its début at the 1928 Olympia Motor Show was the result of lowered suspension, increased steering rake, an MG radiator shell, and minor adjustments to the chassis; the light, boat-tailed, plywood body was covered in fabric. Production actually began in April 1929. Later bodies were metal panelled and when the model ceased some 3,200 cars later, in 1933, other modifications had been made to this '8/33 MG Midget': Hartford friction shock absorbers, new design of inlet exhaust manifold, and larger capacity sump with cooling ribs. By 1930 the Minor-type brake drums with anti-squeal bands had been replaced with ribbed cast aluminium drums with steel liners and a Bowden cable arrangement for front brake operation. In addition, the transmission brake gave way to a system where the hand brake coupled all four wheels via the cross shaft.

That the first Midget chassis carried the Morris Garages telephone number suggests that someone, possibly Cecil Kimber, had a sense of humour when it came to numbers. The '33' of the designation '8/33' meant nothing and if the prospective MG owners like to think it referred to brake horsepower that was their business. Interestingly, the later Morris Eight was sold to the Australian market as the '8/40' with this cryptic designation appearing on the early radiator badges. Certainly the '40' was not bhp, if anything it could have been 40mph.

Despite what some historians may say, the Morris Minor was undoubtably a success when one considers that a fifth of all cars produced by Morris Motors Ltd in 1929 were Minors. The following year this proportion approached a quarter of total Morris production.

For the 1930 season a coachbuilt saloon with 'Kopalapso' folding head was added to the existing fabric saloon and tourer models. In addition, a 5cwt van, based on the same chassis, made its appearance. This must surely rate as one of the few small commercials to be powered by an overhead camshaft engine. The wider choice of body styles was not reflected in the cellulose colour finish as both open and fabric models were now only available in Niagara blue and the new folding-head saloon in brown. Morris Motors announced that '. . . in order

1930 coachbuilt ohv Minor saloon. (Morris Motors Ltd)

to speed-up production — for despite a factory working day and night the demand for these fine little cars is still outstripping the supply — it has been thought advisable to concentrate on one colour only.'

The new season saw the general introduction of chromium plating to external bright parts (which hitherto had been nickel plated) and Triplex safety glass as standard. Other changes were subtle and often unannounced, such as the stiffeners added to the upper panels of the bonnet to ensure rigidity, and a change over from Smiths to Armstrong friction shock absorbers.

Study of the body construction for the Morris Minor is fascinating, especially when related to present day mass production methods. Much manual work was involved, particularly for finishing, and although these notes specifically refer to the early 1929-31 bodies, they can generally be applied to the later models. Ash was used in the construction of the body frame where it was subject to stresses, while parts not under stress used whitewood or pine to reduce the total body weight. For the floor a birch plywood of nine laminations was utilised. All parts of the body such as the sides, doors, saloon roof, etc were assembled in master jigs. The completed frame then received the outside press-formed steel panels which were of .028in sheet on the tourer, a thicker .036in material being employed for the coachbuilt saloon body. In the case of the latter, seven seperate pressings made up the complete body: the scuttle panel, the windscreen frame panel, door panels, rocker panels (beneath the doors), and the tonneau panel which covered the whole back of the body and was itself a sub-assembly of three pressings welded into a whole. All the belt mouldings and wheel housings were formed as part of the panels, and captive nuts inside the wheel housings allowed the rear wings to be fitted or removed without disturbing the interior trim. The fabric saloon had metal panels fitted on the corners of the rear quarters, roof, and scuttle to hold the fabric to shape, and elsewhere a close-weave hessian canvas covered the frame, being stretched over each section of the body. This was covered with a thin layer of wadding and then, finally, the outside fabric. When the steel panels had been fitted on the coachbuilt saloon frame, and the doors fitted, the body was sprayed with a rustproof primer and the inside woodwork sprayed a lead colour and the flooring with black. These coats were dried hard by passing the body through a kiln, followed by a number of coats of filler dried in the same manner. The body was then rubbed down with wet quartz-grit paper by hand until smooth, again heat dried, and then sprayed with coats of cellulose colour, dried by heat, and finally 'sanded' by hand using fine grit paper lubricated with special rubbing compound dissolved in water. Thus 'flatted' a mist coat of cellulose thinners left the body ready for polishing.

Many coachbuilders used the Morris Minor chassis for their special bodies. Gordon England built a two-door saloon and Merlyn Motors Ltd of Bristol produced their coupé in fabric over a frame of ash. Like Gordon England, Merlyn Motors had already had success with the Austin Seven chassis. George Maddox and Sons, of Huntingdon, produced a drop-head coupé with dickey seat which, at £185, topped the basic Morris model by £60, a large sum in 1930, and other specialist coachwork was produced by Sunrayn Ltd of Bedfont, Jarvis and Sons Ltd of Wimbledon, Arrow Coachworks of Merton, Hoyal

Typical of the many special body builders using the ohv Minor chassis was the Coventry Motor & Sundries Co Ltd. This is their 1930 advertisement.

Body Corporation of Weybridge, and others. At least one Swallow-bodied Minor saloon was made, but research suggests that this was a one-off. Almost so was the Boyd-Carpenter 'BC Junior' Special as only two of these were made, to individual orders. Few of the specials appear to have survived the ravages of time, but at least four CMS bodied two-seaters, a Stewart & Ardern Calshot sports model, a Duple convertible van/tourer, and a Gordon England two-seater are known to exist.

A drastic drop in the production figures from over fourteen thousand ohc Minors for the 1930 season to about half this for 1931, is explained by the introduction early in the latter year of the side-valve version of the Morris Minor. For 1931 the overhead camshaft variations included a new two-seater which, called the 'Semi-Sports Two Seater', was virtually a copy of the Gordon England two-seater mentioned previously. It may well even have been Mr England's design, considering that the Morris Isis coupé body of 1930 (listed in Morris Motors' catalogue) was the product of his design flair. The new two-seater, priced at £125, was an attractive little car finished in black and red fabric with red 'Karhyde' upholstery to match. It was listed for one season only. As with all Minor models for this year (1931) the transmission hand-brake arrangement was replaced by cable braking, where front and rear shoes were applied by both foot and hand application.

The coachbuilt saloon and four-seat tourer continued to be listed with a single-colour, although this was now blue for the tourer and dark maroon for the saloon. Later in the season this closed Minor was also offered in blue. Colour for the fabric saloon was not dependant on the busy cellulosing shop at

*Overhead camshaft engined commercial, the 1930 Minor
5cwt van.* (Bonallack & Sons)

Morris Minor 1931 fire engine. (Morris Motors Ltd)

Cowley so the customer had the choice of blue or black fabric.

One other model, in addition to the 5cwt van, was the '8hp Fire Tender'. Contemporary catalogues tell us that this was in great demand by villages, schools, institutes and factories, but in truth very few were actually made. Unlike the slightly earlier Gwynn with its pump and hose equipment, the Morris fire tender was merely a carrier of some nine large chemical fire extinguishers and a folding ladder. At least three of the machines are known to have existed, and possibly four. One was used by Morris Motors Ltd for their own works fire brigade. This particular model, registered UD2682 was made late in 1929 (so it can be presumed to have been a one-off that inspired the listing of the contraption for the 1931 programme) and remained in use until 1950 when a young draughtsman in the engineering office at Cowley, Bernard Walker, bought it and converted it into a general purpose vehicle; in which form it was last registered in 1956. In March 1958 it was sold to Queen Street Motors of Geddington in Northamptonshire and thereafter probably scrapped. A second Minor fire tender was presented to the Oxford Fire Brigade (then an entirely voluntary organisation) by Sir William Morris in late 1930. This machine was registered JO743 and by 1935 had been fitted with an inelegant roof and canvas side panels, being in regular use for chimney fire calls, etc, up to about 1944 when that too appears to have been scrapped. A third known model was purchased by the Maharaj Rana Bahadure of Jhalawar in November 1930 from Everard & Ellis Motors Ltd of Slough — the Maharaj had selected £7,000-worth of cars, including the Minor fire tender, from a display of fifty cars arranged by the distributor for his benefit in a paddock at Brooklands! The possible fourth is one said to be on the estate of the Earl of Macclesfield at Shirburn Castle in Oxfordshire.

William J. Collett of Aston Rowant, one time employee of Morris Motors Ltd, recalls driving a demonstration model from Oxford to Yorkshire for a Bradford firm. The bell fitted on a bracket attached to the near-side of the scuttle clanged out in sympathy with every bump in the road until he hit upon the idea of tying his handkerchief around the clapper.

The Minor's role as a fire fighting instrument was again evident in mid-1934 when a fleet of that season's side-valve chassis were modified as fire-engines for use in Chinese up-country stations. These later machines were more practical as the equipment included a water pump mounted forward of the radiator in addition to a longer extending ladder, hose fitments, and a spot light mounted high above the windscreen.

It is not surprising, in view of the successes by drivers of the 'M' type MG, that the ohv Minor was occasionally used for sporting events. An example was a standard Minor which had previously covered some 10,000 miles yet won first place in the 1100cc catagory in the Belgium 24-hour Grand Prix, held over the gruelling Francorchamps circuit in July 1931. Its average speed throughout the race was 43.47mph even when taking the severe hills in second gear a speed of 35.26mph was attained. The car made 112 complete circuits of 14.9 kilometres (9.25 miles) each, a total distance of 1,036 miles consistent running. Morris distributors R.A. Barthels & Co, of Antwerp, were responsible for entering the car which was driven by Mm Geomans and Blin d'Orimont. In later years modified side-valve Minors were to be entered in sporting events with some measure of success. Examples include William Sullivan of Belfast in the International TT on the Ards circuit, Michael McEvoy of Derby who entered his supercharged Minors in many events, such as the Junior Fifty at Phoenix Park, and the Skinner family with their Skinner Specials at Shelsley Walsh.

One of the first £100 side-valve Minors with Sir George Kenning at the wheel.
(Kenning Motor Group)

Because of the low profit margin on the overhead camshaft Minor, and the general knowledge within the motor trade that Fords were developing a small car, Morris was prompted to produce a car to sell at £100, a magic figure which it was felt would have the maximum sales appeal. The problem was what today is called 'cost analysis' and it became obvious that the area to yield the maximum savings would be the engine. In both machining and assembly costs, considerable savings could be made if a side-valve engine was used. Work on a new side-valve model went ahead and various other devices such as dispensing with the bumpers, combining the head and side lights into one shell, paint finishing the single-panel windscreen surround and radiator shell, and fitting smaller section 19in tyres on a new, presumably cheaper, wire wheel, brought the manufacturing costs down so that the vehicle could be sold for £100. (It is worth noting that after about November 1930 these new type wire wheels were fitted to the ohv models, although with the larger 4.00-19 tyres.)

There had been attempts to market a car at £100 before, but the sponsors either never got beyond the prototype stage or they were produced in very small numbers. Amongst these was the 'Gillett' made by British Ensign Motors Ltd which, with its four-cylinder overhead valve 8hp engine, made a brief appearance in the mid-1920s, but only twenty-five were made. In 1926, the Seaton-Petter was advertised as 'The Real £100 car at last' by The British Dominions Car Co Ltd of Yeovil, with a specification that included a two-cylinder two-stroke engine. Waverley Cars Ltd, of London, preceded both with their £100 car in 1925, which was propelled by a rear mounted flat-twin 900cc engine driving through a friction disc transmission, but this one did not go into production.

After Lord Nuffield's death in 1963, some notes came to light which obviously referred to a 400 mile test drive of the Morris Minor in which he remarks that the engine was very rough at certain speeds, especially when pulling hard on hills, although the engine seemed to have plenty of power. Other notes refer to a bad knock in the engine at one time, noisy when idling, and a heavy oil consumption of one gallon to 400 miles. At 8mph, the engine, when pulling, was very jerky. The transmission came in for some criticism as having 'too much play'. He was satisfied with the steering and footbrake but added that the front brake came on when taking corners (an obvious reference to the cable brakes which involved the use of pulleys). Of the springs, these were 'OK' when fitted with snubbers on the front. Petrol consumption for the test was a surprising 43½mpg and top speed 55mph.

At the New Year party held for Stewart & Ardern employees at Acton, they rang in the New Year by wheeling onto the ballroom floor the first grey finished £100 Morris Minor side-valve two-seater. (The fact that the early Bullnose cars were finished in grey may have influenced the choice of this colour.) The new cheaper Minor had already had its full share of publicity and the forthcoming introduction was an open 'secret' many months previously, for instance the 1931 catalogue published in September 1930 said that the side-valve Minor models were listed in a separate catalogue. It was announced that deliveries of the £100 Minor would commence in February 1931.

Ironically, the £100 car did not produce the extra sales expected. The magic figure failed because Morris forgot that in the early 'twenties his cheap, but high quality cars, were more attractive to buyers than the cheaper Ford. Those people for whom motoring was possible in the depression did not choose the cheapest available, when for another £20 they could buy a car that was better

Following the Minor £100 model, Morris introduced a
side-valve four-seater tourer version at £112 10s.
(Morris Motors Ltd)

looking and with a better performance.

The publicity people at Cowley found some consolation when in late 1931 a specially built single-seater side-valve Minor, with supercharger, reached a speed of over 100mph on the Brooklands Track driven by Adolf Van der Becke. This accomplishment was followed shortly afterwards by a road run of the car, now unsupercharged, around a selected road circuit in the Midlands. Although coasting was allowed and the speed kept to a low average, 107.4 miles were covered on a single gallon of petrol. Morris was then able to boast 100mph, 100mpg and £100. To the company's dismay some owners of the two-seater lined up outside the factory demanding that their cars be made to give a similar performance!

In addition to the basic £100 two-seater, saloon versions in both fabric and coachbuilt form were soon available. In February 1931 some two hundred Morris agents were assembled at Cowley to see the new models which were disclosed when Sir William Morris (as he then was) drew aside curtains on a stage. That they were not just prototypes was demonstrated when the dealers were also shown over 200 of the new cars ready for despatch. The fabric saloon was listed at £114 and the coachbuilt saloon, with a folding-head fitment, £5 more. In producing these new models Morris used the same body and chassis pressings as used on the overhead camshaft Minor, so little additional tooling had been required. In March 1931 there followed a four-seater tourer priced at

Underbonnet arrangement, 1931 side-valve Morris Minor.
(Morris Motors Ltd)

£112 10s. The 5cwt van was not forgotten either, this small commercial appearing around about February.

For the 1931 season (when the output from Cowley was lower than any year since 1924 and the pre-tax profits less than the 1923 figure) the prospective Morris Minor purchaser had a choice of eleven different variations, some with side-valve engines and some with the original overhead-valve power unit. In the 1931 model year 7,696ohv and 5,434sv chassis were produced.

Traditionally the new season's cars were introduced at the Society of Motor Manufacturers and Traders Motor Show. In October 1931 at Olympia, the public was able to see for the first time the new Morris Minors for the 1932 season. There were some styling changes and redesign, and with the exception of two new long-wheel-base cars called the 'Family Eight' and 'Eight Sports

Overhead-valve 1932 Eight sports coupé showing side-hinged locker at rear. (Morris Motors Ltd)

Coupé', all the Minors now had the side-valve engine. A new radiator design brought back chromium plating to the bright parts on all models, including the two-seater which still sold at £100, and with it other improvements which included the repositioning of the petrol tank at the rear (an SU Petrolift moving the

1932 two-door Minor saloon; the light coloured fillet on the door was a feature of this year's models. (Morris Register)

fuel from tank to carburetter). The saloon body had rounder features and the new petrol tank position allowed for a slightly roomier body with larger doors; an interesting styling feature on the saloon was the light coloured fillet beneath the windows. On top, the folding-head of previous seasons was replaced by an optional Pytchley sliding-head and this, together with new Magna-type wheels, considerably altered the appearance of the saloon.

Wartime photograph showing the obligatory masked ARP headlamps. The car is a 1932 model overhead-camshaft engined 'Family Eight'. Note the use of the deeper radiator shell which, although correct for this four-door saloon, was generally introduced on the following season's Minors. (May Hemmings)

With the exception of the long-wheel-base chassis of 1932, all Morris Minors had cable brakes until the 1934 season. The arrangement of the pulleys on the front braking system is clearly illustrated on this 1932 chassis.

The other side-valve Minor was a four-seater tourer which followed the general lines of previous models although the original flat sides took on a gentle curve.

Strictly speaking the overhead camshaft four-door 'Family Eight Saloon' and 'Eight Sports Coupé' were not officially called Minors. Morris literature of the period referred to them as Morris Eights and it may come as a surprise to some Morris enthusiasts to realise that the Morris Eight was not first introduced in 1934! However, the chassis numbering was included within the ohc Minor series and the basic mechanics closely followed those of the Minor. Perhaps the outstanding feature of these larger cars was the use of Lockheed hydraulic brakes which *The Autocar* (Road Test, March 1932) described as '. . . really powerful and capable of meeting every emergency', quoting a braking distance of 37ft from 30mph compared with 56ft for the cable-braked Minor saloon. Specification for the sports coupé included a side hinged boot at the back, dummy hood irons, louvres over the windows, rear blind and roof lamp. Doors were '. . . fitted with remote control for handles, each have a rope pull and there are also rope pulls on the pillars just behind the doors'.

An increase in petrol tax from 6d to 8d just prior to the Motor Show no doubt increased the interest in all small cars, and Morris Motors Ltd were competing in a market against other small models, notably the Austin Seven but also including the Jowett, the Singer Junior, Standard Little Nine and the Triumph Super Seven. Later, in February 1932, Ford's 8hp car made its debut at an exhibition in the Albert Hall, London. This was a prototype and the new car was not generally available until about July that year, too late to affect sales of the small Morris. The 1932 season proved to be the best for Minor sales with the total output (including 4,487 Family Eight saloons and coupés) reaching 19,251.

The Minor chassis was finding other uses about this time. Ever since 1926, when a fleet of Morris vehicles had been used experimentally in Birmingham for mail delivery purposes, the GPO had become increasingly interested in Morris vehicles for their fleet. In early 1932 six Morris Minor vans with special bodies, produced by W. Harold Perry Ltd in collaboration with the Post

McEvoy Special Morris Minor of 1932. Bodywork for these sports versions was made by Jensen Motors of West Bromwich.
(A.A. Burgess)

Office, were being evaluated for telephone engineers' use. These forerunners of the many hundreds of Morris Minors to be used by the GPO were quaint vans with oil side and tail lamps (complementing the electric head lights) and observation windows fitted above the driver's seat.

The War Office too used the Minor and special open bodied versions of these were equipped with radio apparatus for work as scout cars in pre-war Britain's poorly mechanised army. Similar to, and in the company of, an Austin Seven version, these vehicles had a role with the 1st Divisional Signals of the Royal Corps of Signals as part of the International Supervisory Staff for the Saar Plebiscite in 1935.

By 1933, the 12-15hp class of car accounted for less than a quarter of all sales and small cars such as the Morris Minor took over the popular market. To first time buyers, 'baby' cars had long since lost their frail cyclecar image, now they were simply less powerful versions of larger types, with all the necessary refinements and less expensive in upkeep. A good illustration of this was provided by the proprietor of a taxi firm in Manila (Phillipine Islands) who had a fleet of forty long-wheel-base Minor saloons plying for hire in a livery of black and yellow. The 1933 season Morris models for this market now had only the side-valve engine, although the variations remained the same with a two-seater, saloon with optional sliding-head, and a tourer providing the main range. In addition a cheaper two-seater, which was basically a 1932 model carrying the earlier radiator surround, a three-speed gearbox, and worm and wheel steering (all features now changed in the 1933 specification), provided for the third year in succession a '£100' model and, it would appear, a convenient outlet for the remaining stock of earlier components.

The longer wheel-base saloon and coupé had been superseded by side-valve versions which had lost their 'Eight' appellation and were now called the 'Minor Family Saloon' and 'Minor Special Coupé'. In retrospect it can be seen that the new features on the more expensive Morris models in the range became standard on the following season's cheaper models. For example, on the long-

Side-valve version of the 1933 season coupé. (Morris Motors Ltd)

A trio of 1933 Minor 5cwt vans for newspaper delivery.
(Oxford Mail)

wheel-base models for 1933 the boxed mudguards were replaced by one-piece
domed pressings with side-shields which, later, become the normal fitting for
the 1934 range. In the same manner, the overhead-camshaft long-wheel-base
saloon and coupé in the 1932 season was fitted with a longer radiator shell
which overlapped the starting handle hole, necessitating the provision of an
aperture in the shell which was covered with a neat disc carrying the 'M'
insignia; this new radiator design became standard for the 1933 season. One
exception to this rule was the reversion to cable brakes on the long-wheel-base
cars, which (as the 1933 models had been reduced in price) may have been the
result of a cost cutting exercise.

Mechanically, the latest Morris Minors (with the exception of the £100 two-
seater already mentioned) were improved by the addition of a so-called 'twin
top' four-speed gearbox and Bishop cam-type steering box.

For the motor industry generally, 1933 was a bad year and for Morris
Motors in particular both production figures and pre-tax profits were down.
Despite dropping the overhead-camshaft version, the company sold fewer side-
valve Minors than in 1932, predictably perhaps as attendance figures at the
1932 Olympia Motor Show, where all the new 1933 season's models were
shown, reached an all-time low.

For its final season (1934) the Morris Minor received a complete face lift and
many refinements such as a synchromesh four-speed gearbox, new type
Lockheed hydraulic brakes, Armstrong hydraulic shock absorbers, Moseley
pneumatic seats upholstered in real leather, battery master switch, automatic
advance and retard ignition, smaller wheels with larger section tyres, and

*1933 model side-valve Morris Minor 5cwt van. Note the
combined head/side lamps, scuttle mounted petrol tank,
and Magna wire wheels. (C. George)*

*Catalogue photograph of the
1934 Minor 5cwt van. The
petrol tank is now at the
rear. Unlike the Minor cars
of the same season, the
brakes on the van were still
cable operated.*

semaphore direction indicators on all models. The new body design had added
curves which complemented the domed wings, and this rounding off of sharp
corners had been applied even to the hitherto right-angle edging of the running
boards. On the two-seater the rear now curved down to the chassis line, with
chromium-plated bars upon which rested the hood when folded. A clever
addition, which completely altered the appearance of the radiator, was a

*1934 four-seater tourer. Hydraulic brakes were one of the
features of this year's Minors.* (Morris Motors Ltd)

*Rear view of a 1934 model Minor 5cwt
van. The twin rear lights and reflectors
are a concession to modern traffic
regulations.* (Dave Clarke)

separate framed stone-guard tapering
outwards at the bottom and fitting
over the previous year's shell.
Internally, the gearbox area was
draught-proofed with a rubber cover,
while the handbrake lever took on a
more robust appearance having finger
grips and a central release button.
Even the 5cwt van version had been
updated with some rounding off of
the sharp corners and repositioning of
the petrol filler cap and tank to the
rear, although under the skin the van
continued to be constructed with
many of the older components and it

*Hundreds of hybrid Morris Eight/Minor vans continued to
be made long after the Morris Minor's demise. These were
known by the GPO as 'Minor, 35 cu ft'.* (T.S.W. Tice)

retained to the end the original square radiator shell and the boxed front
mudguards.

Few, if any, changes were made to the special coupé still listed for 1934. The
duotone colour arrangement, which as the name suggests consisted of two
shades of the same colour, remained as before with the purchaser having a
choice of four schemes. The only other model on the long-wheel-base chassis
(7ft 7in), the Minor four-door saloon, had become quite a roomy car despite
the engine size and for the ex-works price of £150 the owner had a car with
leather upholstery (the driver's seat on runners and the passenger's seat
adjustable), interior woodwork of polished mahogany, 'Pytchley' sliding-head,
winding windows in the doors, interior driving mirror, indicators, rear blind
with remote control, roof lamp, interior visor, pile carpets, and pneumatic
cushions. Separate side lights and bumpers were never fitted as standard on
any of the side-valve Morris Minors, and the 1934 models were no exception.

Towards the end of 1934 the new 'Specialisation Built' Morris Eight made its
debut and the name 'Morris Minor' was shelved for fourteen years (except that
the GPO continued to use a hybrid Morris Eight/Minor van which was still
called the Morris Minor) to return as the type title for one of the most successful Morris models of all time, the Issigonis-designed Morris Minor in 1948.
As for the pre-war Minor, by the time the last model left the assembly line at
Cowley, 39,083 overhead-valve and 47,227 side-valve engined chassis had been
produced between late 1928 and 1934.

Morris Minor. Specification
Overhead camshaft models
Engine: four cylinder, overhead camshaft, overhead-valve, 57mm bore × 83mm stroke, three-ring aluminium pistons, capacity 847cc, 8.05hp, 20bhp at 4,000rpm. SU carburetter. Three-speed gearbox. Morris single dry-plate clutch. Scuttle-mounted petrol tank 1929-31. Rear-mounted petrol tank with SU Petrolift 1932. Cable foot brakes with transmission hand brake 1929-30. Cable hand and foot brakes 1931. Hydraulic brakes 1932. Small centre wire wheels 1929-31, Magna wire wheels 1932, wheel size 19 × 3in, Dunlop Cord 27 × 400 (400-19) tyres.
Side-valve models
Engine: four cylinder, side-valve, 57mm bore × 83mm stroke, three-ring aluminium pistons. Capacity 847cc, 8.05hp, 19bhp at 4,000rpm. SU carburetter. Three-speed gearbox 1931-2, and 1933 '£100' two-seater. Four-speed gearbox 1933-4. Morris single dry-plate clutch. Scuttle-mounted petrol tank 1931. Rear-mounted petrol tank with SU Petrolift 1932 - mid-1934. Later 1934 models (chassis 39026 onward) SU petrol pump. Cable brakes 1931-3. Hydraulic brakes (except vans) 1934, and on 1932 long-wheel-base chassis for Family Eight and coupé. Tyres 350-19 in 1931-3, except Family Eight and special coupé with 400-19 tyres. 1934 models 400-18 tyres. Small centre wire wheels 19in × 3in 1931. Magna wire wheels 1932-4.

Morris Minor		
Model	Chassis numbers	Notes
1929 Overhead camshaft	M101-M2738	Some vehicles exported to Australia carried the prefix 'Y'.
1930 Overhead camshaft	M12739-M27002	
1931 Overhead camshaft	M27003-M34699	
1932 Overhead camshaft	M34700-M39187	
1931 Side-valve	SV101-SV5535	
1932 Side-valve	SV5536-SV20300	
1933 Side-valve	20301-33778	Short-wheel-base chassis 'MMS8' Long-wheel-base chassis 'MML8'
1934 Side-valve	33779-47331	Short-wheel-base chassis '34/MS' Long-wheel-base chassis '34/ML' Van chassis '34/MV'

Morris Minor. Overhead camshaft engine models. Body Colours and Upholstery

	1929	1930	1931	1932
Semi-sports two-seater			Black & red fabric body with red Karhyde.	
Four-seater tourer	Brown cellulose with brown Rexine.	Niagara blue cellulose with blue Karhyde (black hood).	Blue cellulose with blue Karhyde (black hood).	
Fabric saloon, two-door	Blue fabric body with brown Rexine.	Blue fabric body with blue Karhyde.	Blue fabric body with blue Karhyde. Black fabric body with red Karhyde.	
Coachbuilt saloon, two-door, folding-head		Dark brown & brown cellulose with brown Karhyde.	Dark maroon or blue cellulose with brown Karhyde.	
Family Eight saloon, four-door, sliding-head				Blue cellulose with brown Karhyde. Black cellulose with green Karhyde.
Eight sports coupé, sliding-head				Blue & black, black, or beige & brown cellulose with brown Celstra leather. Grey & dove cellulose with blue Celstra leather.

Morris Minor. Side-valve models. Body Colours and Upholstery

	1931	1932	1933	1934
Two-seater	Naval grey cellulose with red Karhyde (khaki-grey twill hood). Optional colour scheme after about June 1931, blue cellulose with red Karhyde.	Blue cellulose with brown Karhyde. Green or black cellulose with green Karhyde.	Blue cellulose with brown Karhyde. (3- & 4-speed models). Green or black cellulose with green Karhyde (4-speed model only).	Blue or green cellulose with matching leather. Black cellulose with brown leather.
Four-seat tourer	Naval grey cellulose with blue Karhyde.	Blue cellulose with brown Karhyde.	Blue cellulose with brown Karhyde.	Green or black cellulose with green leather.
Fabric saloon, two-door, fixed-head	Black fabric body with red Karhyde.			
Coachbuilt saloon, two-door	Naval grey cellulose with red Karhyde. After about June 1931, blue cellulose with red Karhyde. (Folding-head)	Blue cellulose with brown Karhyde. Green or black cellulose with green Karhyde. (Fixed & sliding-head versions available)	Blue cellulose with brown Karhyde. Green or black cellulose with green Karhyde. (Fixed & sliding-head versions available)	Blue or green cellulose with matching leather. Black cellulose with brown leather. (Fixed or sliding-head versions available)
Family four-door saloon, sliding-head			Blue cellulose with brown Karhyde. Black or green cellulose with green Karhyde.	Blue cellulose with blue leather. Black or green cellulose with brown leather. (1934 model called Minor four-door saloon)
Special coupé, sliding-head			Duotone cellulose: Green with green leather. Red with red leather. Grey with blue leather. Black with brown leather.	Duotone cellulose: Green with green leather. Red with red leather. Grey with blue leather. Black with brown leather.

Note: duotone here indicates a darker shade of the colour for the superstructure.

Morris Minor. Magazine Bibliography

Model Year	Subject	Source	Date	Pages
1929	Introducing the ohv Minor	The Morris Owner	7/28	p570
1929	Decarbonising the ohv Minor	The Morris Owner	8/29	p726-30
1929	Minor Saloon & Tourer details, ohv	The Morris Owner	9/28	p840-2
1929	Minor Tuning and Maintenance, ohv	The Light Car & Cyclecar	21/3/30	p468-70
1929	Minor Tuning and Maintenance, ohv	The Light Car & Cyclecar	28/3/30	p493-5
1929	Advance details of 1929 Minor, ohv	The Autocar	31/8/28	p442-5
1929	Minor Saloon Road Test, ohv	The Autocar	30/11/28	p1221-3
1930	Morris Minor, Detailed Information	The Autocar	4, 11, 18, 25/4/30	
1930	Special Coachwork	The Autocar	9/5/30	p882-9
1931	Special Coachwork	The Motor	10/2/31	p43-5
1931	£100 sv Minor Road Test	The Morecambe Guardian	23/1/31	p.....
1931	£100 sv Minor Description	The Motor	30/12/30	p1079
1931	£100 sv Minor Road Test	The Motor	6/1/31	p1115-17
1931	Decarbonising the side-valve Minor	The Morris Owner	6/31	p429-32
1931	Minor sv Saloon, Description	The Morris Owner	3/31	p35
1931	£100 sv Minor, Description	The Autocar	9/1/31	p72-3
1931	£100 sv Minor, Description	The Morris Owner	2/31	p1480
1931	Minor sv Tourer, Description	The Morris Owner	3/31	p165
1931	Two-Seater Semi-Sports, ohv, Road Test	The Autocar	31/10/30	p925-6
1931	Two-Seater Semi-Sports, ohv, Description	The Motor	12/8/30	p52
1931	Minor sv Saloon Road Test	The Autocar	5/6/31	p1013-14
1932	Family Eight, ohv, Description	The Autocar	28/8/31	p.....
1932	Morris Programme for 1932	The Motor	1/9/31	p199-206
1932	Morris Programme for 1932	The Autocar	4/9/31	p391-5
1932	Family Eight Impressions	The Morris Owner	2/32	p1366-72
1932	Minor sv Road Test	The Motor	9/2/32	p51-2
1932	Family Eight Saloon ohv, Road Test	The Autocar	18/3/32	p462 & 464
1933	Minor Family Eight Saloon	The Autocar	2/8/32	p.....
1933	Morris Programme for 1933	The Autocar	2/9/32	p411-14
1933	Morris Programme for 1933	The Motor	6/9/32	p214-20
1933	Morris Minor Family Saloon, sv	The Morris Owner	1/33	p1160-1 & 1169
1933	Morris Minor Family Saloon, sv, Road Test	The Autocar	2/9/32	p418 & 420
1933	Minor Family Saloon, sv, Road Test	The Motor	28/2/32	p121-2
1934	Minor Saloon, sv, Road Test	The Autocar	10/11/33	p927-8
1934	The Minor as I found it	Morris Owner	3/34	p22-3
1934	Saturday afternoon with the Minor	Morris Owner	4/34	p122-5

Morris Oxford Six 1930~5

The six-cylinder Oxford, popularly known as the 'LA' from its engine designation, was introduced in August 1929 against a background of a fickle market and management at Morris Motors Ltd who were not, it would seem, sure of which direction to head. In his book *William Morris, Viscount Nuffield* (1976) R.J. Overy explains: 'Although the products had improved considerably, the basic designs [four-cylinder Oxfords and Cowleys] were over ten years old and the market was declining; 54,000 were sold in 1925 but only 32,000 in 1927. The difficulty lay in deciding what to put in their place.A new Oxford and Cowley were produced, different in virtually everything but name. On Morris's insistance the attempt to get into the large-car market was continued, for Morris felt convinced that the larger American style car was to help in the boom of the 1930s. This led to a frantic search for the right formula, in which "models were feverishly shuffled, and quests for holes in the market led to some incomprehensible . . . permutations". Worse still, this policy, so different from the early 1920s was specifically claimed by Morris as his own. At the Forth Annual General Meeting in 1930 he announced that "the policy of the Company has always been such that a proper degree of flexibility of production is available, so the variations in demand for different types of cars as necessitated by public taste can be accommodated".'

The Morris programme for the 1930 season reflected the trend towards the production of larger models with the introduction of two new six-cylinder cars, the Isis-Six and the Oxford.

The four basic body types standard on the new Oxford Six chassis were a panelled saloon with sliding-head, a fabric-bodied saloon, a coupé with sliding-head, and the only open model, a four/five seater tourer. The familiar artillery wheels of the four-cylinder Oxfords had been replaced with small hub five-stud wire wheels carrying 28 × 5.25 (ie 5.25-18) Dunlop tyres, with a spare mounted on the off side, except on the fabric saloon and four-light coupé which carried the fifth wheel on the rear built-in luggage container. Mechanical specification included Lockheed hydraulic brakes operating in 14in diameter drums, Luvax hydraulic shock absorbers, rear-mounted 12-gallon petrol tank with pebble guard, three-speed gearbox, multi-plate cork insert clutch running in oil, and enclosed propeller shaft to a spiral-bevel rear axle. A minor precedent was the use of a winged badge on the chromium-plated slightly tapered, radiator shell carrying the words 'Morris Oxford'. Black finished Lucas R160s Biflex head-

Oxford-Six coachbuilt saloon with sliding-head for 1930.
Note the off-side mounted spare wheel.
(Morris Motors Ltd)

lamps, mounted on a lamp support tube spanning the front wings, were complemented by Biflex side lamps (R370s). The combined stop-tail lamp, horn, coil ignition equipment, and spin-start MT1/6 windscreen wiper motor were also by Lucas.

The Oxford Six chassis frame was simple, yet sturdily constructed. The simplest form of chassis from a fabrication point of view consists two straight main side-members like that of the early Bullnose frames (although even on the early Morris cars the dumb-iron section curved downwards). On the Oxford Six the designers had achieved something approaching this by the use of two parallel sides which only changed level to cater for the rear axle, then continued on in a flat formation. On early Bullnose cars the need for this was avoided by the use of $\frac{3}{4}$ elliptic springs and an underslung axle. With few exceptions the usual forgings used in the construction of a Morris chassis were replaced with pressings, while the normally separate cross reinforcement of the chassis was performed solely by the underslung running board supports and a substantial pressing on the straight rear section, which functioned also as spring hanger and petrol tank support. Considerable use was made of the enormous presses recently installed at Cowley. Bracing between the front ends of the chassis comprised a tubular member which doubled as hangers for the ends of the front half-elliptic springs; all four road springs on the Oxford were on the outside of the frame.

All models for the 1930 season had leather upholstery, including the tourer where the front seat was of the full width adjustable type, while the other variations had bucket seats; two independent 'occasional' seats were provided at the rear of the coupé body. On two of the models, the tourer and the panelled saloon, a specially designed luggage container with two sliding suitcases could be supplied as an extra (£10) to match the coachwork finish. Windows on the panelled saloon were operated by a winding mechanism; on the coupé and fabric saloon these were of the sliding type.

The new side-valve six-cylinder engine with a 63.5mm bore and 102mm stroke, rated at 14.9hp, was a clean, compact unit with some convenient servicing features such as the two detachable tappet blocks, each held by three bolts, which allowed six tappets at a time to be removed. With the object of ensuring the absence of fumes within the car body, an elaborate air filter and fume consuming unit was fitted on the flat cylinder head consisting of a horse-hair filled air filter connected to the SU carburetter intake, to 'consume' any oily vapour expired from the engine. Despite early problems of overheated engine oil (which did nothing to prolong the life of the cork clutch inserts) said to be caused by insufficient water space around the bores, the 'LA' engine proved to be a success in later modified forms as used to power subsequent Oxfords as type 'LC', as well as the Major ('LB', 'LE', & 'LF'), Cowley Six ('LG' & 'LO'), Fifteen-Six ('LJ' & 'LL'), and as the 'LK' in the Morris Commercial G2S Junior Six Taxicab.

Some interesting innovations were to be found on the Oxford Six as originally presented. One such was the high-pressure oil filter bolted to the side of the engine cylinder block. This unit consisted of a series of fixed and moving filter plates (in the manner of a radio tuning capacitor) with a capillary space between each, the moving plates being connected by a push rod to the clutch pedal with the object of imparting a partial rotation to the filter pile to perform a cleaning action every time the clutch was used. Unfortunately, as time and usage revealed, it was an impracticable device and a major source of oil leakage; a simple cylindrical gauze filter with more substantial fixings replaced it at engine number LA24997. Far more functional was the Calorstat set into the underside of the header tank to control the radiator temperature thermostatically. The device, filled with ether, transformed a change in temperature into a mechanical movement to open or shut black-finished shutters arranged vertically in front of the radiator. Failure of this system was not in its design, but more due to neglect on the part of the owner to ensure the free movement of the shutter hinge pins and connecting rod.

No shortage of air-conditioning on this 1931 model fabric-bodied saloon version of the Morris Oxford Six. Sliding windows in the doors, roof and scuttle ventilators, and opening windscreen illustrates the point. (Chris Hudson)

The Morris Oxford Six certainly was a popular car in its first season for the 15,445 Oxfords assembled at Cowley constituted more than a quarter of the total vehicle-build for the 1930 model year. Assuming a similar proportion of the vehicles exported, as chassis or complete cars, then it can be estimated that approximately 1,400 went abroad that year, the remainder joining the million or so private cars on the roads of the United Kingdom. Of the exported chassis it is known that Holdens of Adelaide, South Australia, produced thirty-one saloon ('Sedan'), twenty-six two-seater ('Roadster'), and fifty-one tourer bodies on the Oxford Six chassis during 1930.

For the 1931 season the Oxford was basically unchanged, although there was the almost obligatory cosmetic treatment to justify it as a 1931 model. These additions included the window louvres of Triplex glass on the coachbuilt saloon, a chromium-plated decorative metal band between scuttle and bonnet, and the Magna-type wire wheels with a press-in hub cap decorated with the 'M' insignia which were standard equipment on the coupé and sliding-head saloon. The fixed-head saloon was a new and cheaper variation for the 1931 season; on this and the other models the Magna wheels were optional extras (£2 10s). For the convenience of rear seat passengers a folding arm rest was added to the saloon specification, and all types now had an organ-pedal accelerator and group chassis lubrication system located over the rear axle banjo.

Four/five seater tourer. A 1931 model in the more leisurely days of motoring. (Richard G. Gray)

A study of early photographs suggests that the spare wheel, where side-mounted, changed from the off-side on the 1930 model to the near-side in 1931. (At least one prototype 1930 saloon is photographically recorded with the spare carried at the rear.)

Some of the 'bugs' in the LA engine were sorted out during the 1931 season. For example there had been some difficulty in keeping the original cone seating on the water pump watertight, so this was modified. Clutch spin had also proved to be a problem, traced to the binding of the clutch springs before the full travel of the pedal had been obtained; here the replacement of the round-section springs with square-section springs, together with a shortening of the mechanical link, appears to have proved the solution. Modification undertaken at the time also allowed the chain case to be removed without the necessity, as

1931 Morris Oxford Six coupé outside The Harrow,
Headly. (Eric Jackson)

hitherto, of raising the forward end of the engine.

For the following season the Morris programme listed something like twenty-one different models, including a new version of the Morris Oxford Six. The Cowley publicity machine, although it was not considered as such in September 1931, put their case thus: 'To say that the excellent cars of the current year have been improved out of all recognition for 1932 is a bold statement to make. But such are the advances which have been accomplished during the past twelve months that it is difficult to believe that a range of cars selling at a strictly moderate price could be embellished to the standard represented by the 1932 Morris models. Without exception every model has been improved, not merely in detail but constructionally, and the appearance of the new models, together with their added refinements, must definitely place them at the forefront of their respective price classes for sheer value, good looks and all round performance.' Prices had dropped considerably, with, for example, the Oxford Six saloon down from the 1931 figure of £285 to £265 in 1932. Although this was a reasonable percentage reduction on what today appears to be a ridiculously low price, it is worth putting into perspective as it represented over two years' earnings for a young man working in the motor industry.

For the 1932 range of cars Morris had introduced for each model, from Minor to Isis, a sports coupé body of similar design but, naturally, varying in

Absence of splash guards (or what the Americans called 'fender skirts') date this Oxford Six saloon as a 1932 model. (Morris Motors Ltd)

*Two coupé variations were
listed in the 1932 Oxford Six
programme, the sports coupé
and this four-light coupé.*
(Morris Motors Ltd)

*Two coupé variations were
listed in the 1932 Oxford Six
programme, the sports coupé
and this four-light coupé.*
(Morris Motors Ltd)

size and appointments depending on the chassis. These sports coupé bodies
with sliding-head were jig-built of ash and panelled in 20-gauge steel pressings,
and despite the existence of hood stretchers on the leathercloth rear superstructure the coupé heads were fixed; the stretchers served only as a decoration.
At the rear, integral with the body, was a side-hinged boot. A four-light coupé
with sliding-head was also still available on the Oxford Six chassis. The
remaining representative of the Oxford Six was a saloon, which for the 1932
season was in sliding-head form only; the open tourer of the previous year had
been dropped from the lists.

The new appearance and construction of the 1932 Oxford Six involved the
introduction of a new type of rounded radiator shell with chromium-plated
vertical radiator shutters and a shield-shaped badge. To this superfluity of
bright work was added a chromium-plated shell housing the horn in the centre

*Sports coupé on the Oxford
Six chassis of 1932. The
following season, for some
reason, the same body design
was re-named 'special coupé.*
(Morris Motors Ltd)

of the horizontal lamp support bar, similarly plated headlamp rims, twin-bar bumper and the Wilmot Breeden calormeter with wings. The standard Magna wire wheels were of a new design with the hub caps now attached by means of a central screw. The general line of the bodywork was considerably altered by use of the so-called 'eddyfree' front, designed to lessen wind resistance and thus decrease noise and ensure maximum performance. Window louvres on the saloon and sports coupé provided ventilation without draughts.

Being typical examples of the bodywork and trimming technique of Morris Motors Ltd in the early 'thirties, the Oxford Sixes are perhaps worth looking at in greater detail from the inside. Costs apart, certain restraints were put on the designers where the shape, position, size, etc was often dictated by the use of proprietary components. Whereas in more recent times component manufacturers will supply parts specifically created for an individual model, the drawing office of the 'twenties and early 'thirties tended to design around off-the-shelf components, using handles, instruments, lamps, brake systems, shock absorbers, bumpers, ventilators, sliding roof mechanism, etc, identical to those found on many other makes, albeit this has proved to be a blessing to present-day vintage car enthusiasts. Additionally, a manufacturer would offer, on generous terms, a new device or gadget which would be made standard equipment, often for a single season, such as the 'Rolvisors' made by G. Beaton & Sons Ltd of Willesden and fitted to 1935 Morris cars. At the other extreme Wilcot (Parent) Co Ltd received an initial order of 56,000 sets of 'Wefco' leather spring-gaiters in the 'twenties and these continued to be standard equipment on Morris cars, including the Oxford Six, for many years.

The traditional method of building up the cushions and squabs was used; this consisted of rows of coppered steel springs held together by clips at the bottom and sides, and covered above by metal lace formed of coils of fine-gauge interlocked tinned spring-steel wire. This provided the foundation for horse hair and felt with which the cushions were padded. Upholstery was done in Vaumol leather except on the sports coupé where Celstra leather was used. The rear seats on the latter also deviated from the traditional by the use of an inflatable 'Float-on-air' cushion. The bucket seat frames were fabricated from sheet steel with edges rolled over heavy-gauge wire. Door panels, complete with leather pockets, and other trimming was arranged to be easily detachable, with either screws in cup washers around the edges or what was described as slot-screwed; in the latter case, it was only necessary to remove the bottom screws to allow the panel to be lifted slightly to be released. The driver's seating position could be set to suit the individual by combining movement of the front seat, setting the pedal height position and angle of steering column rake, all of which were adjustable. Passenger comfort on the saloon was enhanced by the addition of a central folding arm and two head rest cushions to the rear seats.

Considerable changes were made to the mechanical specification of the 1932 Oxford Six, not least of which was the introduction of a four-speed 'twin-top' gearbox and the use of a single plate in the cork-insert clutch arrangement. The engine, while basically the same unit, now designated 'LC', had minor modifications including a new type of camshaft (commencing engine 26228) and a redesigned manifolding system with a drip tray on the SU carburetter. The substantial chassis frame of 0.128in-thick pressed steel channel was reinforced by an extra cross tube just forward of the rear axle and an additional cross member below the engine. The curious spring-controlled anchorage for the rear end of the off-side front spring had the object of precluding the possibility of wheel 'shimmy'. A similar device was also fitted to the contemporary Isis chassis.

During the production run of the 1932 model teething troubles included the tendancy for top gear to slip out of engagement, solved by the introduction of stepped splines on the shaft where the top gear rested. Oil leakage in various places appears to have been top of the list of complaints. The change in gearbox oil capacity in period data sheets is explained by the action taken to prevent oil disappearing into the rear axle. An oil return groove was added to the universal rear fork together with a special cork washer cemented to the torque tube ball and, as an added measure, the gearbox oil level

Anti-wheel shimmy device fitted to the 1932 season Oxford Six. (Tom Bourne)

of 'normal' was re-marked 'high'. Oil was also prone to leak from the gearbox into the clutch, which is the case of the Morris unit was not serious; but to rectify this fault a return hole was drilled in the gearbox front wall. Oil which, in its turn, leaked from the engine into the gearbox called for a spring-ring oil-thrower which was added to the gearbox drive gear in front of the bearing shroud.

By coincidence the 1932 season saw the introduction on all Morris cars of a quart 'topping-up' can of oil mounted in a bulkhead mounted clip. In 1931 Alexander Duckhams & Co Ltd, with the support of Morris Motors Ltd, produced a special oil they called 'Morrisol' and promptly upset Matthew Wells & Co of 'Wellsaline' fame, who had been marketing a special oil called 'Morrisol' since 1926! Morris Motors Ltd and Wolseley Motors (1927) Ltd issued a statement to the effect that the word 'Morrisol' was being used as the name of an oil without authority or approval of either concern. 'Sirrom' (Morris spelt backwards), they added, had been registered as a brand name and all future containers of Duckham's Morrisol would bear the words 'Sirrom, regd Brand'. Morris Motors Ltd obviously considered that 'Morrisol' was a name that should only have been used with their authority, whereas Matthew Wells & Co maintained, with some justification, that it was their property because they had been using the name for five years. If the correspondence columns of the trade press of the period is any guide, the oil was supplied to the motor manufacturer at a knock-down price and the cans supplied free of charge; all the more ironical that present day dealers in old motoring curios charge exorbitant prices for these containers.

1933 special coupé Oxford Six. Note the introduction of splash guards on the front mudguards.
(Morris Motors Ltd)

Performance figures for the Oxford Six saloon of the 1932 season as recorded by *The Motor* showed a maximum speed of 59mph in top gear and an acceleration time of 36.2 seconds from 0-50mph. Petrol consumption was given as approximately 24mpg. The braking distances from 40mph of the Isis, Oxford and Cowley (hydraulic brakes), compared with the Minor (mechanical brakes) are interesting. The Isis stopped within 64ft, Oxford 66ft, Cowley 75ft and if you were 94ft away from the cable-braked Minor when it commenced its retardation it was still moving when it hit you!

To the uninformed the 1933 model Oxford Six appeared identical to its 1932 predecessor; certainly it looked the same and the choice of models remained as before, ie sliding-head saloon, sliding-head coupé, and the sports coupé of 1932, now renamed 'Special Coupé'. Nevertheless, there were minor changes in appearance, such as splash guards to the front mudguards, twin-bladed windscreen wipers, scuttle side ventilators and the Wilcot direction indicators (subsequently replaced as mentioned elsewhere). Other minor changes or additions were out of sight, such as the check straps fitted to the rear axle, and a central mounted handbrake.

Inside the saloon the trim was improved by the use of polished walnut woodwork; this, together with the real leather used for the brown upholstery and two corner cushions supplied as standard, made for a luxurious interior.

The major change in the 1933 model Oxford Six was the substitution of a larger 15.9hp side-valve six-cylinder engine for the earlier 'L' series power unit. This new 2,062cc design, designated 'QA', took into consideration the earlier overheating problems by increasing the water capacity of the block by $8\frac{1}{2}$ pints and sump oil capacity by 4 pints. At 3,000rpm the new engine gave an output of 39bhp, an increase of 8.5bhp. The distributor drive came from the pump-shaft instead of the dynamo drive. (One of the surprising statistics revealed by Morris Engines Branch was that the making of a single engine of this type involved 19,000 machining operations and 555 inspection processes.) Rear of the single plate cork insert clutch and four-speed 'twin-top' gearbox, the propeller shaft was now of open tubular type with Spicer needle roller bearing universal joints.

A complete redesign of body and chassis alike was apparent when the 1934

1934 model Morris Oxford Six saloon.
(John R. Millburn)

Folding arm rest and other appointments for passenger comfort in the 1934 Oxford Six. (J.R. Millburn)

season's Oxford Six cars were on display at London's Motor Show at Olympia during 12-21 September 1933. Body options were limited to a saloon and a coupé which, although still retaining the name 'Oxford Six Special Coupé', bore no resemblance to its earlier namesake; gone was the perpendicular style with decorative 'hood irons' and the seemingly separate rear luggage container was replaced by a four-light two-door body having luggage accommodation in the rear overhang. The new saloon body had its four doors hinged to bring the handles together by the centre pillar, which housed a slim concealed Lucas trafficator. Although the Oxford Six of 1934 continued to be based on a 114in wheelbase with a 56in track, the saloon body was slightly longer and fractionally lower than the earlier models. Internally the

*15.9hp 'QB' engine for the
1934 model. The flasher unit
is a modern addition to this
Oxford.* (J.R. Millburn)

saloon was equipped with a single full width front seat broken by a folding arm rest in the centre. At the rear was a similar arrangement with the addition of two head-rest cushions, a folding foot rest, pillar pull-cords, and a folding table arranged along the back of the full width front seat. Rear seats in the special coupé were of the Moseley 'float-on-air' pneumatic cushion type.

The new chassis frame was a cross-braced fabrication with the centre of the cross brace opening out into a tunnel for the propeller shaft. Contemporary Morris publicity described it as a frame within a frame, with bracing units running diagonally from the front cross member to the rearmost cross member, the vice-like cross grip preventing frame distortion and consequent body stresses that so often resulted in loose or rattling doors. Earlier chassis frames had relied in part on the solid mounting of the engine to help stiffen the chassis; such reinforcement was not present in the 1934 design as the power units were cushioned on rubber. The new approach, using flexible suspension, was called 'Equipoise' and aimed at cutting out vibration at its source.

Another device to reduce vibration was the Wilmot Breeden harmonic stabiliser bumper fitted to all Morris 1934 six-cylinder cars. The bumper was based on the theory that however meticulous the calculations as to stress, weight, and frame dimensions, at high speeds over bumpy roads certain vibrations are inevitably set up. However, it is possible to absorb nearly all this vibration by means of a carefully balanced flat spring running transversely in front of the car in conjunction with the bumper, which has weight adjustable dashpots at both ends. A similar bumper was fitted as standard on contemporary Rover, SS, Lanchester, Daimler Straight Eight, and Alvis cars.

For the export market the designers of the 1934 model made provision for both left-hand and right-hand-drive versions. Considering the rougher treatment expected in certain overseas countries, both front and rear road springs had additional leaves on the export cars.

As with other vehicles in the 1934 Morris range, the Oxford Six was equipped with the latest design of radiator shell, a much more rounded shape of chromium-plated shell with an immitation honeycomb and tilted slightly.

The radiator proper was mounted in a vertical position behind. Equipment elsewhere for both saloon and coupé models included a Wilmot Breeden metal cover for the side-mounted spare wheel, fog lamp, interior visor for the driver, paired rear and reversing lamps at opposite ends of the rear number plate, Startex automatic starter (details of which are given on page 157), handbrake lever with shaped finger grip and button release, and rectangular pedals.

An outstanding feature was the provision of a synchromesh four-speed gearbox combined with a Bendix automatic clutch control and free-wheel. The release of pressure on the centre accelerator allowed the pedal to travel back beyond the position where the throttle closed. This over-travel opened a valve which connected the power cylinder to the induction manifold, thus creating a vacuum behind the piston. Thus the piston and rod were forced inward by atmospheric pressure, disengaging the clutch. Changing gear simply involved lifting the foot from the accelerator and moving the gear lever into the desired position. When, for example, the driver wished to utilise the engine for retardation on a hill, the automatic clutch could be rendered inoperative by means of a lockout knob on the facia, this control carrying a stern warning not to attempt to lock the free-wheel on the over-run.

The same bodies were used for both Oxford Sixteen and Twenty in the last season, 1935. Apart from the motif on the radiator honeycomb, they were externally indistinguishable. (E. Miller)

For its final year, 1935, the Morris Oxford was listed with the option of the 15.9hp 'QD' engine as the Oxford Sixteen, and with the 'QF' engine rated at 19.9hp as the Oxford Twenty. Bodywork, mechanics, and prices, were identical and, except for the engine, both followed the general lines of 1934 and were

1935 Morris Oxford Six, Sixteen or Twenty special coupé with sliding-head. (Morris Motors Ltd)

available as saloon and coupé. Clearly the knowledge of forthcoming changes in motor vehicle taxation was responsible for the option of a larger engine. Since January 1921 the Road Fund Tax for a 16hp car had been £16 per annum, but, with the new rates which would come into force on 1 January 1935, the average motorist would only pay £12 tax for his Sixteen and a 20hp car would still cost less to tax than the earlier lower-powered Oxford Six.

The most obvious changes made for the 1935 season were the reversion to single bucket seating at the front on the saloon, a right-hand accelerator pedal, and the wider reduced diameter tyres on the Magna wire wheels, changed from 5.25-18 to 5-50-17; export models had special 700-16 tyres on 4.5 × 16 wheels. Minor changes included the standard fitting of a 'Rollsvisor' to provide protection for the driver against sun glare, and an interesting innovation anticipating the general availability of car radios, a built-in radio aerial in the fabric roof with the downlead terminating at the dash.

Salmons & Sons of Newport Pagnell made use of both the 1931 and 1932 season's Morris Oxford Six chassis for their 'Tickford Sunshine Coachwork' which incorporated the patented Tickford folding roof. By means of an external handle inserted in the rear quarter of the body it was possible to simply wind open the roof. The mechanism consisted of a train of gear wheels in each back quarter of the body between the upholstery and panel, operating a light steel framework. The roof itself was kept taut by a dozen light flexible steel cables running from the back to the front, sandwiched between the outer waterproof material and the cloth roof lining. The operating handle was detachable and the winder hole was covered with a plated snap cover. As originally offered around September 1930, the Morris Oxford Six Tickford Sunshine Coachwork was fabric covered with a real leather hood, priced at £365; the following month the price dropped to £335. In March 1931 the substitution of a cellulose finish to the aluminium panelled body and a hood of waterproof material enabled Salmons & Sons to reduce the price of the special (now called Tickford 'Abbey' bodywork) to £310. The 1932 version was the same price until about May 1932, when it was increased by a further £15. Apart from those details mentioned, the specification included an adjustable front windscreen and soft waterproof leather upholstery on sliding bucket seats at the front and the wide rear seat.

On stand 72 of the Olympia Motor Show in 1930, Maltby's Ltd, the Morris distributors for East Kent, had on display examples of their special bodywork on the Minor, MG, and Oxford Six chassis. The Oxford was a sports two-seater which, to be expected on such a large chassis, had an extremely roomy body, while for occasional use a dickey seat was arranged at the rear. This dickey seat was unusual in its design, the 'lid' came away bodily and functioned as the back rest when located by pegs into holes in the floor; protection for the painted outer surface was by two rubber buffers which registered with a pair of chromium-plated metal discs. For the convenience of the rear passengers, small side pockets were included for small items of luggage. Another interesting feature was a special golf bag locker. 'Golf clubs are unreservedly one of the most awkward cargoes to stow even on a five-seater car' so Maltby provided a small locker behind the near-side door where clubs could be housed in the boot without trouble, yet out of the passenger's way when the dickey seat was in use. With the hood up, driver's visibility was improved by the provision of small celluloid lights in the hood corners. Maltby's offered a choice of colours, including an attractive combination of cream and black; the whole of the upper surface, a space of an inch down the bodywork, and wings having the darker colour. Another scheme was black cellulosed superstructure with ivory sides and wheels, a green waist moulding and green upholstery. The pneumatic seats were covered with red leather, and the spare wheel was at the rear. This model was listed at £305 but there appears to have been, in addition, a drop-head coupé and a 'Speed Model' for an extra £20, but what concessions were made to justify the title is not known. Perhaps the following season's Maltby Oxford Six based on the 1932 chassis provides a clue, for this was described as a two/three seater body of racy lines with a long bonnet and cycle-type wings, the doors being recessed to take a combined arm rest, ash tray and pocket, while in the tail was a dicket seat with quick-opening lid. Bodies were panelled in aluminium. A similar style with the standard Morris wings and running boards brought the price down from £315 to £298. In 1933 the Maltby coach-work on the Oxford six-cylinder chassis continued to be listed.

The Cunard Motor & Carriage Company Ltd of Acton, part of the Stewart & Ardern organisation, produced what they termed standardised coachwork which they applied to several chassis including the 1933 version of the Oxford Six. The lines of this all-weather four-seater 'Calshot' (or 'Calshott' — even they were not sure of the spelling to judge by contemporary advertisements!) body was spoilt by the bulbous rear end forming a huge boot. With the hood up, it was no place for claustrophobic passengers, enveloping, as the hood did, the complete car from front seat rearwards with no concession to visibility (unlike the Maltby vehicle) apart from the minimal rear window-panel. The lines improved with the hood folded for this was arranged to disappear entirely into the coachwork. Within the car the interior woodwork was of black walnut and the upholstery was of leather. Conventional running boards were absent, in their place curious oval step-boards were fitted on outriders beneath the doors. Colours were left to the preference of the customer, but one scheme listed was black and grey with the darker colour covering the top of the bonnet and the uper part of the body following the panel line. Far more graceful was Stewart & Ardern's offering on the 1934 Morris Oxford Six chassis called the 'Lansdowne Continental Coupe' which retained the standard Morris front end arrangement of headlamps on a tubular frame with complementing horn and spot lamp, and the 'Harmonic Stabiliser' bumper, blending in with Stewart & Ardern's rounded wings, which tapered smoothly into the running boards.

Morris Oxford Six. Magazine Bibliography				
Model	Subject	Source	Date	Pages
1930	Decarbonisation on the Oxford-Six	*The Morris Owner*	4/30	p164-6
1930	Morris Programme for 1930	*The Morris Owner*	9/29	p840-5
1930	Morris Oxford Saloon, Road Test	*The Autocar*	11/10/29	p701-2
1930	New Morris Six	*The Autocar*	30/8/29	p401-9
1931	Morris Programme for 1931	*The Autocar*	5/9/30	p445-8
1931	Morris Programme for 1931	*The Morris Owner*	9/30	p847-51
1931	Once over the Oxford	*The Morris Owner*	8/31	p676-8
1932	Oxford Six Saloon Road Test	*The Motor*	9/2/32	p48-9, 51
1932	Morris Programme for 1932	*The Autocar*	4/9/31	p391-5
1932	Morris Programme for 1932	*The Motor*	1/9/31	p199-206
1932	Morris Programme for 1932	*The Morris Owner*	9/31	p787-92
1932	On the youngest Morris Oxford	*The Morris Owner*	4/32	p144-5, 148
1932	Oxford Six Saloon Road Test	*The Autocar*	11/3/32	p418-19
1933	Oxford Saloon Road Test	*The Motor*	28/2/33	p119-22
1933	Morris Programme for 1933	*The Autocar*	2/9/32	p411-14
1933	Morris Programme for 1933	*The Motor*	6/9/32	p214-20
1933	Morris Programme for 1933	*The Morris Owner*	9/32	p709-16
1933	Impressions of 16hp Oxford	*The Morris Owner*	6/33	p370-1, 401
1933	Oxford 16hp Saloon Road Test	*The Autocar*	2/9/32	p416-17
1934	Morris Programme for 1934	*The Morris Owner*	9/33	p816-26
1934	Morris Oxford Road Test	*The Morris Owner*	2/34	p1314-16
1935	Morris Programme for 1935	*The Morris Owner*	9/34	p655-62
1935	Morris Oxford Twenty	*The Morris Owner*	11/34	p834-6
1935	Morris Oxford Twenty, Road Test	*Good Housekeeping*	1/35	p92, 94

Specification

Morris Oxford Six, 1930 model.
Chassis numbers: LA101 - LA15545

Engine type LA. Lockheed hydraulic four-wheel brakes. Handbrake, cable to rear wheels. 14in diameter drums. Centre accelerator pedal. Throttle, ignition, dynamo, and lamp controls on 18in diameter steering wheel. Lucas R160s headlights 'cut & dip' system. Bishop Cam steering. Luvax single-acting hydraulic shock absorbers. Rear mounted petrol tank with pebble guard, 12 gallons, Autovac feed to SU 'under-float feed' carburetter. Enclosed torque to spiral-bevel rear axle. Crownwheel and pinion 5.27:1. Wire wheels, small hub, 5-stud fixing, 3.25 × 18 with Dunlop Cord 5.25-18 tyres. Spring gaiters. Rectangular chromium-plated radiator surround with black shutters operated by Calorstat. Winged badge. Calormeter and wings. Triplex single-panel windscreen (closed models) with electric spin-start windscreen wiper motor, single blade (all models). Twin blade bumpers. Luggage grid standard on tourer and coachbuilt saloon. Wheelbase 9ft 6in. Track 4ft 8in.

Morris Oxford Six, 1931 model.
Chassis numbers: LA15546 - LA23746

Engine type LA. Lockheed hydraulic four-wheel brakes. Handbrake, cable to rear wheels. 14in diameter drums. Centre 'organ pedal' accelerator. Throttle, ignition, dynamo, and lamp controls on 18in diameter steering wheel. Lucas R160s headlights 'cut & dip' system. Bishop Cam steering. Grouped chassis lubrication. Luvax single acting hydraulic shock absorbers. Rear mounted petrol tank with pebbleguard, 12 gallons, Autovac feed to SU 'underfloat feed' carburetter. Enclosed torque tube to spiral-bevel rear axle. Crownwheel and pinion 5.27:1. Magna wire wheels standard on coupé and sliding-head saloon. Small hub wire wheels standard on other models with option of Magna type extra. Wheels 3.25 × 18, Dunlop Cord 5.25-18 tyres. 5-stud fixing. Spring gaiters. Rectangular chromium-plated radiator surround with black shutters operated by Calorstat, winged badge. Calormeter and wings. Triplex single-panel windscreen (closed models) with electric spin-start MT1 windscreen wiper motor, single blade (all models). Twin-blade bumpers. Luggage grid standard on tourer and coachbuilt saloon. Wheelbase 9ft 6in. Track 4ft 8in.

Morris Oxford Six, 1932 model.
Chassis numbers: LA23747 - LA28746

Engine type LC. Lockheed hydraulic four-wheel brakes. Handbrake, cable to rear wheels. 12in diameter drums. Centre 'organ pedal' accelerator. Throttle, ignition and lamp controls on steering wheel. Lucas R165 'dip & cut' headlights. Combined stop and reverse lamp. Bishop Cam steering. Luvax single-acting hydraulic shock absorbers. Rear mounted petrol tank with pebble guard, 12 gallons. SU Petrolift to SU 'underfloat feed' carburetter. Enclosed torque tube to spiral bevel rear axle. Crownwheel and pinion 5.27:1. Magna wire wheels 3.25 × 18. 5-stud fixing, Dunlop Cord 5.25-18 tyres. Spring gaiters. (Front springs decreased from seven leaves plus rebound to six leaves at chassis LA25245.) Chromium-plated radiator surround and Calorstat operated shutters. Shield shape badge. Horn built into lamp bar between mudguards. Calormeter and wings. Triplex single-panel windscreen with electric wiper motor. Twin-blade bumpers. Luggage grid on saloon. Instruments with black dials, white lettering, on mottled aluminium (position of clock and speedometer reversed to 1931 model). Wheelbase 9ft 6in. Track 4ft 8in.

Morris Oxford Six, 1933 model.
Chassis numbers: 28747 - 32382 'MO16'

Engine type QA. Lockheed hydraulic four-wheel brakes. Handbrake, cable to rear wheels. 12in diameter reinforced drums. Centre 'organ pedal' accelerator. Throttle, ignition and lamp controls on steering wheel. Lucas LBD165s 'dip & cut' headlights. Combined stop and reversing lamp. Bishop Cam steering. Luvax double action hydraulic shock absorbers. Rear mounted petrol tank with pebble guard, 12 gallons, SU Petrolift to SU 'overfloat feed' carburetter. Tubular propeller shaft with Spicer needle-roller bearing universal joints. Crownwheel and pinion 5.27:1. Magna wire wheels 3.25 × 18, 5-stud fixing, Dunlop Cord 5.25-18 tyres. Chromium-plated radiator surround and Calorstat operated shutters. Shield shape badge. Horn built into lamp bar between mudguards. Calormeter and wings. Triplex single-panel windscreen with self-starter electric wiper motor and twin blades. Early models originally fitted with Wilcot flashing traffic indicators, later replaced with Lucas semaphore type on arms. Twin-blade bumpers. Luggage grid on saloon. Wheelbase 9ft 6in. Track 4ft 8in.

Morris Oxford Six, 1934 model.
Chassis numbers: 34/032832 - 34/035637

Engine type QB. Bendix automatic clutch control and free-wheel. Lockheed hydraulic four-wheel brakes. Handbrake, finger shaped grip with centre button release, cable to rear wheels. 12in diameter drums. Centre 'organ pedal' accelerator. Rectangular brake and clutch pedals. Throttle, indicator, horn and lamp controls on steering wheel. Headlights 'cut & dip' type. Paired rear and reversing lights at opposite ends of rear number plate. Concealed Lucas Trafficator direction indicators (automatic cancel). Bishop Cam steering. (Left-hand drive optional. Additional spring leaves on export models.) Luvax double action hydraulic shock absorbers. Rear mounted petrol tank, 14 gallons, SU electric petrol pump to SU 'overfloat feed' carburetter. Tubular propeller shaft with Spicer needle roller bearing universal joints. Crownwheel and pinion, 5.27:1. Magna wire wheels 3.25 × 18, 5-stud fixing. Dunlop 525-18 tyres. Chromium-plated radiator shell with imitation honeycomb. Shield shape badge. Calormeter and wings. Horn and fog lamp at front. Triplex single-panel windscreen with self-starter electric wiper and twin-blades. Harmonic stabilised front bumpers. Luggage grid on saloon. Wheelbase 9ft 6in. Track 4ft 8in.

Morris Oxford Sixteen, 1935 model.
Chassis numbers: 35/035640 - 35/039140.
Morris Oxford Twenty, 1935 model.
Chassis numbers: 35/OT35638 - 35/OT39140

Engine: 16hp home, QD. 16hp LHD models, QE. 20hp home, QF. 20hp LHD models, QG. Bendix automatic clutch control and free-wheel. Lockheed hydraulic four-wheel brakes. Handbrake, cable to rear wheels. 12in diameter drums. Right-hand accelerator pedal. Trafficator and lamp switches on steering wheel. Automatic advance/retard. Headlight 'cut & dip' type. Concealed Lucas Trafficators with automatic cancel. Battery master switch. CVC dynamo. Radio aerial built into sliding-head. Bishop Cam steering. Luvax double action hydraulic shock absorbers. Rear mounted petrol tank, 14 gallons, SU electric pump to SU 'overfloat feed' carburetter. Tubular propeller shaft with Spicer needle-roller bearing universal joints. Crownwheel and pinion, 5.27:1. Magna wire wheels, home models 3.25 × 17, export models 4.5 × 16. Tyres, Dunlop, home models 5.50-17, export models 700-16. 5-stud. Chromium-plated radiator shell with imitation honeycomb. Shield shape badge. Calormeter and wings. Thermostat in cooling system. Horn and fog lamp at front. Triplex single-panel windscreen with self-starter electric motor and twin-blades. Harmonic stabilised front bumper. Luggage grid with removable centre section rear bumper on saloon. Wheelbase 9ft 6in. Track 4ft 8in.

Morris Oxford Six. Engines

Type LA, 1930 & 1931-3 models.

14.9hp. 1,938cc. 6-cylinder, side-valve. 63.5mm bore, 102mm stroke. 4-bearing crankshaft. 3-ring aluminium pistons. Compression ratio 5-6:1. Bore/stroke ratio 1.6:1. Valves, inlet & exhaust 27.8mm diameter. Centrifugal water pump. Coil ignition, distributor driven from rear of dynamo. 30.5bhp at 3,000rpm. Combined air filter, pre-heater and fume consumer head. 3-speed gerbox, top 1:1, second 1.72:1, first 3.2:1, reverse 3.88:1. Multi-plate cork insert clutch running in oil.

Type LC, 1932 models.

14.9hp. 1,938cc. 6-cylinder, side-valve. 63.5mm bore, 102mm stroke. 4-bearing crankshaft. Aluminium pistons. Compression ratio 5.6:1. Bore/stroke ratio 1.6:1. Valves, inlet & exhaust 27.8mm diameter. Centrifugal water pump. Coil ignition, distributor driven from rear of dynamo. 32.25bhp at peak 3,400rpm. Combined air filter, pre-heater and fume consumer head. 4-speed gearbox, 'twin-top', top 1:1, third 1.479:1, second 2.28:1, first 4:1, reverse 5:1. Single-plate cork insert clutch running in oil.

Type QA, 1933 models.

15.94hp. 2,062cc. 6-cylinder, side-valve. 63.5mm bore, 102mm stroke. 4-bearing crankshaft. 4-ring aluminium pistons. Compression ratio 5.65:1. Bore/stroke ratio 1.557:1. Centrifugal water pump. Coil ignition, distributor driven from oil pump shaft. 41.75bhp at peak 3,600rpm. Combined air filter, pre-heater and fume consumer head. 4-speed gearbox, 'twin-top', top 1:1, third 1.479:1, second 2.28:1, first 4:1, reverse 5:1. Single-plate cork insert clutch running in oil.

Type QB, 1934 models.

15.94hp 2,062cc. 6-cylinder, side-valve. 65.6mm bore, 102mm stroke. 4-bearing crankshaft. 4-ring aluminium pistons. Compression ratio 5.65:1. Bore/stroke ratio 1.557:1. Centrifugal water pump. Valves, inlet & exhaust, 30mm diameter. Valve lift, inlet & exhaust, 7.5mm. Coil ignition distributor driven from oil pump shaft. 41.75bhp at peak 3,600rpm. Combined air filter, pre-heater and fume consumer head. 4-speed gearbox, top (synchromesh) 1:1, third (synchromesh) 1.479:1, second 2.28:1, first 4:1, reverse 5:1. Single-plate cork insert clutch running in oil. Bendix automatic clutch.

Type QD, 1935 home models.
Type QE, 1935 export models.

15.94hp. 2,062cc. 6-cylinder, side-valve. 65.5mm bore, 102mm stroke. 4-bearing crankshaft. 4-ring aluminium pistons. Compression ratio 5.65:1. Bore/stroke ratio 1.557:1. Centrifugal water pump. Valves, inlet & exhaust, 30mm diameter. Valve lift, inlet 7.7mm exhuast 7.8mm. Coil ignition distributor driven from oil pump shaft. 41.75bhp at peak 3,600rpm. Combined air filter, pre-heater and fume consumer head. 4-speed gearbox, top (synchromesh) 1:1, third (synchromesh) 1.479:1, second 2.28:1, first 4:1, reverse 5:1. Single-plate cork insert clutch running in oil. Bendix automatic clutch.

Type QF, 1935 home models.
Type QD, 1935 export models.

19.82hp. 2,561cc. 6-cylinder, side-valve. 73mm bore. 102mm stroke. 4-bearing crankshaft. 4-ring aluminium pistons. Compression ratio 5.65:1. Centrifugal water pump. Valves, inlet & exhaust 33mm diameter. Valve lift, inlet 7.7, exhaust 7.8. Coil ignition, distributor driven from oil pump shaft. 52bhp at peak 3,700rpm. Combined air filter, pre-heater and fume consumer head. 4-speed gearbox, top (synchromesh) 1:1, third (synchromesh) 1.479:1, second 2.28:1, first 4:1, reverse 5:1. Single-plate cork insert clutch running in oil. Bendix automatic clutch.

Morris Oxford Six. Body Colours and Upholstery

	1930	1931	1932
4/5 seater tourer	Deep maroon or blue cellulose with matching leather.	Black cellulose with brown leather. Blue cellulose with blue leather.	
Saloon	*Sliding-head* Deep maroon or Niagara blue cellulose with matching Vaumol leather.	*Sliding-head & fixed-head* Blue cellulose with blue leather. Blue cellulose with brown leather.	*Sliding-head* Blue & black, black, brown & black or wine & black cellulose all with brown leather.
Fabric saloon	Black or dark red fabric body with matching Vaumol leather.	Black or red fabric body with red leather.	
Coupé (sliding-head)	Deep maroon or blue cellulose with matching leather.	Black cellulose with brown leather. Blue cellulose with blue leather.	Blue & black, black or wine & black cellulose all with brown Celstra leather.
Sports coupé			Blue & black, black or beige & brown cellulose all with brown Celstra leather. Grey & dove cellulose with blue Celstra leather.

	1933	1934	1935
Saloon (sliding-head)	Blue & black, black, green & black or wine & black cellulose all with brown leather.	Blue & black cellulose with blue leather. Green & black cellulose body with green leather. Brown duotone or black cellulose with brown leather.	Blue & black cellulose with blue leather. Black cellulose with brown leather. Green & black cellulose with green leather. Red & black cellulose with red leather.
Coupé (sliding-head)	Blue & black, black, green & black or wine & black cellulose all with brown leather.		
Special coupé	Green duotone cellulose with green leather. Red duotone cellulose with red leather. Grey duotone cellulose with blue leather. Black duotone cellulose with brown leather.	Green duotone cellulose with green leather. Brown duotone cellulose with brown leather. Cream & green cellulose with green leather. Black cellulose with brown leather.	Green duotone cellulose with green leather. Brown duotone cellulose with brown leather. Cream & green cellulose with green leather. Black cellulose with brown leather.

Morris Isis and Twenty-Five-Six 1930-5

It was in July 1929 that the trade had its first sight of the new Morris six-cylinder model when a motor dealer's banquet was held at the Holborn Restaurant in London. This new model, called the Isis-Six and generally introduced to the public the following month, was the successor to the Morris-Six. It utilised the same 17.7hp overhead camshaft engine although the body design was entirely different. A change of name was obviously

1928 Dodge Victory Six used Budd body pressings identical to the early Morris Isis.

desirable as the appearance of another six-cylinder Morris car, the Oxford-Six, made the original name 'Morris-Six' somewhat ambiguous. The Isis (named after the river Isis, which is the name given to the Thames near Oxford) for the 1930 season came as a large open tourer, an all-steel bodied saloon and following the precedent of the Morris-Six, a Gordon England-designed 'Club Coupé' with sliding-head.

The all-steel body chosen for the Isis, produced from Budd dies and tooling by The Pressed Steel Co at Cowley, had already been in production in Britain for a year and used for the Wolseley version of a six-cylinder saloon, while in

America the same body was being used on the short-lived Ruxton Sedan and the Dodge. The Ruxton, which made its debut at the September 1929 New York Motor Show, raised a few eyebrows in motoring circles for the manufacturers, New Era Motor Company, had introduced front-wheel-drive — several months before the L29 Cord, three years before the first traction-avant Citröen and almost five years before the famous 810 model Cord. it had a short history and production ceased late 1930 or early 1931, after about 200 of the sedan versions had been made with bodies imported from England. Bodies from Budd dies were also used by Dodge for their Victory Six model. The first Victory Six (131 series) was introduced in November 1927 and after six or seven month's production, a slightly longer body on the same chassis was used for the 130 series Victory Six. This second model shared an identical body with the Morris Isis. When Dodge became a division of the Chrysler Corporation, the Victory Six was renamed the 'DA' and a clever re-style ensured that minimum

*Front view of the 1930 model Morris Isis-Six illustrating
the twin spare wheels supplied on export versions.*
(Morris Motors Ltd)

changes were needed to the dies; as an example, the doors from a Victory Six would fit the DA model with only the belt moulding out of alignment. The Wolseley car which used the Isis bodies was the 21/60hp 'Messenger' introduced in September 1928 and primarily intended for the export market. Seven of these cars were provided for the Prince of Wales for use on his safari in East and Central Africa in 1930. Although Triumph had pioneered hydraulic brakes in Britain, the Wolseley Messenger was the first car from the Morris empire to make use of the Lockheed system manufactured by Automotive Products (who held the sole license to make the components outside the United States). Both Austin and Ford looked upon hydraulic braking with suspicion, preferring a mechanical arrangement.

The footbrake on the Isis was also of the Lockheed hydraulic type, operating internal expanding aluminium shoes in 14in drums on all four wheels, and for the hand-brake (or more correctly, the parking brake) the designers had borrowed a feature from the smallest Morris of the period, a transmission brake of the caliper-shoe type operating on a drum behind the three-speed gearbox. Also common to the Morris Minor was the use of an open propeller shaft, although for the Isis the universal joints were, of necessity, the more robust Hardy-Spicer all-metal units combined with a splined sliding section at the forward end of the shaft to allow for movement of the back axle due to springing (this was not necessary on the Minor due to the flexibility of the fabric discs used). For the home market the large five-stud wire wheels were

This 1930 Isis saloon came to light when an elderly
spinster died in January 1981 at Ilkley, Yorkshire.
(Ken Martin)

*Decorative winged badge on
the early Isis.*

shod with 29 × 5.50 (5.50-19) Dunlop tyres, while export models were fitted
with similar size tyres on steel artillery wheels, as well as two side-mounted
spares. Ironically Morris commercials for export were lacking even a tyre on
their single spare wheel!

The basic Isis chassis, which was constructed with excessively deep side
channels joined by four heavy pressed-steel cross members and a tubular cross
member between the dumb-irons, became even more rigid when fitted with the
four-point mounted engine at the front, and the 15-gallon petrol tank at the
rear. Onto this chassis the all-steel saloon body was secured by twenty-five
bolts. Not only did the body design contrast with its predecessor, but the
chassis differed dimensionally, having a wheelbase shorter by 3in, although the
4ft 8in track was retained. A closer look at the details of the body used on the
saloon and tourer reveals that the 'sill-less', 'mono-piece', or 'all-steel' body (all
terms applied to this type of construction in the 'thirties) was made-up of
various parts formed in shape by presses, complete with the necessary mouldings, recessed panels and wing housing, and then welded together. In the case
of the saloon the whole side of the body from the screen line to the join at the
rear was pressed as one piece. Reinforcing pressed steel angle plates were added
in the corners and large areas of panel were supported by U-section struts spot-
welded to the framing. The sides of the finished body had flanges turned-in at
the bottom of the doors, and these flanges provided the support to allow the
whole to sit on the chassis and the body sides to hang below and be bolted to
the frame sides; a hollow moulding then concealing the fixing bolt heads.
Timber was still used for the floor panels and the roof frame.

The Isis saloon must surely rank as the most colourful model ever produced
by Morris Motors Ltd, not only because of the number of colour alternatives
offered but also the way these colours were applied. The saloon body had
indented panels and mouldings (ostensibly for decoration, but actually a means
of strengthening the huge pressings) which tempted the colour stylist to infill

the panels with contrasting shades. For example the export version of the royal blue and ivory scheme had the basic bodywork of royal blue with the panels beneath the windows, bonnet top, and the artillery wheels cellulosed in ivory, while stove-enamelled black was used on mudguards, aprons, headlamp shells etc. The black rubber mats on the running boards had 'Isis-Six' moulded on them. Similarly colourful was the Morris Motors' catalogue description of the Isis saloon: 'Assuredly the zenith of quiet magnificence, the Morris Isis Saloon is a fully accredited luxury vehicle, in the manufacture of which no expense has been spared. With artistic interior fittings, toning with the rich hides of the resilient upholstery, an interior refinement is offered to which the printed word hardly can do justice. The comfort and convenience of both driver and passenger is studied to the smallest detail; handily disposed smoker's and lady's companions, large pockets on the backs of the forward seats and a plated footrail, among other items, being furnished for the rear occupants.'

Soon the Isis saloon was offered in a de luxe form where, for an additional £10 (on the basic price of £385), the bodywork finish was royal scarlet and ivory cellulose with red leather upholstery; or royal blue and ivory, already mentioned, with complementing blue upholstery and ivory wheels. Unspecified 'special interior equipment' was included which, presumably, referred to the provision of two loose cushions, window louvres with chromium-plated frames, and a folding arm-rest in the centre of the rear seat.

The most expensive of the three 1930 Isis models was the Isis-Six 'Club Coupe' with sliding-head (£399) which followed the lines of the previous year's Isis cars, the Club Coupé had an entirely new American-styled chromium-plated radiator shell with 'Calorstatically-operated' shutters in addition to a fan (fitted to all but the earliest Isis engines, but which had not been standard on the same engine when employed in the Morris-Six) and a new badge which was effectively the normal disc motif with wings added. Surmounting the radiator was a Calometer temperature gauge provided with decorative wings, the hallmark of the larger Morris models of that period. This was the season when chromium plating of external bright parts and the use of Triplex safety glass was standardised for all Morris cars. Although not publicised by Morris Motors Ltd, there was a second Gordon England design based on the Isis chassis, the 'England Isis Four-Door Saloon' displayed in the Carriage Work Section at the 1929 Olympia Motor Show. Equipment for this light fabric-bodied version included a special Gordon England ventilator above the single-panel adjustable windscreen, a rear blind operated from the driver's seat, loose cushions and a special holder for a woman's umbrella!

Perhaps it was the need for a gimmick that manifested itself in the special eye-catching, vehicles used by the various newspapers before World War II. This is not so evident now perhaps with other means of effective advertising. In the 1930s the *London Evening News* used at least one Morris Commercial 'Tonner' chassis with a special body equipped with illuminated news panels on the roof, and about the same time the *Daily News & Star* had a number of Isis chassis fitted out as special vans with coachwork by Stewart & Ardern Ltd. Based on the standard car chassis, with strengthened springs to carry a 15cwt load, the vans had a fixed panel embodying a sliding window on the off-side of the cab, while a half-door was provided on the near-side. A sliding door separating the driver from the van area, and roller shutters at the rear allowed the loading and unloading to be affected from both ends. These special Isis vans were finished in the group's standard livery of red, gold and black, and ran on export-type artillery wheels.

For the second year of production (1931) the Isis returned to the Morris stand at Olympia in much the same form as before, although the ostentatious colour schemes had been toned down a little. There were some small changes beneath the shell but some of these, like the replacement of the transmission handbrake to a more conventional arrangement operating on the rear wheels, and a redesigned rear axle, would not have been evident on the show cars as the first 300 or so 1931 models still retained the earlier system. Conversely, the Magna wire wheels offered as an option on the export (normally artillery wheels) and home models (small-centre wire wheels) of the 1931 Isis became standard on the last 540 cars of that season. Unfortunately for the historian, the equipment specification on Morris cars and commercials did not always change at a convenient model year datum.

Variations as applied to export models have already been mentioned. There were other differences not immediately obvious, such as the two-piece sliding seat at the front of the 1931 saloon in lieu of the home model's two individual pneumatic front seats which were independently adjustable; export models had

Found in Malta in the 'sixties and restored. The direction indicators fitted to this 1931 Isis saloon are later additions. (K. Knowles)

a coil-spring interior. Triplex louvres which were provided for all six side windows on the home version of the 1931 saloon were not fitted to export cars, and special to the export model was the crown-wheel and pinion on the spiral-bevel rear axle with 9/49 teeth giving a 5.44:1 ratio (compared with the home models 11/53 teeth with a 4.82:1 ratio), extra leaves in the rear springs, and the optional extra of a second spare wheel for fitting forward of the near-side running board.

For 1932, the third season for the Isis, the model was completely redesigned. It lost its transatlantic look and the body construction was back to a traditional jig-built ash frame with pressed steel panels. The chassis frame, as a consequence of the increased overall length (now 6in longer with a corresponding increase in the wheelbase), reverted to a more conventional depth of $6\frac{1}{2}$in, but was fabricated of thicker (0.128in) pressed steel channel. With the new body came a new radiator shape carrying a shield-shape badge which was to become familiar on almost all Morris motor cars for the next seven years, while in the case of the Isis the chromium-plated radiator shell enclosed vertical shutters with a similar plated finish. The 'eddyfree' fronted body on these new models was in effect a protrusion-free radiused surface above the windscreen to reduce air disturbance and thus decrease noise; its aerodynamic characteristics must have been negated by a frontal aspect almost 6ft square. The front view presented an imposing appearance with the large $6\frac{3}{4}$in diameter Lucas 'Biflex' chromium-finished headlamps mounted on a bar which also provided

Spotlight and mascot on the horn, etc, are later additions
to this 1932 model Isis saloon.
(Birmingham Post & Mail)

additional support for the large mudguards; the leading edge of the mudguards was now almost down to bumper level and fitted with so called 'splash guards' or what the Americans (who had introduced them on the Graham Blue Streak) called 'fender skirts'. Another feature of the headlamp bar was the centrally mounted horn in a special plated casing.

The two models available on the Isis chassis for 1932 were the sports coupé and a four-door saloon. Equipment to be found on both styles included the Pytchley sliding-head, leather upholstered seating with adjustable front seats and arm rests and pull cords for the rear passengers (Moseley 'Float-on-air' rear cushion in the coupé), door pockets, fitted ashtrays, remote operated rear-window blind, roof lamp and winding windows. Saloon owners also had a central folding arm rest on the rear seat, blinds on the rear quarter-lights and special head rest cushions. For the driver's convenience, finger-tip controls (ignition, throttle and lamps) were grouped in the centre of the steering wheel, concealed lighting illuminated the dash instruments, and an organ type accelerator was positioned between the clutch and brake pedals.

That the driver's door was lockable and fitted with a removable key, and that the remaining doors could be secured from within, was given prominent publicity when extolling the full equipment on the Morris Isis. It may seem strange but in the early 'thirties it was not permitted to lock a car left in a public car park. An exception was in the London Traffic Area after mid-1932, when the Minister of Transport made a concession allowing motorists to lock the doors of their parked cars in areas under their jurisdiction!

A comparison of the mechanical features with the previous season's Isis shows the use of shorter, but wider, road springs, and the introduction of a curious anti-shimmy device fitted to the front off-side spring. The gearbox, in common with other six-cylinder Morris cars of the 1932 range, was now a four-speed unit giving even more flexibility to the same 17.7hp overhead camshaft engine. Experience had shown that the clutch was prone to spin, especially in cold weather due to the higher viscosity of the lubricating oil (compounded oils were not recommended by Morris) at lower temperatures. Modifications had been made by adding additional oil release holes in the flywheel cover plate. Presumably, this was not the complete answer for all 1932 models were fitted with a coil spring on the hub of the front corked plate, with the object of assisting the cork driver plate to leave the driven plates, and the number of separating pins was increased from four to eight with heavier springs.

It was customary for motor manufacturers of the period to supply the basic chassis to coachbuilders, and Morris Motors Ltd were no exception. One such Isis chassis went to the Minories Coachworks, late in 1931, to receive a commodious ambulance body for use by the Newcastle General Hospital. The builders of that ambulance (registered VK5986) could hardly have known that they were anticipating the fate of many Isis and other large cars which were converted for a similar role following the outbreak of hostilities some seven years later. In China a war had already broken out so it is not surprising to find a record of an Isis saloon, fitted with a special interior bullet-proof screen fitted behind the driver for rear passenger protection, being supplied in November 1932 by the Morris Hong-Kong distributors, Dodwell & Co Ltd to the Government of Kwang-si province. A 1932 Isis saloon provided the British Consul-General at Canton, Sir Herbert Phillips, with transportation.

Despite arguments put forward during 1932 advocating the discontinuance of the Motor Show as an annual event (a view probably reinforced after the show, with dropping attendance figures), the 26th International Motor Show took place at Olympia during 13-22 October. On the Morris Motors Ltd stand a last-minute surprise exhibit was a new 25hp car. Indeed the catalogues and press release packages already prepared made no mention of the new model. Although it was hailed as a new model, in practice it was an Isis with a 3,486cc side-valve engine rated at 25.01hp and delivering 60bhp at the maximum 3,000rpm. This same engine was soon to provide the motive power for the Morris Commercial 'C' series of vehicles. Both Isis and Twenty-Five were offered with identical saloon or special coupé bodywork finished in the same colour schemes and, not surprisingly, shared the same chassis number sequence. Later in the season a tourer version appeared; this was listed about March 1933.

Only minor changes had been made in the technical specification for the 1933 season Isis and in the main these features were incorporated in the Twenty-Five. The Wilcot traffic signals, mentioned elsewhere in this book, naturally appeared on the early productions and the electric windscreen wiper featured double squeegees. Joseph Lucas Ltd had just developed a new starter system known as the 'Startix' which comprised a compact box containing

*1933 Morris Isis saloon. Note the ill-fated flashing traffic
direction indicators. (Morris Motors Ltd)*

A surviving 1933 model Isis special coupé. (Ken Martin)

solenoid relays designed to operate the starter motor automatically whenever the ignition switch was on and the engine stationary; thus should the engine stall at any time the Startix automatically re-started it — this system was incorporated in the Isis and Twenty-Five.

Earlier problems with the Isis clutch were finally solved on the new season's car by the introduction of a single-plate cork-insert clutch. This apparently

As found in the Isle of Man. This 1933 Morris Twenty-Five special coupé is now being restored to its original condition. (Michael Pearce)

simple solution involved the modification of the Isis engine bearings, for with the multiple plate clutch the load was taken by the rear main bearing, clutch release pressure being exerted towards the front of the engine. With the single-plate clutch this pressure was exerted towards the rear of the engine and the Cowley engineers had found it necessary to make the front main bearing a flange type and add a case-hardened steel thrust washer to take the crankshaft thrust.

Detail differences between the Isis and the Twenty-Five included larger section tyres (600-19) on the latter, a slightly longer bonnet moving the radiator forward, and, strangely, the retention of the Autovac when the Isis had now been equipped with an SU electric petrol pump. Both models retained the centre accelerator pedal. Externally, the coachwork on the special coupé remained the same (for some unknown reason the coupé version for the 1932 Morris range had been called the 'Sports Coupe', while similar bodies fitted on the range for 1933 were advertised as 'Special Coupe') and at first glance so did the saloon; but the hinges on the rear doors were reversed so that on the 1933 cars they opened at the front, allowing for easier exit and entry.

Out of the combined Isis/Twenty-Five production of 1,038 car for the 1933 season only 228 were Twenty-Five models; the major proportion being saloons. There were eight tourer bodies, forty-four special coupés, and twenty-two chassis — only one chassis went to the home market, the remaining twenty-one being exported to Australia. Export sales accounted for more than one third of the Twenty-Fives produced for the 1933 season.

Of the few coachbuilders to make use of the Isis chassis was Messrs T.H. Gill & Sons Ltd of Paddington. Their 'Special', a convertible saloon finished in deep red, was something of a swan-song for the firm who had specialised in all-

In 1934 the original price of this Morris Twenty-Five
tourer was £385. The body and equipment was almost
identical to that of the same season's Isis tourer. (J. Offer)

weather bodies since 1914. The Isis featured in the company's last appearance at Olympia, in 1932, and by 1935 they had joined the ever increasing number of specialist coachbuilders to go out of business.

The penultimate season for the Isis and the Twenty-Five in perpendicular form was 1934. The models for both this and the 1935 season (before the introduction of the 'Series' cars), can be considered together as, apart from the identification plate carrying the revealing '25' prefix, there was no difference in design or appointments. True, the blue option for the tourer of 1934 was substituted by black in the 1935 model listings, but with this exception the colours were the same. The low number produced over the short period the two models were available for the 1935 season (from about September 1934 to June 1935 for the Isis, and to late 1934 for the Twenty-Five), together with an unaltered engine prefix for both years, leads to the conclusion that the Isis and Twenty-Five, although nominally 1935 models, were in fact left-overs from the previous season, especially as these were the only Morris cars of 1935 to retain a centre accelerator. Combined production figures for the Isis and Twenty-Five (as chassis, tourer, coupé and saloon) totalled 852 for 1934, and a mere 207 for 1935.

For these final models of the Isis and Twenty-Five the thermostatically controlled shutters had been replaced by what became something of a standard for Morris cars of the next few years, a more rounded chromium-plated surround combining a dummy honeycombe, behind which the radiator proper was positioned. Morris held the price of the Twenty-Five to £395 for the saloon and coupé and the tourer at £385 throughout its three-year listing, but the Isis prices had increased by £20 on the closed models and £10 on the tourer for the 1934 and 1935 seasons, at £370 and £350 respectively. However, there can be no doubt that the last models were the most luxurious of the Morris Sixes, in particular the saloon which now had full width leather-upholstered front and rear seats with folding centre arms. In addition the rear passengers were provided with loose seat cushions, an occasional table, folding foot rest and even corner head rest cushions.

A new chassis frame of thinner (0.116in) steel had been designed for the larger horsepower cars in the Morris range for 1934, and this was said to resist any tendency to twisting. A cross-braced frame between the side members gave the chassis more rigidity and as a consequence put less stress on the body. In adding extra cross-members the designers had to provide passage for the transmission shaft and this problem was overcome by a short tunnel.

Concealed traffic indicators were fitted within the door pillars (except on the touring car which retained the boxed type on outriggers) with tiny tell-tale mirrors in the corners of the windscreen; frameless louvres, interior visor, automatic advance/retard distributor, battery master-switch, Armstrong hydraulic shock absorbers, and 'stabilising' bumpers were also fitted. The four-speed synchromesh gearbox was combined with a Bendix automatic clutch control and free-wheel which allowed the driver to make gear changes without using the clutch pedal.

Both Isis and Twenty-Five continued to utilise common bodywork and export models were available in left-hand and right-hand drive versions, with

1926 'Convertible' Bullnose Morris (removable van body).   Photo: Harry Edwards

1927 Morris Cowley Four-seat Tourer.   Photo: David Adair

1930 Morris Cowley Fixed-head Saloon.   Photo: Ray Jenkins

1931 Morris Major Tourer.   Photo: Harry Edwards

1933 Morris Oxford Six Saloon.   Photo: K. Martin

1935 Morris Ten-Four Saloon.   Photo: K. Martin

1929 Morris Minor Fabric Saloon.   Photo: Jim Dalton

1931 Morris Minor Two-seater.   Photo: B. Maeers

1932 Morris Minor Saloon.   Photo: Harry Edwards

1933 Morris Minor Two-seater.   Photo: J. Roles

1930 Morris Minor 5-cwt Van.   Photo: Janet Denton

Series I Morris Eight 5-cwt Van.   Photo: Harry Edwards

1935 Morris Eight Four-door Fixed Head Saloon. Photo: Tony Boaks

Series I Morris Eight Four-seater Tourer. Photo: Kelvin Earlburg

Series E Morris Eight Tourer followed by a Series II Eight Saloon.   Photo: Harry Edwards

Series E Morris Eight Two-door Saloon.   Photo: Paddy O'Riordan

1939 Series E Four-seater Tourer.   Photo: M. Hannigan

Series E Four-door Saloon.   Photo: J. Bates

Series Z Morris Eight 5-cwt Van.   Photo: Finton Foley

Series E Morris Eight Tourer with Australian bodywork by T. J. Richards & Sons, of Adelaide.   Photo: Harry Edwards

Series II Morris Ten-Four Saloon.   Photo: D. J. King

Series M Morris Ten-Four Saloon.   Photo: Brian J. Olner

Series II Morris Ten-Four Saloon. Photo: R. Stansfield

1935 Morris Twelve-Four Saloon.   Photo: N. Beech

Series II Morris Twelve-Four Saloon.   Photo: A. Shelden

1934 Cowley Six Special Coupe.   Photo: Geoffrey Allo

Series II Fourteen-Six Saloon.   Photo: D. A. McKenzie

Series III Twenty Five-Six Saloon.   Photo: Ralf Garratt

Series MM Minor Two-door Saloon.   Photo: Harry Edwards

1950 Morris Six Saloon.   Photo: Harry Edwards

1951 Series MO Morris Oxford Saloon.   Photo: Harry Edwards

750-16 tyres. On the Isis the spare wheel was fitted on the near-side front wing (the 1930 and 1931 models had had this side mount on the off-side), while the Twenty-Five had two spares, one mounted each side. Wilmot Breeden metal wheel covers were standard on both. The other means of identification, apart from the badge, was the provision of two fog lamps on the Twenty-Five contrasting with the single unit standard on the Isis.

Both models ceased to be listed in mid-1935 and it was not until 1954 that the name 'Isis' was re-used on the BMC 2.6 litre Six. By the time production ceased something like 7,607 Morris Isis and Twenty-Five six-cylinder cars had left the lines at Cowley; of this total about half were Isis vehicles with all-steel bodies.

Morris Isis and Twenty-Five. 1930-5 models. Chassis Numbering			
Model	Chassis Numbers	Model	Chassis Numbers
1930 Isis	101 - 2730		
1931 Isis	IS2731 - IS4039		
1932 Isis	IS4040 - IS5523		
1933 Isis	MI 5524 - MI 6648	1933 Twenty-Five	5611-6648 *
1934 Isis	34/I 6649 - 34/I 7500	1934 Twenty-Five	34/TF 6649 - 34/TF 7500 *
1935 Isis	35/I 7501 - 35/I 7707	1935 Twenty-Five	35/TF 7501 - 35/TF 7707 *

* Number series used for both Isis and Twenty-Five models. For the 1934 season the first number used for the Twenty-Five was actually 6655.

Morris Isis. 1930-5. Specification

Engine: 1930-1 models, type 'JB'. 1932 models, type 'JH'. 1933 models, type 'JJ'. 1934-5 models, type 'JK'. Six-cylinder, overhead-camshaft, inclined valves. 69mm bore × 110mm stroke, 2,468cc, 17.7hp. Compression ratio 5.5:1. 49.8bhp at 3,200rpm.
SU carburetter. Multiplate cork-insert clutch running in oil, 1930-2 models. Single-plate cork-insert clutch running in oil, 1933-5 models. Three-speed gearbox, 1930-1 models. Four-speed 'twin-top' gearbox 1932. Four-speed synchromesh gearbox, 1933-5 models. Rear mounted petrol tank using Autovac, 1930-2 models. SU Petrolift, 1933 models. SU electric petrol pump, 1934-5 models. Wire wheels (small centre) with 5.50-19 Dunlop tyres on home 1930 and early 1931 models, with optional extra of Magna wire wheels. Magna wire wheels (3.25in × 19in) with 5.50-19 Dunlop tyres on home models late 1931-5. Export models 1930-1 steel artillery wheels with 5.50-19 Dunlop tyres, Magna wire wheels optional extra for 1931 season. Export models late 1931-5, Magna wire wheels. Tyres on 1934-5 export models known to be 750-16. Steering, Bishop-Cam. Lockheed hydraulic foot brakes to four wheels. Handbrake on 1930 and first few 1931 models operate on transmission, later models by rod to rear wheels.

Morris Twenty-Five. 1933-5. Specification

Engine: 1933 models, type 'OA'. 1934 models, type 'OB' & 'OE'. 1935 models, type 'OE'. Six-cylinder, side-valve, 82mm bore × 110mm stroke, 3,485cc, 25.01hp. Compression ratio 5:1. 74bhp at 3,400 maximum rpm.
SU carburetter. Single-plate cork-insert clutch running in oil. Four-speed synchromesh gearbox. Rear mounted petrol tank using Autovac. Magna wire wheels (3.25in × 19in) with 600-19 tyres. Steering, Bishop-Cam. Lockheed foot brakes to four wheels. Handbrake, rod to rear wheels.

Morris Isis. 1930-2 Models. Body Colours and Upholstery		
	1930	1931	1932
Tourer	Blue/black or maroon/ wine with matching leather. (Black hood material)	Blue cellulose with blue leather. (Buff hood material). Lake cellulose with red leather. (Buff hood material)	
Saloon	*All-steel saloon* Holborn blue cellulose with ivory wheels and blue leather. Olive green & ivory cellulose with green wheels and red leather. Niagara blue & grey cellulose with blue wheels and blue leather. For an additional £10, the following colours were available: Royal blue & ivory cellulose with ivory wheels and blue leather. Sometime during the 1930 season wine & maroon cellulose with red leather was listed. *De luxe saloon* Royal blue & ivory cellulose with blue leather. Deep maroon cellulose with red leather.	*All-steel saloon with sliding-head* Black cellulose with red leather. *All-steel saloon with fixed-head* Black cellulose with green or red leather. Lake cellulose with red leather. Blue cellulose with blue leather.	*Coachbuilt saloon with sliding-head* Black, blue & black, wine & black or brown & black cellulose all with brown leather upholstery.
Coupé	*Club coupé* Black fabric body with blue leather. Blue wheels.		*Sports coupé* Blue & black, black or beige & brown cellulose all with brown Celstra leather. Grey & dove cellulose with blue Celstra leather.

Morris Isis and Twenty-Five. 1933-5. Body Colours and Upholstery		
	1933	1934	1935
Tourer	No record of colours. Leather upholstery.	Blue or brown cellulose with brown leather.	Black or brown cellulose with brown leather.
Saloon, sliding-head	Wine & black, green & black, blue & black or black cellulose with brown leather.	Blue & black cellulose with blue leather. Brown duotone cellulose with brown leather. Green & black cellulose with green leather. Black cellulose with brown leather.	Blue & black cellulose with blue leather. Brown duotone cellulose with brown leather. Green & black cellulose with green leather. Black cellulose with brown leather.
Special	Green duotone cellulose with green leather. Red duotone cellulose with red leather. Grey duotone cellulose with blue leather. Black duotone cellulose with brown leather.	Green duotone cellulose with green leather. Brown duotone cellulose with brown leather. Cream & green cellulose with green leather. Black cellulose with brown leather.	Green duotone cellulose with green leather. Brown duotone cellulose with brown leather. Cream & green cellulose with green leather. Black cellulose with brown leather.

Morris Isis and Twenty-Five. 1930-5. Magazine Bibliography

Model Year	Subject	Source	Date	Pages
1930	Morris Programme for 1930	The Morris Owner	9/29	p840-5
1930	De Luxe Isis Saloon	The Morris Owner	7/30	p608
1930	Morris Isis Six, Description	The Autocar	2/8/29	p211-19
1930	Isis Saloon, Road Test	The Autocar	8/11/29	p1004-5
1930	Isis seeks the Pyramids	The Morris Owner	5/30	p316-19
1930	Isis England Saloon, Road Test	The Autocar	30/5/30	p1057-8
1930	Isis, Road Test	The Motor	3/9/29	p188-90
1930	Isis Saloon, Used Cars Tested	The Autocar	28/8/31	p345
1930	Isis Saloon, Used Cars Tested	The Autocar	16/12/32	p1129
1931	Once Over the Isis	The Morris Owner	3/31	p28-31
1931	Isis. Door panel servicing	The Morris Owner	7/31	p560
1931	Morris Programme for 1931	The Morris Owner	9/30	p847-51
1931	Morris Programme for 1931	The Autocar	5/9/30	p445-8
1932	Isis, the largest Morris model	The Morris Owner	3/32	p22-4
1932	Isis Saloon, Road Test	The Motor	9/2/32	p46-52
1932	Morris Programme for 1932	The Motor	1/9/31	p199-206
1932	Morris Programme for 1932	The Morris Owner	9/31	p787-92
1932	Morris Programme for 1932	The Autocar	4/9/31	p391-5
1933	With the Twenty-Five in Scotland	The Morris Owner	11/32	p948-50
1933	Morris Programme for 1933	The Morris Owner	9/32	p709-16
1933	Morris Programme for 1933	The Motor	6/9/32	p214-20
1933	Morris Programme for 1933	The Autocar	2/9/32	p411-14
1933	Morris Twenty-Five, Road Test	The Motor	28/2/33	p118-22
1933	Isis, Road Test	The Autocar	2/9/32	p419-20
1933	Morris Twenty-Five Saloon	The Autocar	14/10/32	p696
1933	Morris Twenty-Five	The Motor	11/10/32	p431-4
1934	Morris Programme for 1934	The Autocar	1/9/33	p361-4
1934	Morris Programme for 1934	The Morris Owner	9/33	p819-26
1935	Morris Programme for 1935	The Morris Owner	9/34	p655-62

Morris Major, Cowley Fifteen-sixes 1931-5

For the present-day vintage car enthusiast the Morris Major provides the perfect example of the pitfalls encountered when considering the calendar year instead of the model year while searching around for elusive spares. Name apart, the Major for its first season in 1931 was a completely different design to that of the following two years. As introduced in September 1930 the Morris Major was basically a Cowley with the six-cylinder side-valve engine as fitted to the contemporary Oxford Six, and as a six-cylinder block would require extra cooling it was no coincidence that the 1931 version of the four-cylinder Cowley also had a larger radiator; the Major version of this taller radiator had thermostatically operated vertical shutters operated by mechanical linkage to a Calorstat unit housed in the header tank.

A distinguishing feature of the 1929-31 Cowleys was the rounded front wings linked by an inverted 'U'-shaped lamp bar. Apart from the 1929 Oxford, the Major was the only other model to share identical one-piece wings. Components new to the Cowley were, naturally, incorporated into the Major design and included the group lubrication system with a panel of six nipples mounted above the rear axle to carry grease to rear brake-camshafts and spring seats. Both Cowley and Major had an 8ft 9in wheelbase and 4ft track, hence many of the transmission, axle and brake components were common, but not the large 11-gallon petrol tank, which on the Major was situated at the rear of the chassis, supplying the SU carburetter via an Autovac.

The three Morris Majors listed for this first season were all closed models, although the saloon and coupé versions had the advantage of a folding-head.

Side view of the 1931 model Major Six coupé. (Len Barr)

*Rear view of the 1931 model
Major Six coupé.* (Len Barr)

The third variation was a coupé-like car called the 'Salonette', covered in black fabric and upholstered in contrasting red Karhyde. Standardisation of small-hub, five-stud, 19in diameter wire wheels gave the Major a lighter appearance than the contemporary Cowley, which continued to be fitted with the older style steel artillery wheels, although wire wheels had been available for some months as an optional extra on the Cowley, these became standard on the fixed-head version by July 1931.

Although a number of coachbuilders made use of the later Major chassis for their special bodywork, few used the 1931 chassis. One exception was the old established firm of Salmons & Sons of Newport Pagnell (whose Buckingham-shire factory is now occupied by Aston Martin Lagonda Ltd) who produced a

The
**MORRIS MAJOR SIX
SALONETTE**
(Fabric)

*Morris 1931 catalogue illustration of the Major Six
salonette.*

fabric bodied 'Tickford Sunshine Saloon' with their patented Tickford-head in real leather. This folding-head was operated by means of a winding handle inserted in an aperture in the rear quarter enabling the whole of the roof area to fold back, a manoeuvre which the manufacturers and patentees claimed could be done in ten seconds. Adjustable front bucket seats and a wide rear seat were upholstered in waterproof leather. Salmons' price for the Sunshine Saloon of £310 (Triplex safety glass extra) was £85 dearer than the standard Morris saloon.

Almost brand-new, this 1931 model Morris Major saloon came to grief in Wiltshire that year.

Something like 4,000 Majors were made for the 1931 season. The 1932 model had little in common with its predecessor except for the same basic six-cylinder side-valve engine, but with a bore reduced from 63.5mm to 61.25mm, thus bringing the RAC rating down to below 14hp. Although the Morris clutch running in oil was retained, the number of plates were reduced from two to a single disc which, because of its larger diameter and modified splined gearbox attachment, contained fifty-two individual corks as before. A complete redesign of the Major chassis resulted in a wider track and slightly longer wheelbase, with wheels of the five-stud Magna type, while the use of Lockheed hydraulic foot brakes and a cable hand brake to the rear wheels made it possible to dispense with the dual brake shoes in the rear drums, long a feature of the Morris mechanical braking system.

In common with the rest of the 1932 Morris cars the coachwork had become more curvaceous as the facilities available at Pressed Steel Company were increasingly utilised. Gone were the sharp features of the earlier cars and indeed much publicity was given to the 1932 innovation of the so-called 'eddyfree' front to all the closed models. This eddyfree design was possible now that the folding-roof with its built-in peak had been replaced with a Pytchley sliding-head where applicable. Body styles on the new Major chassis were more numerous, with a tourer version carrying the Major name for the first time and a saloon with or without the sliding-head. The Riley-like fabric bodied salonette was dropped in favour of two coupé designs. The first of these had a four-light body which other manufacturers would have described as a close-coupled saloon, while the second, called a 'Sports Coupé', was one of a range

Major Six coupé for the 1932 programme.
(Morris Motors Ltd)

Sports coupé of the 1932 season. An almost identical model for the following year was re-named the special coupé. The hood irons at the rear were for decoration only.

(Morris Motors Ltd)

Surviving example of the 1932 season's Major saloon.
(Chris Creevy)

of similar bodies with sliding-head fitted with false landau-irons and offered on
the season's Isis, Oxford, Major, Cowley and Minor chassis. The generous
plating of external bright parts on the Major now included the temperature-
controlled radiator shutters and the twin-bar bumpers. A more rounded
radiator shell lost its winged badge in favour of a shield-shaped emblem.

For reasons which are not clear, changes were made to the size of the sliding-
head guides on the early 1932 season bodies. The first 375 coupé bodies had
49$\frac{7}{8}$in roof-slide guide channels, later changed to 53$\frac{1}{4}$in. On sliding-head
saloon bodies the first 800 were fitted guides measuring 60$\frac{1}{2}$in, with 65$\frac{1}{8}$in
channels on subsequent bodies. These figures provide an interesting estimate of
the ratio of saloon to coupé bodies in the early part of the season. Another
change in specification came after the first 793 cars of the 1932 model when the
Lucas headlamps type R140 were changed at chassis 4919 to the L140 type.

Salmons & Sons also used the 1932 Major chassis. Their offering for that
year was again the Tickford Sunshine Saloon, with cellulose-finished aluminium panel construction, Tickford roof (no longer real leather) and leather
upholstery. At a price of £285 complete it represented an increase of £70 on the
cost of the standard Morris version, but they also offered a service — on a six-
day turnover — where a Morris owner could have the roof of his car converted
to Tickford folding-head from £19 10s. Maltby's Ltd of Folkestone listed a
1932 'Major Special Two-Seater' with dickey seat and cycle wings, panelled in
aluminium at £269, or for £10 less the same model with standard Morris wings
and running boards. Cunard Motor & Carriage Co Ltd — by 1932 part of the

1933 season Morris Major four-seater tourer.
(Morris Motors Ltd)

Stewart & Ardern organisation — used the Major chassis for their 'Calshot' four-door semi-sports four-seater with fold flat windscreen. An example of this model, with body and wings in aluminium, was displayed at the Olympia Motor Show in October 1931, described as being resplendent in two shades of blue with blue leather upholstery. Research also suggests that a Cunard drop-head coupé body was also available on the Major chassis, for such a model in dark blue with fawn leather seating was offered by a Hampstead motor dealer in 1937 for less than £20, representing a prodigious depreciation over five years. T.H. Gill & Sons Ltd of Paddington also appear to have produced special bodywork for the Major.

The 1933 season's Major was little changed externally, in either appearance or body style (the same design of 'Sports Coupé' was for some reason renamed the 'Special Coupé') although there were the usual changes in colour schemes; the front wings had added side panels described as 'splash guards' and, as on all the 1933 models except the Minor, scuttle mounted direction indicators were standard equipment. A subtle change to the Magna wire wheel design was a male instead of a female fixing on the hub covers. Internally the changes involved a new dash panel with the rearrangement of many of the controls and the replacement of the Lucas 'MT' spin-start windscreen wiper motor with the later 'DW' self-starter type. Elsewhere the Autovac was superseded by a SU Petrolift while the Armstrong hydraulic 'pear shaped' shock absorbers were substituted for the same manufacturer's adjustable friction type. At the rear, a stop-light was incorporated with the tail lamp.

The six-cylinder engine for the normal home market was the 13.9hp type 'LE' giving something like 28.5hp at maximum 3,400rpm. At home the motorist had to contend with the Road Fund Tax which was based on £1 per RAC horsepower (and £1 for the part unit, hence the curious engine ratings of the time such as 11.9, 13.9hp etc, to avoid spilling over into the next taxable unit). Such constraints did not affect the export market where, for example, a Major saloon weighing approximately a ton and a half when laden, driven by a 1,803cc engine, was considered underpowered. A bored-out version (63.5mm) of the six-cylinder engine for the 1933 export Majors, the 'LF', went some way towards answering this criticism, giving an additional 3.7bhp at maximum revs. For the record, an average load was considered by the Cowley designers to be two 12-stone people in the front and two 11-stone passengers in the back seat.

Some Morris Majors briefly made news in the 'thirties, such as a 1933 Major saloon purchased by the Emir of Katsina on the occasion of a VIP visit to the Cowley plant and shipped off to Nigeria. A 1932 Major saloon went to serve His Highness Raja Sahib of Poonch — the problem of getting the car to the remote Himalayan state was solved by dismantling and having the sections carried by native porters for the last forty miles over narrow mountainous paths! A Mr and Mrs Huddle and a Major Grilles purchased a second-hand 1932 Major saloon and a Rice folding caravan from Appleyards of Leeds in 1933 and journeyed with the ensemble to Kenya via the Sahara Desert. The French authorities at the last post before the desert proper (so convinced were they that the car and caravan would not make the crossing) insisted on the vehicles being pledged as a guarantee in case they had to send out a rescue party. One Major sliding-head saloon (registered OW3365) was won by a Mr F.S. Payne of Durley in a competition organised by the Round Table on behalf of Hampshire hospitals. To win the prize (which had been donated by Wadham Bros of Waterlooville) Mr Payne had to estimate the number of passengers travelling on Southampton trams and buses on Saturday 1 July 1933. Another Major was notable in that it had the first Buckinghamshire three-letter registration number issued, ABH1, taken out in March 1933 by the then Chairman of the Buckinghamshire County Council.

A total of 18,493 Majors left Cowley between September 1930 and August 1933, but the name 'Major', unlike 'Cowley', 'Oxford' and 'Minor' was only used briefly again. In the late 'fifties Wolseley introduced a compact 1½-litre saloon, the 1500, which was basically an expanded Morris Minor with rack-and-pinion steering and a close-ratio four-speed gearbox. This Wolseley when marketed in Australia was known as the Morris Major!

The natural successor to the Major for the 1934 programme was the Cowley-Six. For this new design the engine was a 1,938cc (14.9hp) version of the basic six-cylinder engine in resilient mountings. The new design of both body and reinforced cruciform chassis frame bore little resemblance to the Major of 1933, although the wheelbase and track dimensions was retained. The saloon (fixed or sliding-head) had the lower portion of the rear panel with a relatively gentle curve under, and a mounting was provided in this area for the spare wheel with metal cover and a luggage grid. On the special coupé, which bore no resemblance to the Major, the spare wheel remained a side mount. No open touring models were listed.

Considerable efforts had obviously been made by the management at Cowley to revitalise the sagging business and these efforts were to bear fruit in later years with Leonard Lord (later Lord Lambury) at the helm. However, his predecessor E.H. Blake, who retired in 1933 to make way for Lord, made something of a determined attempt in his programme for 1934 to win back some of the lost sales by updating the existing models and introducing two new sixes (Ten-Six and Cowley-Six) to fill what was obviously considered to be a gap in the Morris range. For the 1934 season Morris Motors Ltd boasted that an easychange synchromesh gearbox, Lockheed hydraulic brakes, hydraulic shock absorbers, concealed direction indicators and real leather upholstery were refinements which would appear on every Morris model irrespective of price. Automatic ignition advance and retard was another improvement made to all but the Minor, and on some models including the Cowley-Six a new type of handbrake lever with centre button release was a new feature.

With Len Lord's massive reorganisation at Cowley underway and the 'Series' models (which were to see the end of the 'new model every Motor Show' tradition) in the pipeline it is probably true to assume that the 1935 models, with the exception of the new Morris Eight, were stop-gap modification of 1934 designs. This was to prove a short model year anyway, ending abruptly with the announcement in mid-1935 of the Series cars.

Certainly the Morris Fifteen-Six was virtually the Cowley-Six given a new name. It was a good sales ploy to draw the attention of the prospective pur-

1935 Fifteen-Six saloon.

A feature of the 1935 Fifteen-Six cars was the lamp bar passing behind the radiator honeycomb. The model shown is a special coupé.

chaser to the horsepower rating, in view of the 1934 Finance Act which had cut the Road Fund Tax by a quarter, operative from the 1 January 1935. This move away from model names to model numbers indicative of taxable horse-power was the reason why the four-cylinder Cowley become the Twelve-Four for the 1935 season. (Similarly the introduction of a flat rate of private car taxation in 1948 saw the general return of model names.) Both saloon and coupé bodies were the same as before, the ex-works prices were identical and even the chassis numbering continued on from the Cowley-Six sequence. The most noticeable modification in styling was in the area of the radiator where in Cowley-Six form a lamp-support tube spanned the front wings in front of the radiator shell, but, as the Fifteen-Six, the lamp support tube passed through ferrules in the side of the shell to be concealed behind the imitation honeycomb.

Less noticeable were the minor changes to the 63.5mm bore × 102mm stroke engine, re-designated type 'LJ', a move from centre to right-hand accelerator pedal, and the crown wheel and pinion changed from a 11/58 to a more robust 9/47 ratio. Both right-hand and left-hand drive versions of the Fifteen-Six were exported and on these export models smaller diameter 4in × 16in wheels were fitted with 600-16 extra low pressure tyres.

Morris Major, Cowley-Six and Fifteen-Six. 1931-5. Engine Data

Type LB, Major 1931 model
Type LE, Major 1932 model
Type LE, Major 1933 model
Type LF, Major 1933 export
Type LG, Cowley-Six 1934
Type LJ, Fifteen-Six 1935

Six-cylinder side-valve. LB, LF, LG, LJ, 63.5mm bore, 102mm stroke, 1,938cc, 14.9hp.
LE, 61.25mm bore, 102mm stroke, 1,803cc 13.9hp.
LB, 3-ring, all others 4-ring, aluminium alloy pistons. 28mm diameter inlet and exhaust valves. Compression ratios: LE, 5.49/5.57; LF, LG & LJ, 5.7. Combined pre-heating and air cleaning device on head of engine. Coil ignition. Distributor drive off dynamo (4½in Lucas). Cooling by water pump and fan, thermostat by-pass on LJ. Automatic advance and retard on LJ. SU carburetter, 1⅛in diameter. Three-speed gearbox on LB, other engines four-speed gearbox. Cork-insert clutch running in oil, multi-plate on LB, other engines single-plate. Sump capacity 1¼ gallons. Engine rubber mounted forward of crankcase and rear of gearbox on LG and LJ. Maximum rpm 3,400. LE, 28.5bhp. LF, 32.2bhp. LG and LJ, 36bhp.

Morris Major, Cowley-Six and Fifteen-Six. 1931-5. Specification and Chassis Numbers

Major, 1931 model.
Chassis numbers: MJ101 - MJ4125

Mechanical four-wheel brakes, foot pedal to all wheels, hand brake to independent shoes at rear. Right-hand accelerator pedal. Plymax dash. Dipping headlights. Bishop Cam steering. Armstrong friction-type shock absorbers. Rear petrol tank, 11-gallons, Autovac feed to SU carburetter. Enclosed torque tube to spiral bevel rear axle; home model ratio 4.75:1, export model ratio 5:1. Group chasis lubrication nipples over rear axle casing. Wire wheels with small hub, 5-stud, Dunlop Cord 500-19 tyres. Spring gaiters. Rectangular chromium-plated radiator surround with black shutters operated by Calorstat, winged badge. Triplex single-panel windscreen with electric windscreen wiper. Single blade bumpers front and rear. Folding luggage grid on saloon. Track 4ft. Wheelbase 8ft 9in.

Major, 1932 & 1933 models
Chassis numbers: 1932: MJ4126 - MJ13061; 1933: MJ13062 - MJ18594

Hydraulic four-wheel brakes, handbrake by cable to rear shoes, 10in drums.
Centre organ-type accelerator pedal. Dipping headlights. Traffic indicators on
1933 models (early versions with Wilcot flashing units). Bishop Cam steering,
adjustable rake. Armstrong friction-type shock absorbers on 1932, Armstrong
hydraulic type on 1933. Enclosed propeller shaft to spiral bevel rear axle, ratio
5.27:1. Magna wire wheels 3 × 19, 5-stud, Dunlop 500-19 tyres. Splash guards on
front wings of 1933 models. Rounded radiator shell chromium-plated with
chromium-plated shutters operated by Calorstat, shield shaped badge. Triplex
single-panel windscreen with electric windscreen wiper. Twin blade bumpers front
and rear. Track 4ft 4in. Wheelbase 8ft 10in.

Cowley-Six, 1934 model
Chassis numbers: 34/CS501 - 34/CS7720

Hydraulic four-wheel brakes, button release handbrake by cable to rear shoes.
Aluminium brake shoes. Centre organ-type accelerator pedal. Cut & dip
headlights. Reverse lamp on coupé. Traffic indicators. Rear roller blind. Driver's
glass sun visor. Bishop Cam steering. Armstrong hydraulic shock absorbers. Rear
petrol tank, 10-gallons, SU electric petrol pump to SU carburetter. Tubular
propeller shaft with Spicer universal joints to spiral bevel rear axle, 5.27:1 ratio.
Crown wheel & pinion 11/58. Additional leaf springs on export models. Magna
wire wheels, 3 × 19, Dunlop 500-19 tyres. Metal cover on spare wheel. 'harmonic
stabilizing' front bumper and matching rear bumper. Chromium-plated radiator
shell with vertical strip, shield shaped badge 'SIX' motif on grille. Lamp support
tube in front of radiator grille with horn in centre. Triplex single-panel
windscreen with electric windscreen wiper. Track 4ft 4in. Wheelbase 8ft 10in.

Fifteen-Six, 1935 model
Chassis numbers: 35/FS7721 - 35/FS15970
(chassis numbers shared with Twelve-Four and 8/10cwt van production. First
Fifteen-Six chassis number actually 7733.)

Hydraulic four-wheel brakes, handbrake by cable to rear shoes. Steel brake shoes.
Right-hand roller accelerator pedal. Traffic indicators with tell-tale mirrors.
Rollsvisor driver's sun blind. Battery master switch. Cut & dip headlights. Stop
light. Compensated dynamo voltage control. Rear roller blind. Pneumatic
cushions on rear seat of coupé. 'Karvel' carpets. Ventilating windows arranged to
operate at the end of the window winder travel. Bishop Cam steering. Armstrong
hydraulic shock absorbers. Rear petrol tank, 10-gallons, SU electric petrol pump
feed to SU carburetter. Carburetter guard and undershield on RHD export
model. Tubular propeller shaft with Spicer universal joints to spiral bevel rear
axle, 5.22:1 ratio. Crown wheel & pinion 9/47. Magna wire wheels, 3 × 19,
Dunlop 500-19 tyres. Export models fitted 4 × 16 wheels with Dunlop 600-16 ELP
tyres. Wilmot-Breeden metal cover on spare wheel. Chromium-plated radiator
shell with vertical strip, shield shaped badge. Lamp support tube runs behind
radiator grille. Thermostat control for cooling system. 'Harmonic stabilizing'
front bumper and matching rear bumper with hinged centre section fixed to
luggage grid. Triplex single-panel windscreen with electric windscreen wiper.
Track 4ft 4in. Wheelbase 8ft 10in.

Morris Major. Body Colours and Upholstery

	1931	1932	1933
Saloon and coupé (1932-3 coupé sliding-head only)	*Folding-head* Lake cellulose with red Karhyde. Black cellulose with green Karhyde.	*Sliding-head & fixed-head* Black or green cellulose with green Karhyde. Blue cellulose with brown Karhyde.	*Sliding-head & fixed-head* Black cellulose with green Karhyde. Blue or brown cellulose with brown Karhyde.
Special coupé		Blue/black, black, grey/ dove or beige/brown cellulose with brown Celstra. (1932 models called 'Sports Coupe')	Green or red duotone with matching leather. Grey duotone with blue leather. Black duotone with brown leather. (1933 models called 'Special Coupe')
Tourer four-seater		Blue cellulose with brown Karhyde.	Blue cellulose with brown Karhyde.
Salonette fabric body	Black fabric body with red Karhyde.		

Morris Cowley-Six & Fifteen-Six. Body Colours and Upholstery		
	1934 Cowley-Six	1935 Fifteen-Six
Saloon	*Sliding-head & fixed-head* Black cellulose with brown leather. Blue cellulose with blue leather. Green cellulose with green leather.	*Sliding-head & fixed-head* Blue cellulose with blue leather. Green/black or black cellulose with green leather. Red/black cellulose with red leather.
Special coupé	Green duotone or cream/ green cellulose with green leather. (green top-half) Brown duotone or black with brown leather.	Green duotone or cream/ green cellulose with green leather. (green top-half) Brown duotone or black with brown leather.
Note: The term 'duotone' was used by Morris Motors Ltd to indicate a darker shade of the paint finish or leather cloth colour for the superstructure.		

Morris Major, Cowley and Fifteen-Sixes. 1931-5. Magazine Bibliography				
Model Year	Material	*Source*	Date	Pages
1931	Morris Programme for 1931	*The Morris Owner*	9/30	p847-51
1931	Major Road Test	*The Autocar*	10/10/30	p702-3
1931	Morris Programme for 1931	*The Autocar*	5/9/30	p445-8
1932	Morris Programme for 1932	*The Autocar*	4/9/31	p391-5
1932	Morris Programme for 1932	*The Motor*	1/9/31	p199-206
1932	Morris Programme for 1932	*The Morris Owner*	9/31	p787-92
1932	What I think of the Major	*The Morris Owner*	1/32	p1278-80
1932	Major Road Test	*The Motor*	1/9/31	p207
1932	Major Road Test	*The Autocar*	25/12/31	p1202-3
1932	Major Servicing Data	*Motor Commerce*	6/32	p50-4
1933	Morris Programme for 1933	*The Autocar*	2/9/32	p411-14
1933	Morris Programme for 1933	*The Motor*	6/9/32	p214-20
1933	Morris Programme for 1933	*The Morris Owner*	9/32	p709-16
1933	Saturday afternoon with the Major	*The Morris Owner*	8/33	p692-5
1934	Cowley-Six Road Test	*The Autocar*	10/11/33	p726, 928
1934	Morris Programme for 1934	*The Morris Owner*	9/33	p819-26
1934	Cowley Six-Cylinder	*The Morris Owner*	12/33	p1110-12
1935	Fifteen-Six Road Test	*The Autocar*	23/11/34	p977-8
1935	Morris Programme for 1935	*The Morris Owner*	9/34	p655-62
1935	The Fifteen-Six	*The Morris Owner*	1/35	p1020-1, 1030

Morris Ten
1933-5

A steady increase in the number of cars registered in Britain in the 10hp class began to show itself in 1931 and all signs were that this trend would continue. As petrol tax had been increased twice in that year, then the reasons for a rapid development of the Morris Ten becomes apparent. The shift in demand continued and by 1935 the 10hp registration figures had risen from a 1931 total of 16,000 to 67,000; the Morris Ten formed a substantial percentage of these new registrations, for it proved to be one of the most successful models of the early 'thirties.

The Morris Ten in perpendicular 'pre-Series' form was first announced in August 1932 for the following season as a saloon (with fixed or sliding-head options) and as a special coupé. A tourer version was announced in December and was the subject of a special leaflet issued by Morris Motors Ltd about March 1933.

Although all Tens are generally referred to as the 'Ten-Four' by present-day enthusiasts, the pedantic would be correct to point out that for the first model year they were simply known as the Morris Ten. 'Ten-Four' was the description given to the model for the 1934 and 1935 seasons to differentiate it from the 'Ten-Six', which by then was part of the range.

The new 10hp from Cowley was powered by a side-valve engine of 63.5mm bore × 102mm stroke, which for taxation purposes was calculated by the RAC formula as 9.99hp. Although a flat-head engine, at first glance it appears to be an overhead-valve unit due to the 'Breather, Fume consumer, Air pre-heater and Filter' added above the cylinder head. Power was transmitted through a

Twin-bar bumpers and a vertical radiator identify this Ten saloon as a 1933 model. (P. Harris)

1933 Morris Ten side-valve engine. (John Lowrie)

Morris Ten, four-cylinder, 1,292cc side-valve engine
showing the air filter-fume consumer fitted on top of the
cylinder head.

four-speed gearbox, with what the Morris publicity people called 'twin-top' ratios (6.9:1 and 4.7:1), and the well tried and tested cork-insert clutch running in oil. The tubular propeller shaft, with fabric universal joints drove to a spiral-bevel back axle.

On the saloon version the seating was upholstered in a type of leathercloth called 'Karhyde', the coupé having real leather, and the interior woodwork was of dark oak with a matt finish. Winding windows were provided, as well as silk rope door-pulls, remote control rear-blind, illuminated instrument panel, a scuttle ventilator and roof lamp. On the fixed-head verion of the saloon a parcel net hung above the front seats. This, the cheapest of the Tens at £165, was devoid of bumpers and luggage grid.

Early production models of all Morris 1933 season cars, except the Minor, were equipped with a direction indication device known as the Wilcot Robot, made by Wilcot (Parent) Co Ltd, but this proved to be something of an expensive embarrassment for Morris Motors Ltd. Direction indicators drew much controversy about this time and as early as March 1929 the Ministry of Transport had arranged with the Royal Automobile Club to hold a demonstration in Richmond Park of the various devices in existence. The 'Trafficator' was a German invention for which Lucas had obtained patent rights from A.H. Hunt (Safetisigns) Ltd and as a consequence a royalty of 6d was payable (2½p) on every pair sold to car manufacturers as original equipment. William Morris, it appears, did not like this arrangement so the Wilcot indicators were chosen. In appearance the Wilcot Robots resembled a miniature set of road traffic lights; various positions of the operating switch caused the coloured lights to flash different combinations. For example, the driver wishing to turn left would show a green light to the off-side and a red and amber would flash on the near-side. This operating switch had a built-in timed return to the off position. Not surprisingly, criticism of the dangers that could be caused by confusion came from many quarters, including Sir Herbert Austin who said he did not think that '. . . a complicated signalling device such as that recently introduced by a certain make of car is a step in the right direction.' Eventually the Minister of Transport refused to sanction their use and in March 1933 Morris Motors Ltd announced that in future semaphore signals would be fitted and any existing vehicles modified free of charge.

The exercise had cost Morris £50,000 and it is said that many of the unused Wilcot indicators were dumped into a large pit behind one of the factories. (This appears to have been the normal way of disposing of unwanted components. It is on record that left-over parts, including chassis frames, of some 500 'F' type six-cylinder cars were used to help fill in thc lake behind the Osberton Radiator Co's premises at Bainton Road, Oxford. Elsewhere, a number of Wolseley 32/80 Straight Eight engines provided part of the infilling for the factory floor at Adderley Park.) Another 2,000 sets were bought by Gamages of Holborn who resold them as novelties for 2s 6d per set — at the same time as the makers were advertising the units for £4 4s. Morris's loss would appear to have been Gamages' gain, but the ministerial decision affected the Holborn store in another way. They had been marketing their own version of Robot Indicators for 25s a set in 1932, a price later reduced to 5s. An

Wartime black-out lighting regulations were not absolutely clear to motorists in October 1939 when this photograph of a five-year-old Ten-Four was taken. The owner had blacked out both headlights and improvised masks for the side lamps.
(E. Hope)

amusing sequel is recorded by Robert Jackson in his book *The Nuffield Story;* apparently the brother of one of Morris Motors Ltd directors (Hans Landstad) who, seeing the devices on sale at Gamages, bought a set and tried to interest the Morris concern in using them!

Noteworthy of the specialist coachbuilders of the day to use the Morris Ten chassis was the main Morris agency Stewart & Ardern, whose subsidiary

Tourer version of the 1934 Morris Ten-Four.

1934 model Morris Ten-Four saloon with sliding-head.

Cunard produced the 'Calshot Drop-Head Coupé' version with two independently adjustable front seats and a special Cunard design of hammock seat in the rear which could be folded up to allow accommodation for luggage. This two-door coupé with 'interior cabinet work in black walnut' and a folding-head of enamelled leather sold at something like £47 10s more than the standard Morris special coupé.

Two new Ten-Four models were added for the 1934 programme: a two-seater with dickey seat (often wrongly described nowadays as the 'Doctor Coupé') and a 'Traveller's Saloon' which was basically a standard saloon with a side-hinged door at the rear allowing removable back seat space to be used for goods and samples, an idea originally exploited by Stewart & Ardern on the Cowley chassis in 1929. Additionally, as already mentioned, a new 'Ten-Six' range was introduced. Although this was something of a misnomer, as the rating of the engine was twelve horsepower, the name is understandable as the six-cylinder models used identical Ten-Four bodies on a longer wheelbase chassis, the extra 6in being absorbed in bonnet length to house the larger engine.

One other Ten-Six that did not make the early editions of the Morris catalogue for 1933 was the sports special four-seater — a car out of character for Cowley. The six-cylinder engine used in this model was a specially tuned version with twin linked SU carburetters and remote control gear change. Despite the single-plate oil filled cork-insert clutch (smooth, but hardly the clutch for tyre-burning getaways) all efforts appear to have been made to promote the sports car image, such as a quick-filler cap on the petrol tank, a revolution counter on the dash along with other large-diameter instruments, spring spoked steering wheel, louvred low side-panels in lieu of running boards, Burgess straight-through silencer, stoneguards to the radiator and headlamps, imitation knock-on wire wheels, grab handle on the dash (which the catalogue described as a 'mechanic's grip'), and leather strap across the bonnet. (Centric Super-Chargers Ltd soon

Arrangement of the twin SU carburetters on the prototype Ten-Six special.

added to their range of standard superchargers one for the Morris Ten-Six, priced at £35 fitted.) These are all features one would associate more with Abingdon-on-Thames, and there perhaps lies the answer!

There was, apparently, no love lost between Leonard Percy Lord who had by 1933 been given the task of re-organising the Cowley Works, and Cecil Kimber who was managing director of the MG Car Company. According to Adolf Von der Becke, who worked for Morris Motors Ltd at the time, Lord intended to show that MG did not have the exclusive know-how on sports car design within the Morris organisation, and this was his project. Unfortunately, for Lord, it was not a success. It may, or may not, be coincidental that the MG drawing office was transferred to Cowley in 1935, and that Lord Nuffield clamped down on all MG design enterprise.

If catalogues of the period are any guide, there was also a version of the sports special making use of the four-cylinder engine which, like the six-

Morris Ten-Six special. The absence of bumpers and the arrangement of the bonnet louvres suggests that this is the prototype.
(Morris Motors Ltd)

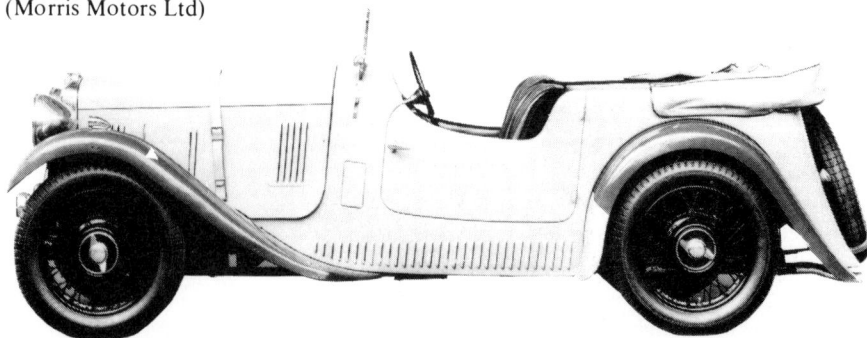

One of a handful of restored Morris Ten-Six specials still in existance. (Chris Creevy)

cylinder, had a high-lift camshaft, twin SU carburetters and special manifolding, straight-through silencer, etc. The 'ME' engine for this gave 36bhp at 4,100rpm against the ordinary unit's 28bhp rating at 3,500rpm. The writer has never seen such a model, nor contacted any one-time owner, but to substantiate its existence the catalogue of the Cunard Motor & Carriage Co

But for the slightly longer bonnet and the 'Six' motif, this coupé version of the 1934 Ten-Six could be mistaken for the Ten-Four. (Geoff Osborn)

1934 Ten-Six.

Ltd issued in 1934 lists their 'International Type Four-Seater' on both Ten-Six and Ten-Four special chassis. Cunard made their special sports models which reinstated running boards and combined 'a charming appearance with comfortable accommodation for four persons'. The finish offered was cream/black with black upholstery, or red, or green, with upholstery to match. Specification for the Cunard versions of these specials built on an ash frame with 18SWG aluminium panelling, included louvres on the bonnet top, Triplex folding windscreen with twin electric wipers and lower cut doors than those to be found on the lower priced Morris-bodied versions. The Ten-Six special was £230 from Cowley and £249 10s from Stewart & Ardern Ltd. Carbodies of Coventry also made use of the Ten-Six special chassis as a basis of a well proportioned fixed-head coupé which was sponsored by W. Watson & Co (Liverpool) Ltd as the 'Watson Special Saloon'.

A Cumberland registered Ten-Six special (ARM64) was among the starters from John o'Groats to compete in the 1935 Monte Carlo Rally. Driven by G.F. Hobley and crew as entrant number 91, this car completed the event successfully if not spectacularly. What became of this car? Or, for that matter, does the Ten-Six Sports engine that was presented to the Bradford Technical College for instructional purposes by Morris Motors Ltd in 1935, still exist?

The major changes to the Ten-Four for the 1934 season were not immediately apparent as they involved the use of a synchromesh gearbox and a new chassis frame described as an 'unusually sturdy deep sectioned down swept frame with generous cross stiffening and resilient engine mounting'. No doubt the need to provide a longer chassis for the six-cylinder model made the incorporation of a stronger one viable for the four-cylinder cars. Apparent changes were minor although the single-blade chromium-plated, black filled, bumper in place of the previous season's twin-blade type and the sloping radiator stoneguard altered the frontal appearance slightly. A design change to the handbrake lever put on moulded finger grips and a central button release, while under the bonnet the distributor was fitted with automatic advance/retard mechanism, even though the manual controls on the steering wheel centre remained. The larger seats provided were now upholstered in real leather, the exception being the dickey seat on the two-seater which was of Rexine. Except on the open models, the boxed Trafficators on outriggers (which were a legacy of the ill-fated Wilcot Robots) were replaced with concealed units between the doors. As the bodies were common to both Ten-Four and Ten-Six, these improvements applied to both series.

From time to time the Morris Ten figured in news items. One instance was the presentation of a cheque by Haslemere Motor Co to a Mr F.K. King of London, the winner in an *Evening News* £100 competition in which the entrant had to write a brief description of a trial run in a Morris Ten. It is not clear if the condition of entry was the purchase of a car, but the press photographs of the time showed the winner posed with a Morris Ten registered BPB942. The Morris Ten is hardly the car one would expect Royalty to use, nevertheless, there was one occasion in early 1934 when Queen Mary (grandmother to the present Queen) was transported in a Morris Ten saloon when journeying back from Sandringham. Apparently Percy Titmouse came across the Royal car in trouble by the roadside and gave her a 'lift' into Cambridge. 'Her Majesty is reported as having remarked upon the comfort and roominess of the Ten.'

Many firms in the 'thirties catered for Morris Ten owners who wished to personalise their car with special bumpers and various other extras. Wilmot-Breeden offered black-enamelled metal spare-wheel covers with a chromium-plated band for 43s 6d, or in all-chromium-plate for an additional 20s. Wheel discs to cover the wire spokes of the Magna wheels were popular in either finish. From Manchester the firm of David Moseley & Sons could convert the seating in the Morris Ten to 'Float-on-Air' pneumatic, using the existing upholstery covers. Other firms such as Millers of Sparkbrook, Weathershields, and Fabram of Yorkshire sold tailored loose covers for the seats; Miller's set included door and side covers to 'harmonise with every colour scheme'. Car mats were another easily obtained extra with The Car-Mat Co of Portobello Road providing fibre, hair, pile, or sorbo rubber types specially designed for the Morris Ten. A special 'Super Easyfit' carrier in black enamel with

A 1934 model Morris Ten-Four sliding-head saloon fording a stream.

rubber inserts to prevent damage to the luggage was the product of Frank Ashby & Sons, while at Willesden G. Beaton & Sons Ltd assembled chromium-plated framed sun visors and window louvres fitted with safety glass. Weathershields Ltd of Birmingham, would even convert the complete roof of the Morris Ten to enable the top to fold right back in open tourer style.

Lord Nuffield bequeathed to Nuffield College, Oxford, an interesting example of the silversmith's craft, a model of the 1934 Morris Ten-Four saloon. This silver model is correct down to the smallest details including opening doors, winding windows, sliding roof, and even a toolbag and jack. Under the opening bonnet can be seen the engine complete with its external fittings, while the underside reveals a faithful reproduction of the silencer, exhaust pipes, hydraulic brake pipes and similar detail. On the plinth an engraved plate records that the model was presented to Lord Nuffield at a gathering of the major Morris Distributors held at Grosvenor House, London, in late 1934, to mark his elevation to the peerage earlier that year. There had been a precedent over a decade earlier when in July 1923 his original agents had presented him with a silver replica of the Morris Oxford car — on that occasion the illuminated address with the model expressed the congratulations of the agents to William Morris on attaining the position of premier light car manufacturer of the United Kingdom.

1935 model Ten-Four saloon with sliding-head.
(John Farmer)

The third and last season for the Ten-Four was to prove a short one of nine months before the new-look Series models made their appearance. The 1935 model saw the trend to standardise the accelerator pedal position, moved to the right-hand side; other refinements were the new pull-up handbrake lever with accessible adjustment, steel brake shoes replacing the earlier aluminium cast type, a round petrol tank to supersede the squarer section one, and an

1935 Morris Ten-Four tourer.
(John Atkinson)

improved rear bumper which incorporated an insert section to fill the gap when the luggage grid was in its folded position. Tyres were increased in section, but reduced by 1in diameter to 4.75-18. The Lucas dynamo (Type C45A) was arranged to give switched 'summer' or 'winter' charging rates, with maximum output when the headlamps were in use at night. A 'Rollsvisor' consisting of a miniature roller blind on a stiff-hinged frame mounted above the driving seat helped prevent glare on sunny days.

With the exception of the traveller's saloon which was no longer listed on either the Ten-Four or Ten-Six chassis, the model choice for 1935 remained the same as for the previous year. However, the trend in the mid-'thirties towards what is now called 'two-tone' was apparent in the new colour schemes for the closed models, with most saloon bodies finished with a black superstructure with colour below the waistline. In accordance with Morris practice dating to very early cars, the wheels and mudguards were always black, as they were stove enamelled and the bodies were cellulose sprayed before being mounted on the running chassis.

Although the 1935 Ten-Six continued to share identical bodies with the Ten-Four, the bumpers on the Six were of a new Wilmot-Breeden 'harmonic stabilising' type.

Exports of Morris cars were gradually increasing again following the drop after the peak year of 1927, and to cater for the demand from some overseas markets the Tens were available (certainly in 1935 and possibly in 1934) with a left-hand drive layout. These export models had disc wheels.

The number of Ten-Fours and Ten-Sixes produced between late 1932 and mid-1935 is only known as a combined figure as both types shared the same chassis numbering series. For the 1933 model year the number of four-cylinder cars topped 14,000 and the combined total for the three seasons reached 49,238.

Morris Ten. Pre-Series models, 1933-5	
Model	Chassis Numbers
Ten, four-cylinder, 1933	101-14280
Ten-Four, 1934	34/T14281-34/T35185
Ten-Four, 1935	35/TN35186-35/TN49338
Ten-Six, standard, 1934	34/TS14281-34/TS35185
Ten-Six, special, 1934	34/TS/SP14281-34/TS/SP35185
Ten-Six, standard, 1935	35/TS35186-35/TS49334
Ten-Six, special, 1935	35/TS/SP35186-35/TS/SP49334

Morris Ten. Pre-Series models, 1933-5. Specification	
Ten, four-cylinder	Engine: type 'MA' 1933, 'MB' 1934, 'MF' 1935. Side-valve, 63.5mm bore × 102mm stroke. 1,292cc. Cork-insert clutch running in oil. SU carburetter. Rear mounted petrol tank using SU Petrolift up to chassis 25088; later models fitted with SU petrol pump. Four-speed gearbox. Magna type wire wheels (except on some export models with disc wheels), 2.5in × 19in on 1933-4 models; 3.0in × 18in on 1935. Tyres on 1933-4 models 4.50-19; 1935 models 4.75-18 (export 5.75-16) Lockheed hydraulic brakes. Rear axle ratio 1933, 4.7:1; 1934-5, 5.22:1.
Ten, six-cylinder	Engine on standard chassis for 1934 season, type 'RA'. 1935 chassis, 'RD'. On the special chassis for 1934, type 'RB'; 1935 models 'RF'. Side-valve, 57mm bore × 90mm stroke. 1,378cc. 12hp. Domed pistons. Cooling system by thermo-syphon except on 'RF' engine which had water pump. Cork-insert clutch running in oil. Rear mounted petrol tank using SU Petrolift to chassis 24987; later models fitted with SU petrol pump. Four-speed gearbox with synchromesh on top and third (remote control lever on special models), SU carburetter (twin on special). Tyres for 1934 models were 4.50-19 on 2.5in × 19in Magna wire wheels (4.75-19 tyres on special). 1935 models had 4.75-18 tyres on 3.0in × 18in Magna wire wheels. (Except some export models with disc. Export size tyres 5.75-16 on 3.5in × 16in wheels.) Lockheed hydraulic brakes. Rear axle ratio 5.55:1.

Morris Ten. Four and Six-Cylinder models 1933-5. Magazine Bibliography				
Model Year	Subject	Source	Date	Pages
1933	Impressions of the New Morris Ten	*The Morris Owner*	10/32	p845-7
1933	Morris Ten introduced	*The Motor*	30/8/32	p194-9
1933	Morris Programme for 1933	*The Morris Owner*	9/32	p709-16
1933	Morris Programme for 1933	*The Autocar*	2/9/32	p411-14
1933	Morris Programme for 1933	*The Motor*	6/9/32	p214-20
1933	Ten Saloon, Road Test	*The Autocar*	2/9/32	p415, 417
1934	Ten-Four Saloon, Road Test	*Practical Motorist*	14/7/34	p385
1934	Ten-Four Saloon, Road Test	*The Motor*	10/7/34	p999-1000
1934	Ten-Six at 10,000 miles	*The Morris Owner*	3/34	p26-9
1934	Morris Programme for 1934	*The Morris Owner*	9/33	p819-26
1934	This Ten-Six	*The Morris Owner*	11/33	p1008-10
1934	Touring with the Ten	*The Morris Owner*	5/34	p222-5
1934	Morris Ten-Six Servicing	*The Morris Owner*	3/34	p26-9
1934	And now Six-Cylinders	*The Morris Owner*	12/33	p1110-12
1935	Ten-Six Sports Tourer, Road Test	*The Autocar*	12/4/35	p635-6
1935	Ten-Six Sports Tourer, Road Test	*The Light Car*	5/35	p . . .
1935	Morris Programme for 1935	*The Morris Owner*	9/34	p655-62
1935	Ten-Four Saloon, Road Test	*The Light Car*	26/10/34	p734-5
All	Maintaining the Morris Ten	*Car Mechanics*	6/61	p52-4, 80

Morris Ten. Four-Cylinder Pre-Series models. Body Colours and Upholstery

	1933	1934	1935
Two-seater, with dickey seat		Green or black cellulose with green leather. (Dickey seat in Rexine)	Green or black cellulose with green leather. Red cellulose with red leather.
Tourer, four-seater	Black or brown cellulose.	Green or black cellulose with green leather.	Green or black cellulose with green leather. Red cellulose with red leather.
Saloon, sliding-head & fixed-head	Green or black cellulose with green Karhyde. Blue cellulose with brown Karhyde.	Green or blue cellulose with matching leather. Black cellulose with brown leather.	*Sliding-head only:* Blue cellulose with blue leather. *Sliding & fixed-head* Green & black or black cellulose with green leather. Red & black cellulose with red leather.
Special coupé	Green duotone cellulose with green leather. Red duotone cellulose with red leather. Grey duotone cellulose with blue leather. Black cellulose with brown leather.	Green duotone or cream & green with green leather. Brown duotone or black cellulose with brown leather.	Green duotone or cream & green with green leather. Brown duotone or black cellulose with brown leather.
Traveller's saloon		Black cellulose with brown leather.	

Note: Duotone indicates a darker shade of colour for the superstructure.

Morris Ten. Six-Cylinder Pre-Series models. Body Colours and Upholstery

	1934	1935
Two-seater, with dickey seat	Green cellulose.	Green or black cellulose with green leather. Red cellulose with red leather.
Tourer, four-seater	Green or black cellulose with green leather.	Green or black cellulose with green leather. Red cellulose with red leather.
Saloon, fixed & sliding-head	Blue or green cellulose with matching leather. Black cellulose with brown leather.	Blue cellulose with blue leather. Green & black or black cellulose with green leather. Red & black cellulose with red leather.
Special coupé sliding-head	Green duotone or cream & green cellulose with green leather. Brown duotone or black cellulose with brown leather.	Green duotone or cream & green cellulose with green leather. Brown duotone or black cellulose with brown leather.
Traveller's saloon	Black cellulose with brown leather.	
Special sports, four-seater	Colour schemes not known. Leather upholstery.	Black or scarlet & black cellulose with red leather. Saxe blue & Oxford blue cellulose with blue leather. Cream & green or green duotone cellulose with green leather.

Note: Duotone indicates a darker shade of colour for the superstructure.

Morris Eight
1935, Series I & II

There can be little doubt that the most successful of all the pre-war Morris cars — and the one which helped to lift Morris Motors Ltd out of the depression years — was the ubiquitous Morris Eight. By July 1935, only nine months after its introduction (during which period the Road Fund Tax on an 8hp car was reduced by 25 per cent to £6pa) over 50,000 Eights of all body types had been sold. This success continued despite competition from other small car manufacturers, including the Ford 8hp saloon which was reduced at the Ford Motor Show in October 1935 from £110 to £100 with the boast that 'This makes the first £100 saloon car produced in Great Britain.' Fords cautiously included the word 'saloon' in this announcement, remembering that a Morris Minor two-seater had been available at that figure during the 1931-3 seasons, and Austin's £100 two-seater, the Austin Seven Opal, was already on the market.

1935 Morris Eight two-door saloon. The wheels would have originally been stove enamelled black.
(C. van Breedam)

The Morris Eight, as announced in August and introduced at the Motor Show in October 1934, gave the prospective purchaser an option of two open models, a two-seater and a four-seat tourer; and two-door or four-door saloons with a Pytchley sliding-head or fixed-head. A 5cwt van version was available with a light body thanks to the generous use of plymax panels which kept the unladen weight of this small commercial below 12cwt, thus qualifying for the £10 annual Road Fund Tax. Colour choice on the cheaper fixed-head saloons was limited to an all-black or a two-tone red and black body with the usual Morris practice of stove enamelled wheels, wings, and aprons in black. To keep the price to £120 for the two-door, or an additional £10 for the four-door body, the trimming was in a red 'Karhyde' leathercloth, while bumpers and Lucas Trafficators were extra. Slightly up-market, the de luxe sliding-head versions must have been a good buy when for an additional £12 10s one had an Eight complete with front and rear bumpers, trafficators, luggage grid, real leather upholstery, and a choice of three two-tone colour schemes. Little did anyone know, back in 1934, that nearly firty years later the most sought after Eight, and the ones to command top bids for Eights at old-car auctions, would be the cheapest of all the Morris models at Olympia — the Eight two-seater. Listed at £118 and £120 respectively, the two-seater and tourer models as originally presented were without bumpers or trafficators (which were extra) and trimming was in Karhyde.

A completely new 918.6cc side-valve engine, type 'UB' with a three-bearing crankshaft, had been designed for the Morris Eight and a proprietary single

Under bonnet arrangement of engine and components on
Series I Morris Eight. (Morris Motors Ltd)

dry-plate clutch, Borg & Beck 6¼in was introduced for the first time. The three-speed gearbox and engine unit was mounted at four points on rubber; the rear mounts consisting of rubber-to-metal bonded attachments. In his biography (*Out on a Wing,* 1964) the late Sir Miles Thomas (Lord Thomas of Remenham) writes: 'One of the reasons why a simple, cheap and dependable side-valve 8 h.p. engine was produced so quickly by Len Lord at the Wolseley factory was that he had no inhibitions about following good examples. Any student of automobile history who cares to compare the 8 h.p. Ford engine of the late nineteen-twenties and the Morris 8 engine of the early nineteen-thirties will find a remarkable resemblance.' Presumably Sir Miles Thomas was referring to the Ford 'Y' type Eight which was introduced not in the late 'twenties, but in February 1932; he was, of course, right and there are many obvious similarities. A Ford Eight was procured by the Morris organisation and data from the completely stripped down engine was used by the assistant chief designer at Morris Engines Branch, Claude Baily, and his team to expedite the design of the new Morris Eight 'UB' power unit which emerged with a smaller capacity and stroke, and a larger bore than the Ford's 56.6mm bore × 92.5mm stroke, 933cc engine. It seems to have been fairly common practice to copy the best of other designs, and certainly Morris Motors Ltd were not averse to doing so. Just before the Morris Eight came on the scene a prototype saloon was built which was decidedly like the Austin Seven Ruby in styling, albeit with hydraulic brakes, hydraulic shock absorbers and semi-

1935 registered Singer Bantam. The similarity to the Morris Eight saloon is obvious. (Maxton Hayman)

elliptic road springs, but nothing came of this. Also when Singer presented their Bantam at the 1935 Motor Show this was undeniably based on the Morris Eight saloon, although its 972cc overhead-camshaft engine ensured better performance; and anyone taking a good look at the 650cc Lloyd produced by Lloyd Cars Ltd of Grimsby, between 1948 and 1951, might be excused if they concluded that they were observing a Morris Series Z van cut down to make a neat four-seater special tourer.

The chassis frame design of the Eight was unconventional for Morris in that the side members were reversed, as it were, so that their open portion faced outwards. Unfortunately, this was to prove a water trap and prone to rusting, especially around the wheel arches. Other departures from normal Morris practice was the positioning of the battery in an accessible container under the bonnet, the use of the (then) small 14mm sparking plugs, transverse mounting of the front Armstrong shock absorbers, and the use of the 'smaller' diameter 17in Magna wire wheels with six-stud fixings — double the number considered sufficient on the previous Minor.

The tendency by designers to move the radiator further and further forward continued, so that on the Eight the dumb-irons were now completely covered by a front apron. The radiator shell design itself, while appearing to follow the general lines of the previous three years, was deceptive, for behind a false honeycomb was mounted the radiator proper and the header tank cap was to

Series I Morris Eight four-seater tourer. This survivor was bought new by Mr E.A. Packer of London and is still in use.

be found in a hidden position under the bonnet. Except on the 5cwt van the two-piece moscot, fitted in place of a conventional radiator cap, was a device said to represent the Morris term 'balanced motoring'. Did the designers deliberately choose the ancient Chinese symbol for 'creation'? This mascot was in fact short lived and replaced by a similar shaped single-piece component carrying a symbolic '8' medallion on each side. From the beginning the van radiator carried a simple sun-dial triangle and a cheaper wire mesh in place of the honeycomb.

From the early stages, the design of the Morris Eight took ergonomic factors into account, with the use of an adjustable skeleton body which allowed the designers to experiment with seating positions, steering rake, etc with average size people *in situ*. The approach of building a chassis to suit the body, rather than the reverse, paid off, but there were minor snags which could not have been foreseen. Readers who are familiar with the control arrangements on the Morris Eight, especially those with an eye to the aesthetic, may wonder why the arm used as a mounting for the horn button, indicator switch and dipping control, has a neat rounded surface on the underside, allowing the various fixings and inspection panel to show on the top surface. The simple answer is that this control arm was originally intended to be mounted with the smooth surface uppermost and the direction indicator switch pointing downwards. Unfortunately it was found that the average driver would knock the protruding switch with his or her knee when entering or leaving the car. When this fact was brought to the notice of Cowley early in 1935, assemblers and Morris agents were instructed to reverse the mounting. Other early modifications were the repositioning of the indicators (when fitted) on the open models from the scuttle to a position behind the doors, and the fitting of a hand throttle control to the lower part of the instrument panel.

Of all the 1935 season Morris models, only the Morris Eight survived the 'Series Model' policy of mid-1935, when it became the Series I. At that time something like 47,000 Eights of all body types had left Cowley and minor changes to the specifications were made, such as the longer-reach Champion L10 sparking plugs, a new design of brake drum with anti-splash rims, and the replacement of the original fabric-disc universal joints by proprietary Hardy Spicer needle-roller units. From time to time small changes were introduced as the demand for the Eight continued unabated. About September 1935 the engine block casting (which required a mould with no less than thirty-one separate core pieces) was modified and this was followed four months later by replacement of the digital speedometer (incorporating a revolving drum which indicated the speed figures in a small aperture) to a needle-type instrument. The mid-months of 1936 saw a breather tube attached to the tappet cover, a change in king-pin bushes to a rolled type, an indent added to the inner edge of the near-side front wing to allow the subsequent fitting of an air silencer on the SU carburetter; a solenoid dipped the near-side headlamp while simultaneously extinguishing the off-side lamp. About the same time the Lucas Altette horn found a new mounting position under the front apron, having originally been mounted on the head of the engine. By September 1936 a more flexible steering wheel of moulded rubber had superseded the earlier hard moulding, and in the

In June 1936 Mrs N. Cosshall of Dartford Heath took delivery of this new Series I Morris Eight two-seater. Forty-seven years later she still drives the same car. (N. Cosshall)

Four-door saloon, Series I Morris Eight. (J. Ratcliffe)

next few months the minor changes continued with a change in the design of the Lucas starter motor (giving a higher ratio), the fitting of light safety guards below the doors on the 5cwt vans, while a self-starting windscreen-wiper motor replaced the 'spinner' type previously used. On the closed models the internal mirror which had hitherto been attached above the windscreen was now incorporated with the adjustable windscreen finger pull on the lower part of the frame. Side-screen fixings on the open models were reversed about May 1937 with a wing nut replacing the previous wing-screw.

The major apparent changes came when the Series II models were announced in September 1937, by which time some 164,000 Morris Eights had been produced; about 15 per cent of these were open cars. Gone was the chromium-plated radiator shell with the imitation honeycomb and in its place a paint finish to match the bodycolour, and an inner of vertical slats brightened by three chromium-plated strips. The Magna wire wheels were replaced with 'Easiclean' disc design, (still stove enamelled, as were the wings, aprons, etc) while the spare wheel mounting had forsaken its brass-buckled leather strap for a studded wooden boss attached to the rear bodyframe. On the sliding-head two-door and four-door saloons the standard luggage grid was now pressed steel. By about December 1937 a new ratio crown-wheel and pinion altered the spiral bevel rear axle from 5.375:1 to 5.286:1. One of the last modifications involved the replacement of the plain hand-brake cable to a covered Bowden type, about April of 1938.

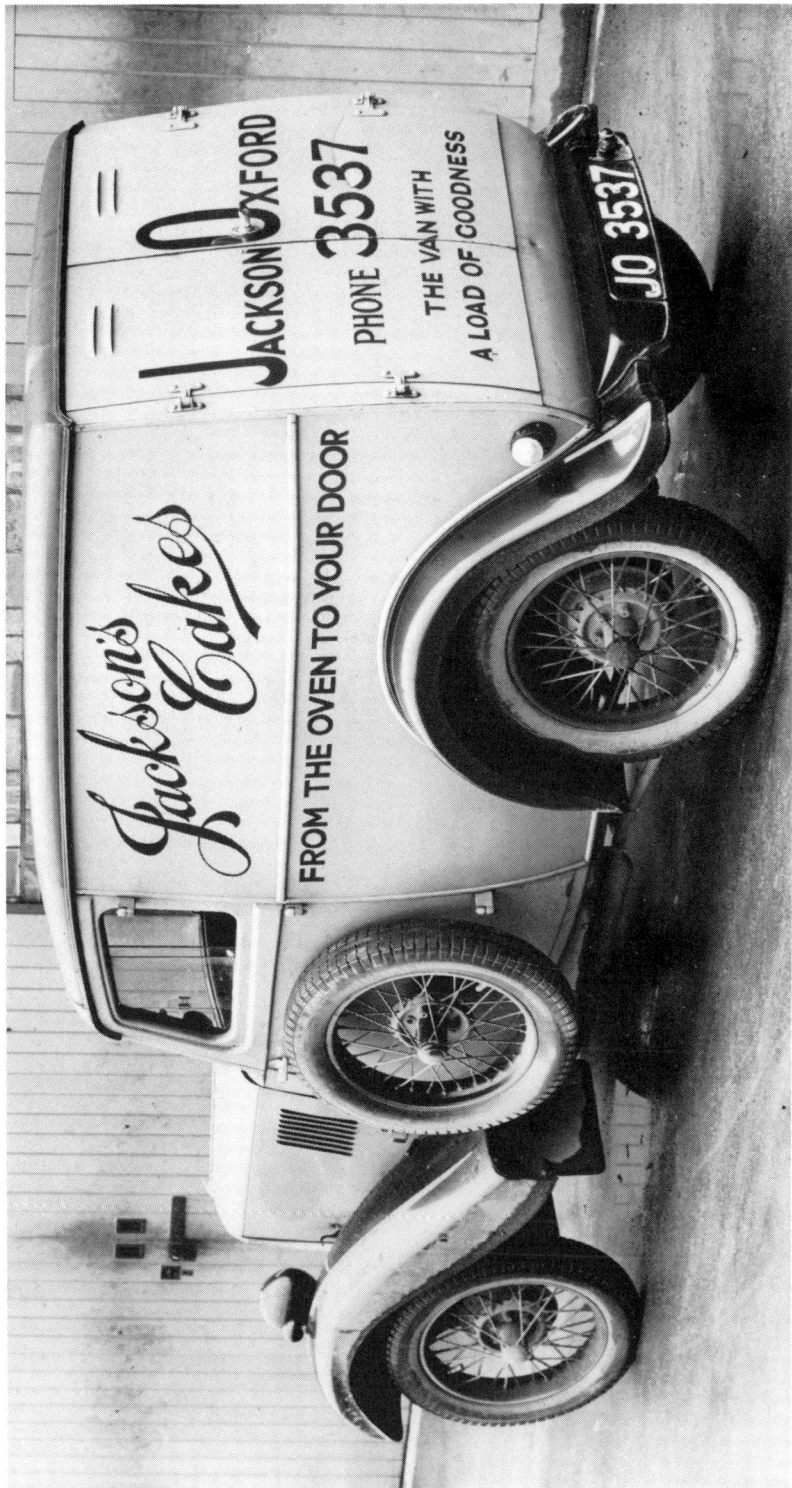

An early Morris 5cwt van used by an Oxford bakery.
(Morris Motors Ltd)

*Posed against the background of the Danish Royal Palace,
this unique Morris Eight 5cwt van chassis with advertising
body in light blue was designed by Mr Wodschow, sales
manager of the Danish Morris agents, Messrs Vilh
Nelleman, and Chr Friis, for the latter's coffee import
business. The bodywork was made in Copenhagen.*

*One of the fleet of specially adapted Morris 5cwt
newspaper delivery vans used by the proprietors of the*
Manchester Evening News *and the* Manchester
Guardian.

It may have been the need to utilise existing stocks of components special to the earlier private cars, but whatever the reason the title 'Series I' and the chromium-plated radiator surround was retained on the 5cwt van right through to the end in 1939. Officially there were no Series II 5cwt vans but the writer has seen a brake version of the Morris Eight carrying the chassis prefix 'S2/EV . . .' (ie Series II Eight van) dating to June 1939, fitted with a professionally built body using the same technique as the post-war Minor 1000 Traveller, even to the method of joining the timber over the rear wheel arches. To this puzzle can be added the cryptic advertisement which appeared in *Motor Sport* in January 1963 offering a '1938 Morris Minor Traveller, 8hp. Rare. One of Six'.

From the outset 'Export' versions of the Eight were available, with both left and right-hand drive options. In the first two years nearly 25,000 Morris cars of various body types were exported, with New Zealand the best overseas customer. Equipment differed slightly on these Eights destined for overseas markets. One example being the Magna wheels of 3in × 16in diameter carrying larger section 5.25 extra low pressure Dunlop tyres on the cars with a similar reduction in diameter on the 5cwt van from 400-18 for the home model to 450-17 on the export model. Some speedometers were marked in kilometres. Other variations were less obvious, such as the increase in the number of leaves in the rear springs. Many of the Eights, either complete or in CKD chassis form, were exported by Morris Industries Exports Ltd to Australia where local body-building firms such as Ruskin Motor Bodies Pty Ltd of Melbourne, and T.J. Richards, turned out some interesting variations of the '8/40' (as the Morris Eight was known in Australia) with fixed-head coupé, four-door tourer, roadster (two-seater) with dickey seat, and 'Ute' (pick-up) bodies.

'Easiclean' disc wheels and a painted radiator shell were features of the Series II Morris Eight.
(Morris Motors Ltd)

That is not to say that Australian coachbuilders had the monopoly of special bodies on the Eight chassis. In the United Kingdom, Stewart & Ardern offered a four-seater drop-head coupé. This model, bodied by Cunard, was of aluminium panelling on an ash frame with the extended rear portion forming a luggage compartment, recessed to cary the spare wheel. In the Midlands Jensen Brothers moved the radiator forward of the cross member, lowering it, and extended the body beyond the rear of the chassis, so as to produce a long bonneted aluminium-bodied four-seat tourer. The Jensen bodied Morris Eight was made in small numbers between 1935 and 1937, and three examples are known to have survived to the present day. One Danish-built variation, a roadster cabriolet with bodywork looking very much American, although stopping short of a dickey seat owing to length limitations, was built by Randers Karosserifabrik in Denmark.

Jensen bodied Morris Eight tourer of 1935.

Some interesting one-off specials were made from time to time on the Morris Eight chassis but perhaps the best known, particularly as it still exists, is the competition car made by William Ashley Cleave. It started as a crashed Series I saloon in 1937 and Cleave rebuilt it using a 1934 Morris Minor four-speed gearbox and supercharged the 8hp engine with a Centric blower. In the hands of Ashley Cleave and his co-director Charles Burleigh it appeared at many pre-war sprints and trials, gaining a substantial number of awards. Since the war it has been extensively modified until today the blue finished aluminium body on small diameter Magna wire wheels hides a 1,250cc supercharged Morris engine which enabled its designer to do a standing quarter-mile in 15.2 seconds and reach a maximum speed of 115mph. As late as 1972 Ashley Cleave, then aged 72 years, was still competing at Shelsley Walsh with the car. Not quite so spectacular are other one-off specials such as the pre-Series tourer upon which a certain Eric D. Clarke transplanted a Series III Morris Ten body in 1948!

Morris Eights of all Series certainly justified the term 'ubiquitous'. A report in the trade press in May 1936 described the use that the Eight chassis was being put to in New Zealand by Perpetual Forests Ltd, where they were equipped with flanged wheels to run on railway lines as fire-fighters where the tracks ran through dense forest plantations near Lake Taupo. The sparks emitted by the wood burning engines which hauled the lumber through these forests represented a source of danger so each train was followed by adapted Morris Eights carrying fire-fighting equipment and a crew of two men. In converting a Morris Eight for the Algemeene Volkscrediet Bank in Batavia (now Djakarta, Java) in 1936, the unknown coachbuilders were really pushing the engine and chassis to its limits for the vehicle had a twelve-seater brake body, used to carry children to and from school. To accommodate its juvenile load the body overhung the rear chassis by something like 18in from the rear wings, and the spare wheel was located on the off-side front wing. What became, one wonders, of the Morris Eight tourer purchased by the exiled Emperor Haile Selassie from Bath Garages for his son, the Duke of Harar, in February 1937, or the Series II two-seater which Alexander Duckham (the oil manufacturer) supercharged in 1938 and affectionately called 'Nipper'? Another Morris Eight, and one which has survived, is a two-seater Series I car which was entered in the Monte Carlo Rally (competing for the Riviera Cup for cars under 1,500cc) in January 1936, driven by Norwegian Bjarne Wist with co-driver Sverre J. Herstad, starting from Stavanger. Of the ninety-two starters in the severe winter of 1936 only seventy-two cars arrived in Monte Carlo, including Wist's Morris. The car still exists in northern Norway.

Use of the tourer version of the Eight by police forces is mentioned elsewhere in this book, these forces including Bradford, Sheffield and Hastings. The War Office also found a use for the Morris Eight and special equipment was installed in the open tourers used by the British Army. Known by the Army as 'Morris Eight, Two-Seater Wireless Cars' these vehicles (like the earlier Morris Minors and Austin Sevens) were fitted with the 'No 1 Set', a Radio Telephone and Morse transmitter-receiver introduced two years before the Eight. The body type was that of the standard four-seater, but the apparent misnomer is explained by the single bucket seat fitted for the driver and accommodation at the rear for the wireless operator; the remaining space contained the wireless equipment, batteries, spare petrol container, aerial gear, etc. The aerial was mounted on special brackets fixed to the off-side rear of the body. The all-over finish was khaki, including those parts normally plated such as windscreen surround, hub caps, radiator shell (and even the radiator mascot). The War Office specification included deep tread Dunlop types similar to the 'Town & Country' type, Autovac in lieu of the normal SU electric petrol pump, a locking cover on the tool box, towing hooks, and electrical interference

*Two-seater Series II Morris Eight used by
the Bradford City Police in 1938.*
(Bradford Constabulary)

*A British Army Morris Eight two-seater wireless car with
the Training Battalion Signals. Despite the four-seater
tourer type body, the Army nomenclature is correct as
only two seats were fitted.*

suppression on all electrical components, which may account for the vacuum-type wipers fitted to the early cars supplied. Bumpers were not fitted, but radiator muffs were.

A few of these ex-Army Eights that have found their way into the hands of present day enthusiasts and are easy to recognise by the chassis number prefix 'S1/EWD . . .'. Because military transport in the 'thirties carried Middlesex civilian registration numbers issued to the RASC Vehicle Reserve Depot at Feltham, they retain their original plates. Many of these Army cars found their way onto the civilian market via a motor agent, T. Scott of Bobbers Mill, Nottingham, so it is not surprising that most of the survivors are in the Midlands. The exact number used by the Army is not known to the writer, but research suggests that the first contract in January 1935 (V2694) was for sixty-six cars, a later contract (V2743) may have been for a similar number, and with the suspicion of a third contract it can be assumed that the numbers used were well into three figures.

It is worth recording that the South African Army also used tourers. Also the side-valve 8hp engine was used to provide auxiliary electrical power for Centurion, Conqueror and Chieftain tanks until the 'seventies.

Morris Eight. 1935/Series I/Series. Specification

Engine: Type 'UB'. Four-cylinder, side-valve. 918.6cc, 57mm bore × 90mm stroke. Three-bearing crankshaft. Aluminium pistons. 5.8:1 compression ratio. 23.5bhp at peak 3,900rpm. Coil ignition. Cooling system thermo-syphon aided by fan. SU carburetter, $\frac{7}{8}$in diameter. Three-speed gearbox with synchromesh top and third. Clutch, single-plate Borg & Beck 6$\frac{1}{4}$in diameter. Lockheed hydraulic brakes. Rear mounted petrol tank, 5$\frac{1}{2}$-gallons. SU electric petrol pump. Spiral-bevel final drive. Bishop-Cam steering. 6-volt electric system, negative earth. Propeller shaft with fabric disc universal joints to chassis 48612, later chassis Hardy-Spicer needle bearings. Armstrong hydraulic shock absorbers. Wheelbase 90in. Track 45in. Wheels, 6-stud. 1935/Series I, Magna wire wheels 2.25in × 17in with 450-17 tyres. Series II, Easiclean spoked disc wheels 2.5in × 17in with 450-17 tyres. Export cars 3in × 16in wheels with 500-16 tyres. 1935/Series I 5cwt vans, wire wheels 2.15in × 18in with 400-18 tyres. Export vans 2.5in × 17in Magna wire wheels with 450-17 tyres.

Morris Eight. Chassis numbers		
Model	Chassis Prefix	Chassis Numbers
Cars, 1935	35/E	901-48612
Series I	S1/E	48613-165000
Series II	S2/E	165001-219000
Vans, 1935	35/EV	901-48612
Series I	S1/EV	48613-221837

Military versions of the Morris Eight Tourer carried the prefixes 35/EWD and S1/EWD.

Morris Eight. 1935/Series I/Series II. Magazine Bibliography				
Model	Subject	Source	Date	Pages
1935	Morris Eight, New model report	*The Autocar*	31/8/34	p358-60
1935	Eight, New Style	*The Morris Owner*	10/34	p759, 760
1935	Morris Eight Saloon, Road Test	*Practical Motorist*	5/1/35	p250-2
1935	Morris Programme for 1935	*The Morris Owner*	9/34	p655-62
1935	Morris Eight Data Sheet 23	*Practical Motorist and Motor Cyclist*	9/56	p231
Series I	Morris Eight Saloon	*The Light Car*	15/11/35	p . . .
Series I	Morris Eight Saloon. Road Test	*The Light Car*	16/10/36	p634, 635
Series I	Morris Eight, Data Sheet	*Practical Motorist*	29/1/38	p432
Series I	Morris Eight	*Motor Trader*	25/8/37	p204-9
Series I	Morris Eight Saloon, Road Test	*The Autocar*	1/1/37	p12, 13
Series I	Morris Programme for 1936	*The Morris Owner*	9/35	p655-8
Series I	Morris 5 cwt. Van	*The Morris Owner*	11/35	p873
Series I	Morris 5 cwt Van, Electrical Test Data D438	*Motor Commerce*	1/45	p119, 120
Series I	Morris Eight Under the Microscope	*The Light Car*	25/6/37	p162-4
			2/7/37	p198, 199
			9/7/37	p230, 231
			16/7/37	p262, 263
			23/7/37	p292, 293
			31/7/37	p . . .
Series II	Morris Eight Tourer, Road Test	*The Autocar*	18/2/38	p288, 289
Series II	Morris Eight Saloon	*The Light Car*	19/11/37	p . . .
Series II	Morris Eight Saloon, Road Test	*Practical Motorist*	25/9/37	p833, 834
Series II	Morris Eight Saloon	*The Light Car*	27/8/37	p . . .
All	Overhauling the Morris Eight	*Practical Motorist and Motor Cyclist*	8/56	p132-4
			9/56	p215, 216
			10/56	p301, 302
All	Overhauling the Morris Eight	*Car Mechanics*	9/59	p45-47
			10/59	p65-7
			12/59	p55, 56

Morris Eight. 1935/Series I/Series II. Body Colours and Upholstery

	1935	Series I	Series II
Saloon, fixed-head, two & four-door	Black or red/black cellulose with red Karhyde.	Black or red/black cellulose with red Karhyde. Blue/black cellulose with blue Karhyde. Green/black cellulose with green Karhyde.	Black or maroon cellulose with red Karhyde. Blue cellulose with blue Karhyde. Green cellulose with green Karhyde.
Saloon, sliding-head, two & four-door	Blue/black cellulose with blue leather. Green/black cellulose with green leather. Red/black cellulose with red leather.	Blue/black cellulose with blue leather. Green/black cellulose with green leather. Red/black or black cellulose with red leather.	Blue cellulose with blue leather. Green cellulose with green leather. Maroon or black cellulose with red leather.
Two-seater and four-seater tourer	Black or green cellulose with green Karhyde. Red cellulose with red Karhyde.	Black or red cellulose with red Karhyde. Green cellulose with green Karhyde. Blue cellulose with blue Karhyde.	Black or maroon cellulose with red Karhyde. Black or green cellulose with green Karhyde. Blue cellulose with blue Karhyde.

Other Series Models Series I & II

Fewer cars were assembled at Cowley in 1933 than in 1925. Indeed, production was the lowest for nine years (with the possible exception of 1931) and with well over half the new cars sold in the under 10hp category, the bread-and-butter Cowley was no longer competitive for the British public. William Morris (by now Sir William Morris) had begun to take frequent trips overseas, particularly Australia, and his personal handling of the business was neglected. Thus the scene was set for the entry of Leonard Lord, 'Specialisation', and the Morris 'Series' cars.

Leonard Percy Lord (later, in the 'sixties, to become Lord Lambury) started his working life as an apprentice draughtsman for Courtaulds at the age of 16. When Morris took over Hotchkiss et Cie in 1926 he had been working for the Coventry firm for two years as a jig and tool draughtsman. After promotion he was made responsible for the purchase and design of machinery at a time when the renamed Morris Engines (Coventry) Ltd was being reorganised. He subsequently, held a similar post at the newly acquired Wolseley Motors Ltd. Lord was a good engineer and organiser, but was not popular. To quote the late Lord Thomas of Remenham:

> He knew what he wanted and knew how to get it. He frightened the living daylight out of them — right from the highest to the lowest. If he wanted a thing done there were no quibbles, that was done. Lord said so and Lord had it done. The lower ones he was more friendly with but they were still frightened of him — if there was one word out of place — Out! Lord was ruthless, yet capable of touching generosity. He was crude in both speech and manner and a victim of a sizeable inferiority complex.

By 1933 Morris decided that a long planned thorough reorganisation of the production side of Cowley was overdue. Over the years the main assembly plant had grown piecemeal to keep pace with output, but the large sums spent had gone mainly on building extensions and maintenance. Ironically, the bad trading year for 1933 made the proposed changes physically possible and finances for the project had been kept in hand. Morris chose the hook-nosed, red-haired, Warwickshire man Leonard Percy Lord for the task of reorganisation, and his predecessor, E.H. Blake, stood aside as managing director of Morris Motors Ltd.

The major new installation was a moving assembly line to replace the old system where a chassis, fitted with slave flanged wheels, was manhandled along a track as each stage of work was completed, the parts necessary for the appropriate assembly being stacked on the nearby floor space. The whole process was now mechanised with the chassis being slowly carried forward by a moving chain to coincide with a supply of components delivered by overhead conveyor direct from the store of the sub-assembly department. Subsequent reorganisation of the sub-assembly departments were made; for example, the trimming shop where the long benches had been piled high with stitched upholstery and filling was fitted with central moving belts. Five new sewing machines were specially developed to enable Karhyde or leather to be sewn direct onto plywood panels at a speed of 700 stitches per minute. The new paint shop additions went into service in March 1935, here the half-mile long line with four double rows of conveyors running its entire length, costing £83,000 to build and equip, was said to have a production capacity of 2,000 bodies in a 40-hour week.

By June 1934 Sir William Morris was able to announce that a sum of £250,000 had been spent on the reorganisation. This figure reached £500,000 two years later, by which time Morris Motors Ltd had purchased an aditional 125 acres to erect new buildings with a floor space of 359,125 sq ft for service, repairs, and a car dispatch department capable of holding 1,620 cars. Included in the complex was a sales school and a showroom to hold the complete Morris range. Electric petrol pumps within the building were fed from storage tanks holding 5,000 gallons of fuel.

Late model of the Series II Ten-Four saloon with Easiclean wheels. The Twelve-Four was identical in appearance.

Some of the methods of material testing, described by the Morris publicity people at the time of the reorganisation, present an interesting picture of the behind-the-scenes activities before a Morris vehicle reached its new owner. The cast cyclinder blocks were not cast by the conventional methods using green sand cores which are damp (resulting in a casting which on the surface is extremely hard while the centre, taking longer to cool, remained relatively soft) but by the use of oil sand cores which were baked dry to avoid local chilling, resulting in an even cooling throughout the casting. The completed engines were electrically rotated on a test bed for two hours to run-in the bearings thoroughly then, under their own power, the units were motored for a further two hours at varying speed while bhp, petrol consumption, oil and water temperatures were monitored. The engines were fitted to a complete test car, and a rolling road machine linked the final power output to an electrical dynamo to ascertain the energy dissipated in transmission losses between engine and road wheels. The same machine (similar to those used for present day MoT tests) tested the braking efficiency, the brake linings themselves having been tested on a machine which effectively simulated continuous braking for a hundred miles. A similar arrangement of rollers vibrating through cams applied a 200 per minute bump test to wheels and suspension. Some of the purpose-made testing gear had curious, if self explanatory, names such as the 'Weather-o-meter' which subjected test pieces of rubber, paint, and other materials to penetrating rays of a lamp many times more intense than brilliant sunlight, while heat and humidity was simulated, compressing two year's life of the samples into a couple of weeks. Another machine, the 'Fugit-o-meter' did similar tests on leather, hood cloth, blind and trimming materials after test pieces of the material had been ripped asunder by the 'Salter improved Dynamometer'. In a cold chamber, large enough to take the largest Morris model, the effects of the lowest known climatic conditions were closely observed. Some of the machines ripped sample metal components apart or put parts as large as cylinder blocks under compression to a point of collapse with the 200,000lb capacity available. Messrs W. & T. Avery Ltd of Birmingham made a machine which delivered giant hammer blows to samples to reveal brittleness, and another Avery machine, with a capacity of 6,000lb, compressed road springs to indicate ratio of deflection and load over their working range. A crude but effective means of giving the chassis a 200 per cent overload was done by placing the car wheels astride an inspection pit and applying force with a block and tackle anchored to the pit floor. In the laboratory an American made moisture meter (basically a test of electrical resistance) ensured that timber used in the bodywork contained just the right amount of moisture. Laboratory personnel also tested the purity of horsehair, used for upholstery padding and air filters, by dissolving samples in a caustic solution for analysis. A machine was even used to simulate 150 miles of foot wear on the rubber inserted pile carpet.

One of the first models to make full use of the new production facilities was the Morris Eight, designed by a team under the leadership of S. Westby. Once the initial production was underway, measures were taken to keep details of the new model secret until the press release was made to the motoring public in

August 1934. For this reason facilities for organised parties of visitors to the Cowley works were withdrawn in June 1934 for three months. The cautious original estimate that 35,000 units would be required during a twelve month period did not allow for the immediate success of this new 918cc side-valve model.

'Mass production' was a term not in favour at Cowley —preferring as they did 'Cars built by Specialisation'. 'We ask motorists to compare the Specialisation-built Morris with its Mass-Produced competition' ran the Morris publicity. Sir William Morris tried to explain the subtle difference in 1934: 'The best cars have always been produced by Specialisation, but at a prohibitive cost. Now Morris brings you the same precision manufacture at a reasonable price.'

Showroom display of Series II cars. The colour extending down to the running-board on the saloon dates this photograph to late 1936. (Stewart & Ardern Ltd)

The success of the Morris Eight continued and by May 1935, the future bright for Morris Motors Ltd, the new 'Series' policy was implemented. At a luncheon at Grosvenor House in London the new Series II 10hp and 12hp models were shown to the press. Len Lord, introducing the new models, said 40,000 Eights had been sold since September (in retrospect this would seem to have been an understatement) and these new cars were the result of 'tearing up many thousands of drawings'. Many ideas that might have improved the design had been tested, including novel suspension and transmission, but few of them had proved their advantage. Lord said that investigations by designers had shown that cruciform chassis bracing was 'not worth having for small cars', but tests had led to the adoption of the boxlike formation of the front members now used on the new models. Every care had been given to providing as much room as possible for the passengers and although the cars did not look large, actual measurements would show that the dimensions inside the centre posts were exactly the same as the 25hp Morris.

Dash and controls on the Ten and Twelve-Four Series II models. (Th. J. de Mooy)

1936 example of the Series II Morris Ten-Four saloon. (J.W. Edgell)

Announcing the new 'Series' policy, Lord stated that in future there were to be no more dated Morris models, they were to be known by Series numbers and the new versions would be brought out as and when desirable. The idea behind this new policy was to avoid a seasonal buying peak in March and April which tailed off dramatically before the annual Motor Show, while prospective purchasers awaited news of the new season's models. The effect was to be seen later in the year when in August 1935, normally a pre-Motor Show slack period, 7,635 cars were delivered from Cowley compared with a previous high for August of 2,380, which included 1,450 cars to be stocked by agents for show time. Three years later, in June 1938, the 'Morris Statistic Department' (presumably to confirm the success of their Series policy) revealed that in taking an average for the 1935, 1936, and 1937 seasons, June, July and August had proved to be the second-best selling quarter of the year. As the Society of Motor Manufacturers & Traders had agreed a rule that no announcements of the new season's programme would be made before 13 August, accusations were made by other manufacturers that Morris Motors Ltd had broken faith by announcing new 'Series' models early in the year. L.P. Lord countered with the statement that this was not the case as the new cars were not described as 1936 editions: 'In any case, Morris reserved the right to run the Morris business in their own way.'

The new Tens and Twelves, identical in all but engine and badge, were classified as Series II. At the same time the Morris Eight, with minor modifications, had ceased to carry the year prefix and was designated Series I. Why only the Eight was known as Series I is something of a puzzle, the theory that because the Eight body style was still somewhat 'perpendicular' in contrast to the new 'streamlined' models, only holds good until late in 1937 when the updated model with the same body was known as Series II. To add to the historian's confusion, gasket manufacturers, such as Payen and Hallite, incorrectly referred to the pre-Series 1935 Ten-Four model as 'Series I'.

As originally introduced the Series II Tens and Twelves had, underneath the shell, much in common with the 1935 Ten-Four. For example the well tried cork-insert clutch running in oil, 18in diameter Magna wire wheels, and a similar side-valve engine. The engine now looked like a side-valve unit, having a cylindrical air cleaner/fume consumer mounted across the head. Various 'new' features had already been introduced in the previous Ten-Four, such as the right-hand accelerator pedal, automatic advance/retard (combined on the 1935 pre-Series model with the steering column control), Tecalemit external oil filter, steel brake shoes, and split-bumper combined with luggage grid. It has further been suggested that the last of the pre-Series Tens had Spicer needle-roller bearing type universal joints; although the writer has not seen any written confirmation of this change in the specification. A feature already tried on the Ten-Six, the so-called 'Harmonic Stabilising Bumper', was also fitted to the Ten-Four.

1,293cc side-valve engine used in the Series II Morris Ten.
(Th. J. de Mooy)

The four-cylinder side-valve engine for the Ten was a 63.5mm × 102mm unit with a capacity of 1,292cc; a similar engine (not a bored-out version of the Ten) rated at 1,548cc with a 69.5mm bore and larger diameter valves provided the power unit for the optional Twelve. During the production run of the model, between May 1935 and the end of July 1937, various changes in specification were made, for example, the gearbox. In its initial form the car was equipped with a three-speed box and (it is pertinent to note) the 'MK' and 'TJ' type engines in the Ten and Twelve respectively, with the oil filter mounted on the oil pump. By November 1936 the option of a three-speed or four-speed gearbox was offered, but the purchaser choosing a four-speed model had to pay an extra £5 to have the built-in 'Jackall' hydraulic jacking system; these devices being standard on the three-speed models. By February 1937 the manufacturers announced that having found 92 per cent of orders had specified the four-speed gearbox, it would henceforth be made standard. (Sceptics may now think, knowing the model to be nearing the end of its production run, it suited the supplies in hand.) Where the four-speed gearbox was fitted the engine types 'MSJM' and 'TSBM' with cylinder-block mounted oil filters were standard for the Ten and Twelve respectively. One of the minor changes (about March 1936 at chassis 74057) was to supersede the little green light in the centre of the steering wheel which indicated to the driver that the flush fitting Lucas trafficators were operating correctly. In its place a pair of small mirrors were fitted, one in each corner of the windscreen, allowing the driver to see the semaphore arm.

Early version of the Series II Fourteen-Six with the colour confined to the bonnet top, styling line and the rear of the body. (John Lowrie)

In the closing months of 1936 the new style 'Easiclean' disc wheels (3.5in ×
16in diameter) appear to have been offered as an alternative to the Magna wire
wheels (3in × 18in diameter) before becoming standard equipment some
months later. Tyre sizes differed, with Dunlop Cord 4.75-18 for the Magna
wheels and 5.50-16 on the smaller diameter Easiclean type. Vehicles destined
for foreign parts continued to be fitted with the customary smaller diameter
and wider width tyres (5.75-16 extra low pressure) on similar steel wheels.

When the press accepted an invitation to a luncheon at Stewart & Ardern's
showrooms in Berkeley Square one day in June 1935 it was to witness the
introduction of further Series II vehicles. The same formula as the Ten/Twelve
was applied utilising common body pressings for the Sixteen-Six and Eighteen-
Six with a 117in wheelbase, while the 121in wheelbase six-cylinder Twenty-One
shared the body with a larger capacity Twenty-Five model. The 2,062cc 'QH'
side-valve engine on the Sixteen became the 'QJ' of 2,288cc when bored-out to
69mm for the Eighteen-Six. Likewise the Twenty-One 'OJ' engine of 2,915.8cc,
in 'OK' form had a capacity of 3,485cc for the Twenty-Five model. Each pair
had their own common gearbox, but the differing rear axles resulted in
dissimilar final drive ratios. The increase in piston size for the larger version of
each engine type resulted in some extra 7bhp.

*The Series II Eighteen-Six
shared the same body with
the Sixteen-Six.*
(Ken Martin)

The odd-one-out in the Series II range was the six-cylinder Fourteen,
apparently introduced to fill the gap between the Twelve and Sixteen models.
This one became available some twelve months after the other Sixes and unlike
the larger models did not share the body with an optional engine size, nor was a
special coupé version offered. The short production run of the Series II Fourteen listed between June 1936 and July 1937 is reflected in the mere thirteen
cars recorded in the hands of present day enthusiasts.

The bodies on all the 'streamlined' Series II saloon models followed a similar
pattern with the hinges to the centre pillar for all four doors and a metal cover
for the spare wheel mounted at the stern, although this was £1 extra on all Ten
and Twelve fixed-head saloons, and with £1 10s for a luggage grid. This last
extra also applied to the smooth top Sixes, including the Twenty-Five. The
major design variation on the six-cylinder models, except the Fourteen, was the

Identical bodies were used on both Ten-Four and Twelve-
Four versions of the Series II cars. This photograph shows
a Series II special coupé. (J. Lowrie)

boot protrusion at the rear, with access from inside the car. The extra sum
asked for the sliding-head version of any of the saloon cars would appear to
have been good value, for with the sunshine roof came real leather upholstery
in place of the leathercloth used with the fixed-head saloons. All horsepower
ratings had a 'Special Coupé' listed except the Fourteen, and here the body
design for the Fours and Sixes differed in that the Ten/Twelve combination
was a two-door, two-light, body. It had imitation pram-irons and an inbuilt
luggage container with a diminutive top lid, indicative of the need for the
additional luggage grid provided.Special coupé six-cylinder cars came with a
four-light body with divided rear windows, a similar deck hinged boot, and
extras such as chromium-plated headlamps, twin fog lamps and twin horns.

The mid-1930s proved a popular period for two-tone colour schemes on car
bodies. Although the term 'two-tone' was probably not in use at the time, it
should not be confused with the Morris description 'Duotone' which referred
to two shades of the same colour. In the case of the Series II saloon cars —
once again exluding the Fourteen-Six — the original colour arrangement
(accepting the long standing Morris practice of stove-enamelled black for
wheels, wings, aprons, etc for saloons, but no longer used on the special coupé)
consisted of a black body brightened by a wide band of colour within the
styling lines and including the bonnet top. After the middle months of 1936
(therefore including the Fourteen) these colours, blue, green, red or grey,
extended down to cover the complete lower half of the bodywork. There were,
in addition, all black models. For the special coupé there was no change; these
continued to be listed as a single all-over colour of black, grey or sports blue.

A guide to colour preference in the 'thirties is given in a Morris press release of 1937 in which it was revealed that 3,000 gallons of cellulose/thinners mixture was used at Cowley each week. An analysis of the quantities used showed that the order of popularity was black (1,500 gallons per week), blue (650), green and maroon (300) and grey (250). Each Morris car had eleven coats of finish, consisting of a primer coat, three fillers, a sealing coat, and finally six coats of 'Bripal' cellulose lacquer.

The gearbox alternatives of three-speed with Jackall, or four-speed without Jackall, for the Ten/Twelve cars, was also available on the Fourteen-Six and Eighteen-Six models in the penultimate month of 1936. A similar option did not apply to the Sixteen or Twenty-One which by then were no longer listed, or to the Twenty-Five cars. Other changes, however, took place during the production period such as the introduction of Easiclean wheels which, on the Twenty-Five, resulted in a change of tyre size from 600-17 to 700-16, bringing the home model into line with the hitherto export-only size. This change in specification dates to chassis S2/TF51593, about November 1936, when chromium-plated lamp shells brightened up the front end of the saloon. On the Twenty-One, as early as November 1935, the running boards were altered to rubber mat type (chassis S2/TO42880), followed in May 1936 when the facia panel bakelite moulding was superseded by a wooden panel (chassis S2/TO46009). The total number of Twenty-Ones built was only 829, of which it can be estimated that less than thirty were special coupés.

Good publicity of all kinds was naturally sought by Morris Motors Ltd for the products of 'Specialisation'. Production had hardly started when a Series II Ten special coupé gained first prize for coachwork at a Bexhill competition. The Scottish Rally Concours d'Elegance of the period gave the first prize for competing closed cars costing up to £250 to a similar Ten, while in the same rally two Tens took second and fourth places for performance against all closed cars up to 1,300cc. Who entered these cars is not recorded, but certainly the Cowley publicity department provided a new Twenty-One and a Twenty-Five to a team of journalists and others (including Cecil Orr, then press officer of the Automobile Association) to cover the route of the first international road, from London to Istanbul, declared open in June 1935. Another Series II Twenty-Five saloon, in November of the same year, made a successful journey by road from England to Nigeria in just over a week, beating the time of the mail steamer and express train by nearly eight days. The Morris used (registered BWL103) was said to be a standard production model apart from the extra tanks fitted with petrol and water. The latter proved its worth when the driver, H.E. Symons, and his two companions came across two well diggers who were dying of thirst in the Sahara. Six months earlier, Symons (who subsequently wrote a book based on his experiences called *Two Roads to Africa,* and was later, as Flt Lt Humfrey E. Symons, to be killed off the Belgian coast during the evacuation from Dunkirk) and his co-driver Leslie Seyd, had made a similar trans-Sahara trip in a Series II Ten saloon (BFC606) in 97 hours, 33 minutes. The car used was a standard model with the exception of larger tyres, a drinking water tank, auxiliary petrol tank, auxiliary cooling radiator, and an Ekco car radio with built-in roof aerial — modifications which, it was said, had

Standard Morris body on the
Series II 10cwt chassis.
(Beck & Pollitzer Ltd)

only been completed two hours before Symons and Seyd drove away from
Cowley at 5pm on Saturday 27 April 1935.

Hardly standard was the Series II Twenty-Five saloon owned by Alexander
Duckham which was fitted with a Perkins 'Wolf' diesel engine for experimental
work in connection with lubrication. The car, registered CUC644, had the
Perkins Diesel motif fixed to the radiator grille, in just over a year it clocked up
25,000 miles, with a saving in fuel costs of £65 — which was twice the cost of
the more expensive diesel engine. Maybe Alexander Duckham was a Morris
enthusiast for he later had a Morris Eight two-seater supercharged for his
personal transport.

One other Series II vehicle which must be mentioned, if for no better reason
that it was the longest running Series II model of them all, was the 8/10cwt
van, listed for over four years following its début in July 1935. Replacing the
more conventional 8/10cwt van on Cowley lines, the model broke new ground
with its semi-forward control driving position and the unconventional off-set
side-valve engine, the 'TK' of 1,547cc based on the 'TJ' version used in the
Series II Twelve. Although the wheelbase was shorter than its predecessor by a
full foot, a precedent was created for Morris by the dissimilar front and rear
track dimensions, the back axle being 4in wider.

The off-set engine of the
Series II 10cwt van is clearly
indicated by the position of
the starting handle.

In the writer's opinion, it was one of the nicest little commercials of the 'thirties, retaining something from the earlier period with its boxed front mudguards, vestigial running boards and a fixed starting handle; yet it was up-to-date with rear wheel spats, a rounded body and, on some later examples, Easiclean wheels. If there was one feature that could be considered ugly it was the home-made appearance of the stark bumper fitted as standard at the front. This apart, it possessed aesthetic neatness lost in its steel bodied successor, the Series 'Y' 10cwt, which will be considered later. For a commercial vehicle the 8/10cwt van was particularly well instrumented with, an ammeter, oil gauge, petrol gauge, hand throttle, inspection lamp sockets and a clock. The steering wheel followed the design of that fitted on the Series II cars in having a trafficator switch (and warning light on early examples) mounted in the centre yet, strangely, was probably the last Morris vehicle to retain the outmoded centre accelerator pedal.

During its currency the Series II 8/10cwt van (or as Morris Motors later renamed it '10 cwt Van') never found favour with the GPO although, or perhaps because, they had a large fleet of the earlier 8/10cwt Cowley-type vans. It was, however, popular with the municipal gas, electric and water undertakings, and older readers will recall many of these vans in Co-op livery. Selfridges of Oxford Street had an interesting fleet of special bodied vans, perhaps remembered better by the store's drivers than the public for the design left the upper portion of the cab doors open to the elements. Another small fleet, based on this chassis, was the thirty loudspeaker vans used by Gordon

A Series II 10cwt van chassis with a special body by Bonallack & Sons. (Freight Bonallack Ltd)

Stewart of Stewart & Arderns to promote his favourite charity, The Children's Safety Crusade.

Announced in June 1938, a variation of the van in small truck form had its first public exhibition at the Royal Show 'Motor Transport in Agriculture' early the following month. In place of the van driver's bucket seat (additional seat extra) the truck cab had a single full width Karhyde upholstered seat in the cab. The truck platform, with removable side panels and tail boards, was of deal tongue and groove boards longitudinally mounted on ash bearers. Ash bends which fitted into 'D' sockets in the sides and a khaki canvas hood and rear roll-up curtain of the same material with a transparent panel were extras to the basic price of £169 10s. One intriguing standard accessory was an additional number plate and tail lamp which, to avoid obscuring the rear number plate when loading or unloading at night '. . . is provided with two hooks that fit over the upper edge of the lowered tailboard, when a lead to the lamp can be plugged into the socket provided.'

By July 1937 the listed range of Series II cars was down to the four-cylinder Ten and Twelve, and sixes were represented by the Fourteen, Eighteen and Twenty-Five, the last two mentioned having now completely broken the Morris tradition with wings the same colour as the monochrome body. Shortly, in October, the Lord Mayor of London, Sir George Broadbridge, opened the 31st International Motor Exhibition at its new venue Earls Court. This coincided with the announcement of the new Series III (and Series II Eight) models from Cowley and, thus, brought Morris back to introducing their new models at the Motor Show.

Series II Models. Chassis Numbers

All six-cylinder Series II cars shared the same chassis series of numbers between 39141 and 57531. It is known that 39141 was allocated to an Eighteen-Six, 39142 to a Sixteen-Six, 39143 to a Twenty-Five-Six, and 39144 to a Twenty-One-Six. The Fourteen-Six started at chassis number 49119.

A separate series running on from the pre-Series Ten-Fours was used for Series II Ten and Twelve cars. Tens were between 49341 and 108704, the Twelves between 49342 and 108659.

Chassis numbering on the Series II 10cwt vans ran from 15901 to 33303.

Each model had its own distinctive prefix thus:

Series II Ten-Four	S2/TN . . .	Series II Twelve-Four	S2/TW . . .
Series II 10cwt van	S2/TWV . . .	Series II Fourteen	S2/FS . . .
Series II Sixteen	S2/SS . . .	Series II Eighteen	S2/ES . . .
Series II Twenty-One	S2/TO . . .	Series II Twenty-Five	S2/TF . . .

Engine Data. Series II Models

Engine Type MK, Ten 3-speed gearbox
Engine Type MSJM, Ten 4-speed gearbox

Four-cylinder, side-valve. 63.5mm bore. 102mm stroke. 1,292cc. 10hp. 4-ring aluminium alloy pistons. 3-bearing crankshaft. Valves, inlet & exhaust, 28mm diameter. Valve lift, inlet and exhaust, 7.31mm. Compresion ratio 5.9:1. Carburetter 1⅛ SU with air cleaner and fume consumer, AC. Clutch, single-plate cork-insert running in oil, spring hub. Distributor Lucas DK4A-A84. Plugs Champion L10, 14mm- Belt driven Lucas C45-NVO dynamo. Coil ignition. Gauze oil filter in sump. Sump capacity 1 gallon 2 pints. Gearbox capacity 1⅛ pints. Thermo syphon & fan for cooling system. Engine mounted at four points on rubber. 30bhp at 3,200rpm. 3-speed gearbox with synchromesh in top and 2nd. Top, 1:1; 2nd, 1.713:1; 1st 3.6:1; reverse, 4.55:1, 4-speed gearbox with synchromesh in top and 3rd. Top, 1:1; 3rd, 1.53:1; 2nd, 2.363:1; 1st 4:1; reverse 5.15:1.

Engine Type TJ, Twelve 3-speed gearbox
Engine Type TSBM, Twelve 4-speed gearbox
Engine Type TK, 10cwt van

Four-cylinder, side-valve. 69.5mm bore. 102mm stroke. 1,547.8cc. 11.97hp. 4-ring aluminium alloy pistons. 3-bearing crankshaft. Valves, inlet and exhaust, 33mm diameter. Valve lift, inlet 7.704mm, exhaust 7.817mm. Compression ratio 5.91:1. Carburetter, 1⅛ SU on TJ and TSBM engines, 1¼ SU on TK engine. AC air cleaner and fume consumer. Clutch, single-plate cork-insert running in oil, spring hub. Distributor Lucas DK4A-A84, later TK engines used Lucas DKH4A-A95. Plugs, Champion L10, 14mm. Belt driven Lucas C45-NVO dynamo. Coil ignition. Gauze oil filter in sump. Sump capacity 1 gallon 3 pints. Gerbox capacity 1⅛ pints. Thermo syphon & fan for cooling system. Engine mounted at four points on rubber. 34bhp at 3,400rpm. 3-speed gearbox with synchromesh in top and 2nd. Top, 1:1; 2nd, 1.713:1; 1st, 3.6:1; reverse, 4.55:1. 4-speed gearbox with synchromesh in top and 3rd. Top, 1:1; 3rd, 1.53:1; 2nd, 2.363:1; 1st, 4:1; reverse, 5.15:1.

Engine Type QSDM, Fourteen 3-speed and 4-speed gearbox

Six-cylinder, side-valve. 61.5mm bore. 102mm stroke. 1,818cc. 14.069hp. 4-ring aluminium alloy pistons. 4-bearing crankshaft. Valves, inlet and exhaust, 30mm diameter. Valve lift, inlet 7.704mm, exhaust 7.817mm. Compression ratio 5.9:1. Carburetter, 1¼ SU downdraught with AC air cleaner and fume consumer. Clutch, single-plate cork-insert running in oil, spring hub. Coil ignition. 43.75bhp at 4,000rpm. 3-speed gearbox with synchromesh in top and 2nd. Top, 1:1; 2nd, 1.675:1; 1st, 3.39:1; reverse, 4.3:1. Engine mounted at four points on rubber. 4-speed gearbox with synchromesh in top and 3rd. Top, 1:1; 3rd, 1.48:1; 2nd, 2.28:1; 1st, 4.08:1; reverse, 5.097:1.

Engine Type QH, Sixteen 3-speed gearbox

Six-cylinder, side-valve. 65.5mm bore. 102mm stroke. 2,062.17cc. 15.94hp. 4-ring aluminium alloy pistons. 4-bearing crankshaft. Valves, inlet and exhaust, 30mm diameter. Valve lift, inlet 7.704mm, exhaust 7.817mm. Compression ratio 5.9:1. Carburetter, 1¼ SU with AC air cleaner and fume consumer. Clutch, single-plate cork-insert running in oil, spring hub. Coil ignition. Distributor Lucas DK6A-A540. Plugs, Champion L10, 14mm. Belt driven Lucas C45-WV1 dynamo. Tecalemit oil filter OF442. Sump capacity 2 gallons 1 pint. Gearbox capacity 1 quart. Water pump in tandem with fan for cooling. Engine mounted at four points on rubber. 48bhp at 3,600rpm. 3-speed gearbox with synchromesh in top and 2nd. Top, 1:1; 2nd, 1.675:1; 1st, 3.39:1; reverse, 4.3:1.

Engine Type QJ, Eighteen 3-speed gearbox
Engine Type QSHM, Eighteen 4-speed gearbox

Six-cylinder, side-valve. 69mm bore. 102mm stroke. 2,288.45cc. 17.7hp. 4-ring aluminium alloy pistons. 4-bearing crankshaft. Valves, inlet and exhaust, 30mm diameter. Valve lift, inlet 7.704mm, exhaust 7.817mm. Compression ratio 5.9:1. Carburetter, $1\frac{3}{8}$ SU with AC air filter and fume consumer. Clutch, single-plate cork-insert running in oil, spring hub. Coil ignition. Distributor Lucas DK6A-A540. Plugs, Champion L10, 14mm. Belt driven Lucas C45-WV1 dynamo. Tecalemit oil filter OF442. Sump capacity 2 gallons 1 pint. Gearbox capacity 1 quart. Water pump in tandem with fan for cooling. Engine mounted at four points on rubber. 3-speed gearbox with synchromesh in top and 2nd. Top, 1:1; 2nd, 1.675:1; 1st, 3.39:1; reverse, 4.3:1. 4-speed gearbox with synchromesh in top and 3rd. Top, 1:1; 3rd, 1.48:1; 2nd, 2.28:1; 1st, 4.08:1; reverse, 5.097:1.

Engine Type OJ, Twenty-One 3-speed gearbox

Six-cylinder, side-valve. 75mm bore. 110mm stroke. 2,915.8cc. 20.92hp. 4-ring aluminium alloy pistons. 4-bearing crankshaft with damper at front. Valves, inlet and exhaust, 35mm diameter. Valve lift, inlet 9.2607mm, exhaust 9.495mm. Compression ratio 5.9:1. Carburetter, $1\frac{5}{8}$ SU with AC air cleaner and fume consumer. Clutch, single-plate cork-insert running in oil, spring hub. Distributor Lucas DK6A-A93. Plugs, Champion L10, 14mm. Belt driven Lucas C45-WV1 dynamo. Coil ignition. Tecalemit oil filter OF362. Sump capacity 2 gallons 2 pints. Gearbox capacity 1 quart. Water pump in tandem with fan for cooling. Engine mounted on four points on rubber. 72bhp at 3,800rmp. 3-speed gearbox with synchromesh in top and 2nd. Top, 1:1; 2nd, 1.675:1; 1st, 3.39:1; reverse, 4.3:1.

Engine Type OK, Twenty-Five 3-speed gearbox

Six-cylinder, side-valve. 82mm bore. 110mm stroke. 3,485cc. 25.01hp. 4-ring aluminium alloy pistons. 4-bearing crankshaft with damper at front. Valves, inlet and exhaust, 35mm diameter. Valve lift, inlet 9.2607mm exhaust 9.495mm. Compression ratio 5.91:1. Carburetter, $1\frac{5}{8}$ SU with AC air cleaner and fume consumer. Clutch, single-plate, cork-insert, running in oil, spring hub. Distributor Lucas DK6A-A93. Plugs, Champion L10, 14mm. Belt driven Lucas C45-WV1 dynamo. Coil ignition. Tecalemit oil filter OF367. Sump capacity 2 gallons 2 pints. Gearbox capacity 1 quart. Water pump in tandem with fan for cooling. Engine mounted at four points on rubber. 79.5bhp at 3,700rpm. 3-speed gearbox with synchromesh in top and 2nd. Top, 1:1; 2nd, 1.675:1; 1st, 3.39:1; reverse, 4.3:1.

Series II Models. Specification

Series II 10cwt van

Engine type TK. Footbrake, Lockheed hydraulic to four wheels. Handbrake, cable to rear shoes. 9in diameter drums. Petrol tank on left-hand side, $6\frac{1}{2}$ gallons. SU electric petrol pump. Open propeller shaft with Hardy Spicer needle-roller bearing universal joints. Rear axle crown wheel & pinion 8/43. Armstrong hydraulic shock absorbers. Bishop Cam steering. Semaphore type indicators 12-volt electrical system, battery under driver's seat. Headlamps Lucas L140 EDS. Stop & tail lamp Lucas ST51A. Side lamps Lucas LD206. Horn Lucas HF935. Windscreen wiper, single blade, Lucas CWR-A95. Front springs 9 leaves, rear springs 10 leaves. Wheels 6-stud, Magna wire 3in × 18in, Dunlop 500-18 tyres. Special orders, Easiclean 3.5in × 16in, Dunlop ELP 550-16 tyres. Wheelbase 90in. Track, front 50in, rear 54in.

Series II Ten-Four and Twelve-Four

Engine type 3-speed Ten MK; 4-speed Ten MSJM; 3-speed Twelve TJ; 4-speed Twelve TSBM. Footbrake, Lockheed hydraulic to four wheels. Handbrake, cable to rear shoes. 9in diameter drums. Petrol tank at rear, 7 gallons. SU electric petrol pump. Open propeller shaft with Hardy Spicer needle roller bearing universal joints. Rear axle ratio, Ten, 5.375:1 (8/43); Twelve, 4.875:1 (8/39). Armstrong hydraulic shock absorbers. Bishop Cam steering. Semaphore type indicators. 12-volt electrical system. Front springs 6 leaves, rear springs 8 leaves. Wheels, home models to chassis 89272, 3in × 18in Magna wire with Dunlop 4.75-18 tyres. Home models chassis 89273 onward, 3.5in × 16in Easiclean with Dunlop 5.50-16 tyres. Export models 3.5in × 16in Easiclean with Dunlop ELP 5.75-16 tyres. Wheelbase 100in. Track 50in.

Series II Fourteen-Six

Engine type QSDM. Footbrake, Lockheed hydraulic to four wheels. Handbrake, cable to rear wheels. 10in diameter drums. Petrol tank at rear, $8\frac{3}{4}$ gallons. SU electric petrol pump. Open propeller shaft with Hardy Spicer needle roller bearing universal joints. Rear axle ratio, 5.33:1 (9/48). Luvax hydraulic shock absorbers. Bishop Cam steering. Semaphore type indicators. 12-volt electrical system, battery under front seat. Headlamps Lucas L146. Stop & tail lamp Lucas ST50. Sidelamps Lucas 1130A. Horn Lucas HF935. Windscreen wiper, twin blades, Lucas A90-CWR. Front springs 7 leaves, rear springs early 10 leaves, later 12 leaves. Export models, early rear 12 leaves, later rear 10 leaves. Wheels, home models to chasis 50733, Magna wire 3.25in × 18in with Dunlop 5.25-18 tyres. Home models from chassis 50734, Easiclean 4in × 16in with Dunlop 5.75-16 tyres, later superseded by 4.5in × 16in Easiclean with Dunlop 700-16 tyres. Export models, 4in × 16in Easiclean with Dunlop 6.00-16 ELP tyres. Wheelbase 106in. Track 56in.

Series II Sixteen-Six and Eighteen-Six

Engine type Sixteen-Six QH; 3-speed Eighteen-Six QJ; 4-speed Eighteen-Six QSHM. Footbrake, Lockheed hydraulic to four wheels. Handbrake, cable to rear wheels. 12in diameter drums. Petrol tank at rear, 10 gallons. SU electric petrol pump. Open propeller shaft with Hardy Spicer needle roller bearing universal joints. Rear axle ratio, Eighteen, 4.75:1 (12/57), Sixteen, 5.273:1 (11/58). Hydraulic shock absorbers. Bishop Cam steering. Semaphore type indicators. 12-volt electrical system. Front springs 9 leaves, rear 9 leaves. Wheels, home models Sixteen-Six and Eighteen-Six, Magna wire 3.62in × 17in with Dunlop 600-17 tyres, later Eighteen-Six, 4.5in × 16in Easiclean with Dunlop 6.50-16 tyres. Export models, Sixteen-Six and Eighteen-Six, Easiclean 4.5in × 16in with Dunlop 700-16 ELP tyres. Wheelbase 117in. Track 56in.

Series II Twenty-One-Six and Twenty-Five-Six

Engine type Twenty-One-Six OJ; Twenty-Five-Six OK. Footbrake, Lockheed hydraulic to four wheels. Handbrake, cable to rear wheels. 12in diameter drums. Petrol tank at rear, 13 gallons. SU electric petrol pump. Open propeller shaft with Hardy Spicer needle roller bearing universal joints. Rear axle ratio 4.454:1 (11/49). Hydraulic shock absorbers. Bishop Cam steering. Semaphore type indicators. 12-volt electrical system, constant voltage control on Twenty-Five-Six after chassis 52556. Front springs 9 leaves, rear springs 9 leaves. Wheels, home models Twenty-One and Twenty-Five-Six to chassis 51592, 3.62in × 17in Magna wire wheels with 600-17 Dunlop tyres. Twenty-Five-Six home models after chassis 51593, 4.5in × 16in Easiclean wheels with 700-16 Dunlop tyres. Export models, Twenty-One and Twenty-Five-Six (known in Australia as '6/25 model') 4.5in × 16in Easiclean wheels with Dunlop 700-16 ELP tyres. Wheelbase $121\frac{1}{2}$in. Track 56in.

Series II Models. Body Colours and Upholstery

	Ten-Four & Twelve-Four	Fourteen-Six	Sixteen, Eighteen, Twenty-One and Twenty-Five-Sixes
Saloon, fixed-head	Black or red/black cellulose with red Karhyde. Blue/black or grey/black cellulose with blue Karhyde. Green/black cellulose with green Karhyde.	Black or red/black cellulose with red Karhyde. Blue/black or grey/black cellulose with blue Karhyde. Green/black cellulose with green Karhyde.	Black or red/black cellulose with red Karhyde. Blue/black or grey/black cellulose with blue Karhyde. Green/black cellulose with green Karhyde. After about January 1937 The Eighteen and Twenty-five-six models were finished in the following colours, including the wings: Black cellulose with brown Karhyde. Blue or grey cellulose with blue Karhyde. Green cellulose with green Karhyde. Maroon cellulose with red Karhyde.
Saloon, sliding-head	Same colours as the fixed saloon, but the Karhyde upholstery was replaced by leather.	Same colours as the fixed saloon, but the Karhyde upholstery was replaced by leather.	Same colours as the fixed saloon, but the Karhyde upholstery was replaced by leather.
Special coupé, sliding-head	Black cellulose body and wings with brown leather. Grey or sports blue cellulose body and wings with light blue leather.		Black cellulose body and wings with brown leather. Grey or sports blue cellulose body and wings with light blue leather.

Series II Models. Magazine Bibliography

Model	Subject	Source	Date	Pages
10cwt van	Service Data	*The Motor Trader*	17/8/38	p170-5
10cwt van	Equipment & Test Data	*Motor Commerce*	9/42	p63
10cwt van	New Model Introduction	*The Morris Owner*	11/35	p873
10cwt van	Truck bodied version	*The Morris Owner*	7/38	p499-500
Ten-Four & Twelve-Four	Model introduction	*The Morris Owner*	6/35	p351-4
Ten-Four & Twelve-Four	Road Test	*The Autocar*	12/2/37	p278-9
Ten-Four	Servicing Data	*The Practical Motorist*	24/7/37	p517
Ten-Four	Servicing Data	*The Practical Motorist*	31/7/37	p542
Ten-Four	Servicing Data	*The Practical Motorist*	7/8/37	p535
Ten-Four	Saloon Road Test	*The Light Car*	22/5/36	p10-11
Ten-Four	Four-Speed Model	*The Light Car*	13/11/36	p798-9
Ten-Four	Saloon Road Test	*The Practical Motorist*	20/7/35	p441-2
Ten-Four	Saloon Road Test	*The Practical Motorist*	5/12/36	p200-1
Ten-Four	Saloon Road Test	*The Practical Motorist*	14/3/36	p734-5, 737
Twelve-Four	Wiring diagram and Test data	*Motor Service*	11/51	p148-150
Twelve-Four	Saloon Road Test	*The Motor*	14/7/36	p1049-50
Fourteen-Six	New model description	*The Practical Motorist*	13/6/36	p278-80
Fourteen-Six	Saloon details	*The Practical Motorist*	1/5/37	p1082-3
Fourteen-Six	Saloon Introduction	*The Morris Owner*	7/36	p438-40
Eighteen-Six	Saloon Road Test	*The Autocar*	28/2/36	p376-7
Twenty-Five-Six	Saloon Road Test	*The Practical Motorist*	14/9/35	p717-8
Twenty-Five-Six	Saloon Road Test	*The Motor*	30/7/35	p1147-8
Twenty-Five-Six	Morris 6/25 Road Test	*The Open Road (Australia)*	1936	p10
All Models	Eight to Twenty-Five	*The Morris Owner*	9/35	p655-8
All Models	Four New Morris Sixes	*The Morris Owner*	9/35	p554a, 554b
All Models	Olympia report	*The Autocar*	16/10/36	p768

Series III Model

The 31st International Motor Exhibition at the new venue of Earls Court opened in October 1937 giving the general public the opportunity of seeing the latest Morris Series models on stand 147. The usual lunch for distributors and dealers had been held about seven weeks earlier when 1,800 guests had been entertained at Grosvenor House, London. Lord Nuffield surprised most of those present when he announced that, with the exception of the Eight, all models of the 1938 range would have overhead-valve engines and four-speed gearboxes, adding, 'And if you cannot sell them you ought to be boiled.' In the event nobody was boiled, not even his advisers who obviously forgot to mention that the 25hp version continued to have a three-speed gearbox!

Now the sole side-valve Morris car was the Eight which had been changed in detail only and became the 'Series II'. The Bradford City Police took delivery of new Series II Eight two-seaters (registered CKU783 to CKU788, and BKY738 to BKY743) to add to the fleet of Series I open models already in service. The Sheffield force also took delivery of thirteen Series II two-seaters. At least one of the Bradford fleet is said to still exist. In the south, the Hastings Police were also using Morris Eights.

The new models, designated Series III, comprised the four-cylinder Ten and Twelve, and the six-cylinder Fourteen and Twenty-Five. Except for the Twelve, the changes were merely cosmetic; for example the Series III Ten body lines

Original price of this Series III Morris Ten-Four special coupé was £220; the 'Jackall' built-in jacking system was £5 extra. (P.W. Jones)

*The showroom introduction of the New Morris Twelve,
Series III.* (Stewart & Ardern Ltd)

remained virtually identical to the model it replaced but there was a painted radiator shell; less noticeable was the reduction in wheel size for the 5.50-16 tyres. This new size of Easiclean wheel had been fitted for some time to the previous Series II Ten/Twelve export models, although with 5.75-16 tyres, a size which continued on the Series II Tens destined for overseas markets.

Three alternatives were offered, starting with the saloon model with fixed-head at a basic price of £185 with the optional extras of a luggage grid for £1 10s, a metal Wilmot Breeden spare-wheel cover for £1 in black or £1 7s 6d in a matching colour, and the Jackall built-in jacking system which added another £5. In sliding-head form the saloon was presumably fitted with the appropriate wheel cover as the only extras to the list price were the Jackall system and luggage grid. The third model was a special coupé at £220.

The engine used in the Ten was a four-cylinder 1,292cc unit, type 'MPJM', with a bore of 63.5mm and a stroke of 102mm driving through the usual cork insert clutch and four-speed synchromesh gearbox. The overhead valves were of dissimilar diameter for inlet and exhaust. Other features included an AC cleaner and fume consumer, SU carburetter, and a pumped system fitted with a by-pass thermostat which provided the cooling for an engine giving 37.5bhp at 4,500rpm. In 'MPJG' form the same engine was used in the contemporary MG 10hp Series II car; and with an appropriate W suffix, in the short lived Wolseley 10/40.

Colour choice for all the Ten models was black, blue, grey, green or maroon but it was not merely a styling feature that dictated a dissimilar shade of body colour from that of the wheels and wings. With the exception of the all-black models (which had stove enamelled wheels and wings and a cellulose finish to the body) the brighter models had a cellulosed body while the wheels and wings were finished in synthetic lacquer.

The Series III Ten had an extremely short production of only twelve months and if the chassis serial numbers (which continue on from the Series II Ten and Twelve) are a guide, something like 13,719 were produced. Of these, and including three special coupé models, thirty-four are known to be at present in the hands of enthusiasts.

FFC330 was the registration number of a Series III Ten sponsored by Morris Motors which entered as number 108 in the 17th Monte Carlo Rally in January 1938. Driven by the *Daily Mail*'s motoring correspondent, W.A. McKenzie, in the company of J.E. Whitehead and F.H.S. Rasch of Morris Motors Publicity Department, the sliding-head saloon model was equipped with additional equipment such as windscreen defrosters, car heater, spot lamps, external mirrors, export-size tyres, together with snow chains and spades. One of five John O'Groats starters to arrive on time and without loss of points, the car was placed fourteenth in the 1,500cc class, gained some welcome publicity, and was later displayed at Stewart & Ardern's showrooms in London.

Of the new overhead-valve four-cylinder models the Series III Twelve-Four had the larger horsepower. In common with all the new Morris cars introduced at the time, the radiator shell and wheels were painted the same colour as the body. The body was a new design and although the same overall length as its predecessor (the Series II Twelve-Four) the wheelbase was some 4in shorter, putting the rear seats over the back axle, rather than forward of it, but still leaving space for the boot with external access. This boot lid, hinged at its bottom edge and fitted with supports, provided an extra platform for luggage in excess of the internal 10cu ft space. Immediately below was a removable

The Series III version of the Twelve-Four saloon continued to use the same body pressings as the Series II cars, but the painted radiator shell and Easiclean wheels considerably altered the appearance. (John Lowrie)

number plate panel to allow access to the spare wheel which was now housed in a separate compartment. Aesthetic balance to the longer overhang at the rear was achieved by giving a more pronounced tail to the rear wings.

Saloon versions only were listed and these came in four-door form with, or without, sliding-head. For £215 the new owner had a car in the choice of black, blue, green, maroon or grey (although the latter was later replaced with beige body/brown wheels and wings) upholstered in leather. Internal fittings included a folding arm rest on the rear seat, large door pockets, parcel shelf under the facia panel, sun visor, pillar pull-cords and a remote control rear blind. Pivoting rear quarter-lights, hinged front windscreen with centre control, scuttle side ventilators and glass louvres over the front door windows helped to provide controllable ventilation. The Jackall hydraulic jacking system came as an extra on both models. Surprisingly, the lack of sliding-head and replacement of real leather by Karhyde only gave a £10 reduction in the listed price of the fixed-head version.

Under the bonnet the 'TPBM' four-cylinder push-rod ohv engine with a RAC rating of 11.978hp (thus cleverly fitting into the £9pa Road Fund Tax bracket) gave a maximum 40.8bhp at 3,800rpm. The capacity of the 69mm × 102mm engine was 1,550cc. A feature not shared with the other Series III cars was the use of a dry clutch, in this case a Borg & Beck A6 type, 8in diameter, with a release bearing of carbon graphite. Because of rationalisation within the Morris Engines Branch the previous year, it is not surprising to find that the same engine in 'TPBW' guise was used in the Wolseley Series III 12/48, and the 'TPBG' version in the MG 12hp Series II car.

Immediately prior to the war some small changes were announced for the Twelve which included improved springing by the introduction of an increased number of thinner leaves to the front springs and a change from six to seven leaves at the rear, new type shock absorbers, bucket-seat back-panels deeper by 2in, and improved waterproofing.

The new Sixes came as the Fourteen-Six and Twenty-Five-Six — gone was the in-between Eighteen — the engine, wheel size and colour presentation had changed, of course, but it was still the same basic body design that carried the new overhead-valve engine. A sliding-head saloon body was the only type listed for the Fourteen-Six and although the price was up to £23 10s on its predecessor many features were so similar that the Morris Oxford Press were able to use some of the original Series II photographs in the new catalogue! Detail improvements were made internally, for example, the figured walnut facia board now boasted lids on the glove compartments and the semaphore arms were self-cancelling. Right-handed smokers almost certainly appreciated the new ergonomic approach which had moved the ashtray to the right-hand side of the sprung steering wheel. The double windscreen wipers had an independent control on the left side while, for the comfort of the rear passengers a folding arm rest had been added; in original form, this must have been a nuisance when lifting the back for access to the luggage compartment.

Perhaps one of the best known Series III Fourteen-Sixes to Morris enthusiasts is that owned by Arthur Peeling of London. Questioned about the anomaly of the chromium-plated radiator shell on his saloon, he explained that

A Series III Fourteen-Six saloon with some non-standard additions.

the vehicle is an ex-ministry car and that he had been told by one of the original ministry chauffeurs that the non-standard parts were the result of a 'costing' exercise. It appears that the normal vehicle issue to eligible officials was the contemporary Wolseley 14/56 saloon which at £275 was more expensive than the £248 10s for the Morris Fourteen-Six, so the Morris version was issued. However, to retain the dignity commensurate with the official's rank,

Fourteen-Six saloon, Series III. (Morris Motors Ltd)

Late models of the Series III Fourteen-Six had external access to the boot. (Morris Motors Ltd)

chromium-plated headlamps and radiator shell, together with an opening boot, at an extra cost of £65, was specified. The result of this was, according to Arthur Peeling, a Morris which cost the taxpayer more than its Wolseley counterpart and, incidentally, appears to have triggered off the provision of an external opening boot lid on the standard models, commencing at chassis number S3/FS60573.

In February 1939, three Halifax business men set out on an ambitious venture: a 5,000-mile journey which included a 3,000-mile trip across the Sahara Desert, always a popular challenge in the 'thirties. The car used was a Morris Fourteen-Six Series III saloon (registered JX7294) towing a small caravan which (to judge by the period publicity photographs) had such narrow section tyres, that it must surely have become bogged down in the desert sand. The three men were Mr George Boulton, Councillor J. Milton Stead, and Mr J.S. Hargreaves.

A landmark for Morris came on 22 May 1939 when the millionth Morris car made drove off the production line at the Cowley Works, making it the first British factory to manufacture a million vehicles (Fords of America had, by contrast, reached this figure as early as 1915 and by 1939 were assembling their twenty-seven millionth car at the Richmond, California, plant. On the other hand it took Austin 40 years, between 1906 and 1946, to reach a million.) Lord Nuffield presented the car, a Series III Fourteen-Six sliding-head saloon, to the Ladies Association of Guys Hospital to be offered as a prize in a special draw. It had taken William Morris 26 years to reach the one million mark and only twelve years later (six of which can be discounted due to the war) the two-millionth Morris left the works.

A regular feature of the day-to-day working at Cowley about this time was the constant passage of groups visiting the factory to see the cars being assembled. Morris encouraged this obvious good publicity for his cars. Indeed, in 1938 and 1939 the Southdown Motor Services ran special excursions to Cowley for a return fare of 15s 6d which included tea! It is interesting to read Southdown's description of what a visitor to the production lines would see: '. . . thousands of men working — thousands of cars are taking shape. From the various specialised Morris factories, the parts arrive at Cowley continuously. Engines . . . bodies . . . wings . . . axles . . . chassis frames . . . radiators . . . wheels . . . a constant stream is arriving, to be instantly sorted out for assembly. Then the chassis frames move slowly along the conveyors, various parts are assembled at different points along the line. Everything happens with clockwork precision. Men and machines are synchronised. From out of the maze of separate parts, complete cars emerge. They grow before your eyes. It is a spectacle worth travelling hundreds of miles to see; a symbol of modernity and progress; an experience you will value all your life.'

Contemporary road tests gave a figure of 8 seconds to 30mph from a standing start for the Fourteen while 'the needle could be taken up to 75mph in favourable conditions.' The engine type 'QPDM' with a 61.5mm bore and 102mm stroke (1,818cc) was rated at 51.25bhp at 4,200rpm maximum. In another form ('QPDC') and fed by a Solex carburetter the engine was used in the Morris Commercial G2-SW, or 'Super Six' taxicab.

With a Road Fund Tax of £18 15s per year, increasing to £31 25s in May 1939 when the Chancellor of the Exchequer raised the tax from 15s to 25s per hp, it is not surprising that cars of the 25hp rating formed a small percentage of sales in Britain. How many of the Series III Twenty-Five-Sixes were made is not known, Morris Motors Ltd included the chassis numbering in with the Fourteen-Six sequence. A total production of 5,419 Fourteen and Twenty-Five cars combined provide the basis of a reasoned estimate of 20 per cent, or approximately 1,000 of the larger vehicles. Certainly of those made, a considerable number were, together with other makes of large car, modified at the start of the war to provide ARP ambulances. At the time of writing only two examples exist within the membership of the Morris Register, while the few known examples of the earlier Series II versions appear to be in Australia where the model was known as the Morris 6/25.

A major new innovation on the Series III Twenty-Five, engine apart, was the provision of a boot lid at the rear, making the luggage grid superfluous. It altered the lines of the body little, but the retention of the spare wheel mounting position made for a weighty bottom-hinged lid. Some time later, about September 1938, a Morris badge incorporating a reflector was added to the boot lid. Most of the changes made in detail and equipment were the same as those mentioned above in relation to the Fourteen, but one new feature not shared was the telescopically adjustable steering column added to the new Twenty-Five. Countering the additional paint finish surfaces at the front were styling strips between the bonnet louvre groups, while inside a luxury touch

*A surviving Series III
Twenty-Five-Six saloon.*
(Robert Lister)

was the provision of corner cushions at the rear and adjustable pneumatic cushions for the small of the back, built into the front seats.

Two models were available. A saloon with sliding-head listed at £320 (which was an increase of £40 on the Series II version) and a special coupé (£5 cheaper than the earlier version!) which in a choice of overall black, grey, or 'Sports Blue', boasted twin fog lamps and horns as standard. With the addition of these electrical components and the twin SU petrol pumps the substitution of the old winter/summer charging arrangement for the constant voltage control system in conjunction with a 60 amp-hour 12-volt battery was desirable.

The 'OPEM' engine in the Twenty-Five (which continued after the war as 'OPEW' to power the long-wheel-base Series III Wolseley 25) contributed considerably (780lb) to the unladen weight of the saloon which weighed over 32cwt exclusive of fuel. This car was the most powerful production Morris of pre-war years, the engine producing 95.4bhp at 3,800rpm.

When introducing the new Series III models in August 1937, Lord Nuffield had some strong words to say about the 'ramp' in the continued advance of steel prices in this country. Of the steel manufacturers he said '. . . an absolute ramp — big cigars and nothing to do!' Despite this, the list price for all Series III models had remained constant throughout the peacetime production period which, in the case of the Twelve, Fourteen, and Twenty-Five, was two years. The smaller Ten-Four, as mentioned earlier, had a shorter life before being superseded by the Series 'M' Ten.

Series III Models. Chassis Numbers	
The chassis numbering series for the Series III Fourteen-Six and the Twenty-Five-Six were shared.	
Series III Twelve-Four	S3/TW101 - S3/TW19565
Series III Fourteen-Six	S3/FS57532 - S3/FS62940
Series III Twenty-Five-Six	S3/TF57532 - S3/TF62936

Series III Models. Body Colours and Upholstery

	Ten-Four	Twelve-Four	Fourteen-Six
Saloon, fixed-head	Black cellulose with brown Karhyde. Blue or grey cellulose with blue Karhyde. Green cellulose with green Karhyde. Maroon cellulose with red Karhyde. (Wheels and wings slightly different shade to body on coloured models due to the use of synthetic lacquer.)	Black cellulose with brown Karhyde. Blue cellulose with blue Karhyde. Green cellulose with green Karhyde. Maroon cellulose with red Karhyde. Grey cellulose with blue Karhyde, later replaced by brown wheels and wings and beige cellulose body.	
Saloon, sliding-head	Black cellulose with brown leather. Blue or grey cellulose with blue leather. Green cellulose with green leather. Maroon cellulose with red leather. (Wheels and wings slightly different shade to body on coloured models due to the use of synthetic lacquer.)	Black cellulose with brown leather. Blue cellulose with blue leather. Green cellulose with green leather. Maroon cellulose with red leather. Beige cellulose with light brown wings and brown leather.	Black cellulose with brown leather. Blue cellulose with blue leather. Green cellulose with green leather. Maroon cellulose with red leather. Grey cellulose with blue leather, later replaced with beige cellulose body, brown wings, and brown leather.
Special coupé, sliding-head	Black cellulose body and wings with brown leather. Sports Blue or grey cellulose body and wings with light blue leather.		

Series III Models. Body Colours and Upholstery	
	Twenty-Five-Six
Saloon, sliding-head	Black cellulose with brown leather. Blue or grey cellulose with blue leather. Green cellulose with green leather. Maroon cellulose with red leather.
Special coupé sliding-head	Black cellulose with brown leather. Sports Blue or grey cellulose with light blue leather.

Series III Models. Magazine Bibliography

Model	Subject	Source	Date	Pages
Ten	Ten-Four Road Test	Practical Motorist	11/6/38	p236-7
Ten	Saloon Road Test	The Light Car	17/9/37	p. . .
Twelve	Twelve-Four detailed description	The Autocar	20/8/37	p335-6
Twelve	Twelve-Four detailed description	The Motor Trader	25/8/37	p197-9
Twelve	Road Test	Practical Motorist	29/7/39	p523-4
Twelve	Twelve-Four detailed description	Practical Motorist	4/12/37	p170-2
Twelve	Twelve-Four detailed description	Practical Motorist	28/8/37	p698
Twelve	Twelve-Four description	The Light Car	20/8/37	p414-16
Twelve	Road Test	The Light Car	3/2/39	p356-8
Twelve	Twelve-Four model description	The Morris Owner	9/37	p648-51
Twelve	Popularity of the Twelve-Four	The Morris Owner	12/37	p956-8
Twelve	Service Data Sheet 57	The Motor Trader	1/2/39	p231-8
Fourteen	Road Test, Fourteen-Six	Practical Motorist	29/1/38	p428-30
Twenty-Five	Specification	The Motor	14/3/39	p217
Twenty-Five	Road Test	The Motor	22/2/38	p143-4
Twenty-Five	Road Test	The Autocar	31/12/37	p1294-5
Twenty-Five	Electrical Test Data D453	Motor Commerce	4/46	p147-8
All Models	Description of new Series III models	The Morris Owner	9/37	p676
All Models	Description of new Series III models	The Morris Owner	10/39	p746

Specification. Series III Models

Series III Ten

Engine type MPJM. Footbrake, Lockheed hydraulic to four wheels. Handbrake, cable to rear wheels. 9in diameter drums. Petrol tank at rear, 7 gallons. SU electric petrol pump. Open propeller shaft with Hardy Spicer needle roller bearing universal joints. Rear axle ratio, 5.25:1. Armstrong shock absorbers. Bishop Cam steering. Semaphore type indicators. 12-volt electrical system (2 × 6). Lucas LD147EDS headlamps. Lucas 1130 sidelamps. Lucas ST508 stop/tail lamp. Lucas HF935 horn. Windscreen wiper, Lucas CW4A84 with dual wiper blades. Road springs, home models and later export models, front 6 leaves, rear 8 leaves. Early export models, front 7 leaves, rear 9 leaves. Wheels, 5-stud Easiclean, 3.5in × 16in with Dunlop 5.50-16 ELP tyres on home models and 5.75-16 ELP tyres on export models. Wheelbase 100in. Track 48⅞in. 'Jackall' built-in jacks optional extra.

Series III Twelve

Engine type TPBM. Footbrake, Lockheed hydraulic to four wheels. Handbrake, cable to rear wheels. 9in diameter drums. Petrol tank at rear, 7½ gallons. SU electric petrol pump. Open propeller shaft with Hardy Spicer needle roller bearing universal joints. Rear axle ratio, 5.25:1 (8/42). Armstrong hydraulic shock absorbers. Bishop Cam steering. Semaphore type indicators. 12-volt electric system, battery mounted on dash. Lucas MD140EDS headlights. Lucas LD109/1 sidelamps. Lucas ST50 stop/ tail lamp. Lucas HF1235 horn. Road springs, chassis 101 - 15257, front 7 leaves; chassis 15258 onward front 12 leaves. Rear springs, chassis 101 - 8840, 7 leaves; chassis 3172 - 16885, 6 leaves; chassis 16886 onward, 7 leaves. Wheels, 6-stud Easiclean, 3.5in × 16in with Dunlop 550-16 ELP tyres. Wheelbase 96in. Track 50in. 'Jackall' built-in jacks optional extra.

Series III Fourteen

Engine type QPDM. Footbrake, Lockheed hydraulic to four wheels. Handbrake, cable to rear wheels. 10in diameter drums. Petrol tank at rear, 8¾ gallons. SU electric petrol pump. Open propeller shaft with Hardy Spicer needle roller bearing universal joints. Rear axle ratio, 5.333:1 (9/48). Luvax hydraulic shock absorbers. Bishop Cam steering. Semaphore type indicators. 12-volt electric system (2 × 6v batteries under front seats). Lucas L147EDS headlamps. Lucas 1130A side lamps. Lucas ST50A stop/tail lamp. Lucas HF935 horn. Road springs, front 7 leaves, rear 12 leaves to chassis 60598, later 10 leaves. Export models, front 7 leaves, rear 10 leaves. Wheels, 5-stud Easiclean 4in × 16in with Dunlop 575-16 ELP tyres. Export models had Dunlop 600-16 tyres. 'Jackall' built-in jacking system optional extra. External boot lid after chassis 60573. Wheelbase 106in. Track 56in.

Series III Twenty-Five

Engine type OPEM. Footbrake, Lockheed hydraulic to four wheels. Handbrake, cable to rear wheels. 12in diameter drums. Petrol tank at rear, 13½ gallons. Twin SU electric petrol pumps. Open propeller shaft with Hardy Spicer needle roller bearing universal joints. Rear axle ratio, 4.454:1 (11/49). Luvax hydraulic shock absorbers. Bishop Cam steering with telescopic adjustable steering wheel position. Semaphore type indicators. 12-volt electrical system. Lucas LD163SEDS headlamps. Lucas LD309S side lamps. Lucas ST50 stop/tail lamp. Lucas RT508 reverse/tail lamp. Lucas SW4-A/104 windscreen wiper with twin blades. Twin fog lamps. Twin Lucas IIE728 horns. Road springs, home and export, front and rear, 9 leaves. Wheels, 5-stud Easiclean, 4.5in × 16in with Dunlop 700-16 ELP tyres. 'Jackall' built-in jacking system. Wheelbase 121½in. Track 56in.

Engine Data. Series III Models

Type MPJM

Four-cylinder, overhead-valve, push-rod. 63.5mm bore. 102mm stroke. 1,292.1cc. 10hp. 4-ring aluminium alloy pistons. Three-bearing crankshaft. Valves, inlet 30.5mm diameter, exhaust 26mm diameter. Valve lift, inlet and exhaust, 8.4mm. Compression ratio, 5.85/6:1. Carburetter, SU $1\frac{1}{4}$in diameter horizontal with AC air cleaner and fume consumer. Clutch, single plate cork-insert running in oil. Distributor, Lucas DK4A-A103. Sparking plugs, Champion L10. Dynamo, Lucas C45YV3, type L. Coil ignition. Cooling system by pump with by-pass thermostat and fan. 37.25bhp at 4,500rpm. Four-speed gearbox with synchromesh in top and 3rd. Ratios: 1st, 4:1; 2nd, 2.363:1; 3rd, 1.53:1; reverse, 5.15:1.

Type TPBM

Four-cylinder, overhead-valve, push-rod. 69.5mm bore. 102mm stroke. 1,547.83cc. 11.978hp. 4-ring aluminium alloy pistons. Three-bearing crankshaft. Valves, inlet 32mm diameter, exhaust 28mm diameter. Valve lift, inlet and exhaust, 8.4mm. Compression ratio, 5.85/6:1. Carburetter, SU $1\frac{1}{4}$in diameter with AC air cleaner and fume consumer. Clutch, 8in diameter Borg & Beck single dry plate. Distributor, Lucas A102-DK4A. Sparking plugs, Champion L10. Dynamo, Lucas C45-YV/3. Coil ignition. Cooling system by pump with by-pass thermostat and fan. 40.8bhp at 3,800rpm. Four-speed gearbox with synchromesh in top, 2nd and 3rd. Ratios: 1st, 3.975:1; 2nd, 2.29:1; 3rd, 1.48:1; top, 1:1; reverse, 3.975:1.

Type QPDM

Six-cylinder, overhead-valve, push-rod. 61.5mm bore. 102mm stroke. 1,818.094cc. 14.069hp. 4-ring aluminium alloy pistons. Four-bearing crankshaft. Valves, inlet 30.5mm diameter, exhaust 26mm diameter. Valve lift, inlet and exhaust, 8.4mm. Compression ratio, 6.2/6.4:1. Clutch, single plate cork-insert running in oil, spring hub. Distributor, Lucas DK6A-A/104. Sparking plugs, Champion L10. Dynamo, Lucas C45-PV3. Coil ignition. Cooling system by pump with by-pass thermostat and fan. Carburetter, SU $1\frac{1}{2}$in diameter with AC air cleaner and fume consumer, fitted with piston damper after chassis 59127. 51.25bhp at 4,200rpm. Four-speed gearbox with synchromesh in top and 3rd. Ratios: 1st, 4:1; 2nd, 2.36:1; 3rd, 1.53:1; top, 1:1; reverse, 5.15:1.

Type OPEM

Six-cylinder, overhead-valve, push-rod. 82mm bore. 110mm stroke. 3,485cc. 25.01hp. 4-ring aluminium alloy pistons. Four-bearing crankshaft with damper at the front. Valves, inlet 38mm diameter, exhaust 34mm diameter. Valve lift, inlet and exhaust, 9.45mm. Compression ratio, 5.9/6:1. Carburetter, SU $1\frac{3}{4}$in diameter with AC air cleaner and fume consumer, piston damper fitted after chassis 59171. Clutch, single plate cork-insert running in oil, spring hub. Distributor, Lucas DK6A-A105. Sparking plugs, Champion L10. Dynamo, Lucas C45-PV3. Constant voltage control. Coil ignition. Cooling system by pump with by-pass thermostat and fan. 95.4bhp at 3,800rpm. Three-speed gearbox with synchromesh in top and 2nd. Ratios: 1st, 3.39:1; 2nd, 1.675:1; top, 1:1; reverse, 4.3:1.

Series Models E & M

The Series M Ten saloon, announced in August 1938, with its 'mono-construction' design of underframe and body in one complete unit, followed the pioneer work by Vauxhall who, with their recently introduced Ten saloon, was the first British manufacturers to exploit this system. In America the Lincoln V12 Zephyr had a form of composite body while on the Continent Lancia were very early exponents. In France the Citroën 7CV front-wheel drive and in Germany the Opel Olympia had this form of construction. This new method of fabrication must have caused a few headaches to the people concerned with accident damage rectification. Morris, aware of the problems, organised special courses of instruction and published a manual, in conjunction with The Pressed Steel Co, to assist the trade make good any damage to a construction where nuts, bolts, rivets, timber, screws and tacks gave way to the modern technology of spot-welds, captive nuts and Phillips-head self-tapping screws. The complete shell comprised some sixty-nine separate pressings welded and fused together, and what had previously

Series M Ten-Four of 1939. An example of the early use of the mono-construction principle where the combined body and chassis was built up from a number of pressings welded together. (Ken Martin)

*On the Series M the boot lid could be folded down and
used to increase the luggage carrying capacity.*

been considered the chassis (with channels of 0.092in thick metal) was now the underframe with each side channel comprising two 0.050in thick pressings welded together in the centre.

Experiments filmed in slow motion were conducted at Cowley in May 1939 to see how the Series M Ten saloon and an earlier Series II Ten saloon with conventional chassis would fare in a crash. As a demonstration for the press, the two cars, electrically controlled, were lined up one each end of a 200yd road. The first test was something of an anti-climax as the vehicles passed each other with a fraction of an inch to spare! The second test was more successful when the two cars hit each other almost head-on at a combined speed of 60mph. The earlier model, some $2\frac{1}{2}$cwt heavier than the 24cwt Series M, overturned and '. . . suffered much more extensive damage than the Series M model, its entire frame was distorted, the front axle badly bent, and damage generally extended as far as the rear springs.' Predictably, the damage to the Series M was confined to '. . . damage in a small area around the front axle and bonnet side which can be quickly repaired by welding in a new section.'

The successful features of the Series III Twelve such as the spare wheel stowage under the boot, the full-width shelf under the wide facia, the tubular-steel-framed front seats, the underbonnet battery location, and the wings devoid of a ribbed edge, were all incorporated in the new Ten design. A continuing trend in body design over the years had been the movement forward of the radiator shell from well back between the dumb-irons, as on the Cowley of the late 'twenties, to the position it now took on the Series M in line with the forward

Flying Officer M.N. Mavrogardato and W.A. McKenzie with the Series E Morris Eight saloon used in the 1939 Monte Carlo Rally.

1939 Series E Morris Eight tourer driven and owned by Kurt Christensen competing in a hill climb event near Copenhagen in 1947. Christensen's successes with this car included first prize in 10km race at an average speed of 127km/hr, and five first prizes in ice races on different lakes in Denmark.

edge of the front wings. The radiator shell almost touched the plainest of bumpers, which provided the anchor points for the standard jacking device supplied with the car, although the Jackall system was still available as an extra.

Suspension on the Series M was the work of Alec Issigonis (better known for his post-war Morris Minor and Mini designs) and an ex-MG draughtsman Jack Daniels. In the development stages Issigonis and Daniels experimented with coil spring and wishbone independent suspension at the front, combined with rack and pinion steering, but the management resisted these innovations. Working within the restrictions imposed the designers realised that in order to obtain comfortable suspension over rough roads the front springs would have to be more flexible than usual. The answer was to use considerably longer springs and to interpose rollers at the tip of each spring leaf. This extra flexibility then posed another difficulty, for the car had a tendency to roll badly on corners and the front axle would twist on the springs under braking torque. These problems were solved by Hubert Charles, one-time head of design and development in the MG drawing office, who came up with the idea of a torsion anti-roll bar laid out with parallel arms connecting it to the front axle beam

Juxtaposition of the Morris Series E and Wolseley 8hp models illustrate the use of common body pressings.

with the addition of radius rods to counter the axle twisting. The original experimental work on the coil and wishbone suspension was not wasted for the system finally evolved was to be fitted in the post-war MG 'Y' type saloon of 1947.

A four-door saloon was the only body type available for the Series M, but this was offered in fixed-head style with Karhyde upholstery, or sliding-head with leather upholstery. The colour schemes as originally introduced were common to both models with an all-over finish in black, blue, green, or maroon, and a more colourful version in beige with light brown wheels and wings.

With a bore of 63.5mm and a stroke down to 90mm, the overhead-valve engine 'XPJM' used in the Series M was rated fractionally under 10hp and gave 37.2bhp at 4,500rpm. A single dry plate (Borg & Beck 7$\frac{1}{4}$in) clutch was used in conjunction with a four-speed gearbox with synchromesh in 2nd, 3rd and top gears. The engine must have been considered a very successful design, for in 'XPJW' form it was used in the Wolseley Ten and bored-out versions went on to power the mid-fifties Wolseley 4/44 as 'XPAW' and MG cars which included the TC, TF, Y, and YB ('XPAG'); and the MG TF1500 as 'XPEG'.

Just before the war some small changes were announced for the Ten which included refinement to the interior trimming and the arm rest pads. The bucket seat backs had been deepened to give an added 2in, in addition to having the back panels entirely redesigned. Mechanically, there were a new type of rear shock absorbers, improved fixing clamps for the battery and exhaust pipe, and

Restored example of the 1939 Series E saloon on display at the 'Calvalcade of Morris Cars' organised jointly by Stewart & Ardern Ltd and the Morris Register at the former's Berkeley Square Showroom in 1970.

redesigned mud flaps. With the commencement of hostilities in 1939, car manufacture for the civilian market virtually ceased, although, presumably to cater for those dealers and distributors still holding unsold stock, the Series M Ten, Series III Twelve, Fourteen and Twenty-Five; and Series E Eight continued to be listed as late as August 1940, by which time legislation prohibiting the sale of new cars without a licence had been passed. From April 1940 the price of the cheapest Eight, Ten and Twelve had gone up by approximately £20.

Some weeks after the announcement of the new Ten came the Motor Show on 13 October 1938, and with it the début of the new Series E Eight. A completely new design had replaced the successful 1935/Series I/Series II Eight which had by then reached a production figure of some 218,000 and would total, in 5cwt van form, another 3,500 before it was completely finished. It was the success of this and other early Series models, the outcome of Len Lord's reorganisation at Cowley, that resulted in the rift between him and William Morris. Lord's idea was that the person who made a business profitable should have a percentage of those profits. Morris would have none of this, and in August 1936 Leonard Lord left Cowley to eventually become head of the Austin Motor Co Ltd.

Following that Motor Show of 1938 and the introduction of the Series E Eight, Kenning's Ltd of Clay Cross, Derbyshire, purchased the entire remaining stock of the earlier Eight models from Morris Motors Ltd. This consisted of approximately 500 Series II two-door and four-door saloons. There had been a precedent in 1927, and on an even larger scale, when George Kenning made a deal with William Morris and bought the complete surplus stock of 'Bullnose' models after Cowley changed over to the flat radiator. To

Open models of the Series E Eight were only available during the pre-war years. (Ken Martin)

this day the cheque for £66,924 13s 3d, dated the 28 September 1927, is preserved at Kenning's headquarters at Chesterfield.

It was in the Series E range that the only Morris open models were now obtainable and, although the same basic body was used, a two-seater and four-seater tourer was offered. The closed models came as sliding-head and fixed-head versions of the two- and four-door saloon. Plastics were, by then, one of the new materials increasingly used by the automobile industry. On these saloons the dash panel was made of phenol formaldehyde by De La Rue Plastics Ltd using the compression moulding method; the open models had a traditional wooden veneered panel. The outstanding feature of the design, and a complete move away from the (then) conventional method, was the mounting of the headlamps in the front wings in the manner of the later Volkswagen, necessitating oval lamp glasses and escutcheons. Unfortunately, water penetration was a problem. Lucas brought out a conversion kit which put the lamp rims at right angles to the road, but the protruding units did nothing to improve the appearance. Also unconventional was the bonnet lid which, crocodile fashion, hinged at the rear end; to counteract the lack of side access a detachable panel was added in the near-side front wing valance to enable the owner to adjust the tappets, etc. The E was the first Morris car to make use of dissimilar track dimensions front and rear, a feature already used successfully on the Series II 8/10cwt van.

Two-door Series E saloon commandeered by the German forces during the occupation of the Channel Islands. The car is shown turning into St Julians Avenue, by the Royal Hotel, Guernsey.

The same basic engine design continued to be used in the new Eight, a side-valve of 57mm bore × 90mm stroke, although the new head and piston design had increased the compression ratio from 5.6:1 to about 6.6:1. This, and the new counterbalanced crankshaft, resulted in greater maximum revs and an increase in output from 23.5 to 29.6bhp.

Pre-war production of the Series E Eights accounted for 52,919 of the total 120,434 made. Modifications during the pre-war run improved the sealing of the headlamps and repositioned the handbrake lever by 4in to give additional clearance for the gear lever. Mud flaps to the front wings and the battery/tool box were other areas subject to modification.

The Series Z 5cwt van had a production period of almost fourteen years, when more than 51,000 vehicles were made.

Before moving on to the post-war period, mention should be made of two Series models which were in the design pipeline before war was declared, and were announced early in 1940. The Series Z 5cwt van, the smaller of the two, and designed as a replacement for the Series I van, was the commercial equivalent of the Series E car. The larger Series Y 10cwt replaced the Series II version and was initially introduced in van and truck variations, although if the survival rate of the latter is any guide, few were made; they were not listed when production of civilian vehicles recommenced after the war. Both types of van represented a move by Morris Motors away from the coachbuilt type of van body to the steel panelled construction, though the roof fabrication was still traditional with leathercloth over a lath timber frame. The 10cwt Series Y, like its predecessor the Series II, had the same 1,547.8cc side-valve engine mounted off-set to give a semi-forward control cab.

Reference has already been made to the counterbalanced crankshaft in the Series E engine, so it is interesting to record that although the engine used in the Series Z van was, externally at least, the same, the crankshaft used was not counterbalanced and was basically the type fitted in the earlier 'UB' (1935/Series I/Series II) Eight, but with shell bearings. This curious anomaly probably arises from the fact that the earlier Eight three-speed gearbox was used on the vans, rather than Series E's four-speed unit. As the three-speed gearbox would only mate up with the UB engine flywheel (although the bell housing is interchangeable) it seemed to suggest that Morris designers had

Series Y 10cwt van of 1947.
(Beck & Pollitzer Ltd)

decided to incorporate this flywheel and its compatible crankshaft into the
Series E 'USHM' engine for use in vans. However, the machining cost of
converting both crank and flywheel to suit would have been minimal, in fact,
the UB flywheel assembly was probably the more expensive due to the addition
of a self-aligning bearing for the first-motion shaft and an oilite bush in the
USHM crank.

Other questions remain unanswered. Why, for instance, did the Series Z not
have the headlamps set into the wings? Why the continued use of the earlier
Morris Eight instruments; and a car-type steering wheel on the export but not
the home models?

The GPO, which was a keen user of Morris vehicles, placed large orders for

*Little red mail vans were once to be seen everywhere. The
Series Z 5cwt van was used for both mail and telephone
engineer duties.* (GPO)

both Series Z and Series Y vans. Although they were almost standard production vehicles, there are sufficient special requirements on these vans to make identification of an ex-GPO van easy. The most curious requirement was the railway carriage type door handle. On the Z vans such extras as a lockable tool box, absence of chrome strips down the radiator grille and bonnet lid, separate side lamps, wood panelling adjacent to the dash panel and on the doors, metal panels inside the front doors, stone-guard on the petrol tank, body lifting hooks at the rear, and integral hub cap Easiclean wheels, were specified. An early batch of 150 'Z' chassis in 1940 deviated even further from the standard panel van, having an incongruous square-cut coachbuilt van body on the lines of the specially-built hybrid Eight/Minor vans which the Z vehicles gradually replaced. No doubt this accounts for the GPO's terminology for the Series Z vans as '50 cu ft Minor Van Type 1'. Similar special equipment was fitted to the Series Y vans used by the GPO and designated '100 cu ft'.

Wartime production of these small commercials was confined to orders against Ministry licence only, so few, if any, fund their way to civilian operators until restrictions were lifted at the end of the war. One or two special versions of what would have been normal production models were made during the war years and perhaps the best known is the Morris version of what service personnel nicknamed the 'Tilly'. These light utility vehicles were also made by Austin, Hillman and Standard, and in all cases were a modification of

Morris ten utility car. Wartime production was based on the Series M Ten. (Morris Motors Ltd)

Wartime special, the Series Y Morris ambulance.
(Les Orton)

a standard civilian model. The Morris version was a truck with canvas tilt based on the Series M, and details of this '4 × 2 Light Utility' were on the secret list until the last years of the war. Mechanical specification was similar to the car although the carburetter used was a Solex and the normal car radiator shell was later replaced by crude wire mesh grille. 4,442 of these Morris Utility vehicles were built between 1940 and 1942.

Another special was an ambulance version of the Series Y vehicle, but on a longer wheelbase chassis and driven by a larger 13.9hp engine, the 1,805.5cc 'TSDM-A'. How many were made is not known, but the first contract for the Royal Air Force was for 500. Bert H. Vanderveen in his publication on World War II vehicles states that the 14hp Standard ambulance superseded the Morris Y type. Presumably its use as an ambulance made the provision of large section 700-16 tyres advisable and in addition to the increase of 10in on the wheelbase compared with the Y van (which had 500-18 tyres) the differences included a four-speed gearbox and a larger 8-gallon petrol tank. In common with many military ambulances the body was truck-like with a canvas tilt. The spare wheel was on the cab roof.

The shortage of light commercial vehicles in the early days of the war prompted Jarvis & Sons of Wimbledon to turn their attention to good second-hand Morris Eight two-door saloon cars and convert these to delivery vans or pick-up trucks for their customers. The van version followed the general lines of the normal Morris 5cwt van while the truck body, built of ash, could be supplied with detachable hoops and a tilt of khaki material, similar to the ambulance just described.

After the war attempts were made to get back to normal, but fuel and steel were in short supply. Some pre-war models were introduced as an interim measure and the saloon version of the Series E & M went back into production in September 1945. In those austere times most motor car production was earmarked for overseas markets to earn foreign currency and the consequence was a rapid inflation in second-hand car prices. It is probably true to say that many old cars now restored and in the hands of enthusiasts would, but for this boom in second-hand cars, have long since gone to the scrap yards. For the lucky chap who finally managed to obtain a new car after August 1946, he had to sign a covenant between himself, the Motor Trade Association, and the

Although the open version of the Series E Morris Eight was confined to the 1939 season in the United Kingdom, post-war chassis continued to carry tourer bodies in Australia. This is a Richards bodied example. Note the integral windscreen, unlike the chromium-plated frame normal to the standard Cowley product.

dealer, undertaking not to re-sell the car within six months without approval of the MTA. Until the beginning of 1946, the sale of a new vehicle had been governed by a Ministry of Supply licence, which included a similar restriction on disposal. When this was discontinued in January 1946, chaos and inflation was caused by many people who placed orders on the waiting lists of a number of different dealers. Following delivery at the list price the speculative new owner immediately re-sold on the black market, resulting in a large tax-free profit. The period covered by the covenant was increased from time to time, eventually restricting re-sale for 2 years, before being relaxed on many models in late 1952.

The Series E model as initially available in the post-war years matched the austerity of the time, a two-door saloon (with fixed or sliding-head) finished in black only, and then at a price almost double that of 1939 — although a percentage of this figure represented 'purchase tax', introduced during war time. Some months later the four-door saloon model was re-introduced and by the end of 1947 both Eights and Tens were finished in moonstone grey, dark green, or black. Probably no four-door sliding-head Series E saloons ever appeared in the new colour schemes for by the end of that year 'In furtherance of model reduction policy' this model was withdrawn, but the four-door fixed-head saloon had 'Improved trimming and polished windscreen channels and hub-caps'.

When the Series M Ten returned in those post-war years it too had been subject to minor changes, such as the anti-corrosion treatment on the floor, continuous anti-draught piping around the doors, improved sealing on the

Series M Ten saloon showing the early type of radiator design.

Later style of radiator design on the Series M Ten saloon. The two spot lamps on this car are non-original fittings.
(C. Wain)

boot, polished fillets for the sliding-head, pillar loops in place of the earlier rope pulls at the rear, new door striker mechanism, new pedal gaiters, and instrument panel finished in duotone colours and door fillets toned to match, in place of the previous mottled effect. Underneath, the suspension system was modified to include double-acting shock absorbers mounted on the rear axle and improved front units, rubber mounted shackles on front and rear springs, and modified rear springs with rubber interleaving. After the first eight thousand or so post-war Tens had left the production lines it took on a new appearance with a re-designed front affecting the radiator shell, incorporating additional bracing and a starting handle guide. Prior to this at chassis 39910 in April 1946, a demonstration model had been made with the new type grille. The familiar winged mascot displaying the '10' emblem was absent from 1,544 cars (chassis 44277 - 45820 inclusive) to incorporate the new radiator design, but this may have been a supply problem.

Indian version of the Series M, the Hindustan Ten saloon.

Perhaps it was not merely a means of updating the Series M Ten that brought about the new style of radiator surround on the post-war models. As early as September 1945 an agreement had been reached between the newly founded Hindustan Motors Ltd in Calcutta and the Nuffield Organisation to ship components for assembly in India. To overcome prohibitive import duties on motor cars shipped complete and similar difficulties that would result from manufacturing what was obviously a Morris car, it was decided to design a different radiator and add a badge with another name. The outcome was the 'Hindustan Ten' which deviated from the Cowley assembled product in some other respects, notably the full-width front seat which resulted in the repositioning of the handbrake lever to an inverted position below the dash, and plain disc wheels. The new radiator was not to William Morris's liking (which is surprising, for in a way the well rounded shape was the nearest thing to his beloved 'Bullnose' radiator that the designers had achieved since its demise in 1926) but someone at Cowley made a decision and it was incorporated in the Series M Ten for the home market.

In researching details of the Series E Morris Eight, a number of interesting stories have come to light which indicate the reliability of these delightful and comfortable little cars. Not surprisingly there are still numbers of these models in daily use. The Monte Carlo Rally of 1939 had a particularly large entry of light cars, including the DKW and the 570cc Simca Fiat. Among the small cars in the forty-one British entries was a Series E two-door saloon (GJO58) driven by Flying Officer M.N. Mavrogardato, who was a racing motor-cyclist as well as a pilot, and W.A. McKenzie, a noted journalist of the 'thirties. Reported as being absolutely standard, this saloon (open tourers were banned by the regulations) started from John O'Groats and covered a distance of 3,624 miles during the four days and nights of the rally, taking the car the length of Scotland and England to Folkestone, and thence through France down to the Alpine section from Grenoble to Monte Carlo, arriving, it was said, too early! W.A. McKenzie's name again appears, in the entry list for the RAC Rally in May 1939, when with his co-driver D. Talfield they used the same Series M Ten

(GWL148) that had set up a record run from Ankara to London the previous year.

After the war there were tales of Morris cars being concealed from the occupying powers. Messrs J.J. Molenaar of Amersfoort, Holland, managed to hide a number of new Morris cars in part of a building. In Batavia a similar subterfuge was achieved by the Mascotte Trading Company, Nuffield Distributors for the Dutch East Indies, when fifty Morris Eights were stripped down and concealed in parts in various warehouses and made to look like scrap. The Japanese were outwitted, and after liberation the cars were rebuilt again. Most of these Series E tourers were later used as taxis. The Dutch appear to have been good at this for on the very day of Holland's liberation three new Wolseley cars were driven out of hiding to Schiphol Airport near Amsterdam by employees of Dirk van der Mark, the Wolseley and Riley Distributor.

In Holland is an unusual stained glass window which depicts a Series M Morris Ten saloon. This colourful window was presented by the employees of The Hague Morris dealer, N.V. Bolland, to the directors as a novel way of celebrating the firm's 30 years of trading, in May 1947. When the dealer's premises in van Slingelandstraat and Willen de Zwijgerlaan were sold in 1967 by the grandson of the founder, an interconnecting door between the two premises contained the stained glass window, and there it remains today.

A long distance journey in a Series E Morris Eight was undertaken by Trevor Webster and Alan Taylor of Essex in 1958. With their families, including an 18-month-old baby, they were emigrating to Australia — not by the normal sea voyage from Tilbury, but overland in two 20-year-old Morris Eights, a saloon and an open tourer. Their route took them through Belgium, Switzerland, Italy and Yugoslavia, onto the Dalmation coast. The mountains of Montenegro proved a formidable barrier for the two small overloaded cars; then came the Serbian Plains and some excruciating roads to the Greek border. The bitumen roads to Athens were a pleasant change before they plunged into the mountainous country heading north to the Turkish territory. A delay because of the Iraq revolution and the petrol shortage there was solved when BP kindly donated some of their small emergency stock. Crossing into Iran bitumen roads became a novelty until Pakistan was reached. Along the Black Sea coast they experienced heavy rain and had to have the cars manhandled through one river crossing. At Trabzan they turned inland along a series of jagged mountain trails, on past Mount Ararat then to the Persian border. 500 miles of dust and bumps took them to Teheran, then a northerly route between the Elburz mountains and the fringe of the Great Salt Desert. After Afghanistan they swung south towards the Baluchistan Desert. The heat was its most intense at Sibi in Pakistan, so the rest of the journey to Lahore was undertaken in the cool of the night. India came next and Madras where they sailed to Penang. Malaya's superb roads allowed for 60mph driving and so onto Singapore where they joined a ship bound for Australia.

Since 1958 many place names have changed, but today, settled in New South Wales, Trevor Webster is an enthusiastic member of the Morris Register of New South Wales, and is about to restore a London registered Series E saloon, FUW13 — the same one, of course, that covered those 17,663 miles twenty-five years ago.

Specification

Series E Eight

Engine type USHM. Footbrake, hydraulic to four wheels. Handbrake, cable to rear wheels. 8in diameter drums. Petrol tank at rear, $5\frac{1}{2}$ gallons. SU electric petrol pump. Open propeller shaft with Hardy Spicer needle-roller bearing universal joints. Rear axle ratio 5.286:1 (7/37). Armstrong hydraulic shock absorbers. Bishop Cam steering. Semaphore type indicators. 6-volt electrical system. Lucas F133-AP-EDS headlamps combined. Lucas ST51 stop/tail lamp. Lucas HF935 horn on early models, later models Lucas HF1235. Windscreen wiper, Lucas SW4-A110 on saloon, Lucas CW2 on tourer. Road springs, front 5 leaves, rear 8 leaves on early models, 7 leaves on later models (home and export). Wheels, 6-stud, 2.5in × 17in with Dunlop 450-17 tyres. Metal spare wheel cover optional extra on tourer models. Wheelbase 89in. Track, front $44\frac{5}{8}$in, rear $46\frac{1}{4}$in.

Series Z, 5cwt van

Engine type USHMV. Footbrake, Lockheed hydraulic to four wheels (type changes at chassis 41707). Handbrake, cable to rear wheels. 8in diameter drums. Petrol tank at side, 6 gallons. SU electric petrol pump. Open propeller shaft with Hardy Spicer needle-roller bearing universal joints. Rear axle ratio, 5.286:1 (7/37). Special export ratio 5.857:1 (7/41). Axle type changes at chassis 41707. Armstrong hydraulic shock absorbers. Bishop Cam steering. Semaphore type indicators. 6-volt electric system. Lucas L133G head/side lamps. Lucas ST38 stop/tail lamp. Windscreen wiper Lucas A114-CW. Road springs, front 5 leaves, rear 7 leaves (home and export). Wheels, 6-stud, Easiclean 2.5in × 17in with 400-17 Dunlop tyres on home models, 450-17 tyres on export models. Wheels on home models up to chassis 41706 and export up to chasis 201, not fitted with separate hub caps. Later chassis with hub caps as the Series E cars. Passenger seat optional extra. Wheelbase 89in. Track, front $44\frac{5}{8}$in, rear $48\frac{1}{4}$in.

Series M Ten

Engine type XPJM. Footbrake, Lockheed hydraulic to four wheels. Handbrake, cable to rear wheels. 9in diameter drums. Petrol tank at rear, 7 gallons. SU electric petrol pump. Open propeller shaft with Hardy Spicer needle-roller bearing universal joints. Rear axle ratio, 5.286:1 (7/37). Luvax hydraulic shock absorbers on early models, Girling on later models. Bishop Cam steering. Semaphore type indicators. 12-volt electrical system. Lucas MD140EDS headlamps. Lucas LD109 sidelamps. Lucas ST51L stop/tail lamp. Lucas HF1235 horn. Windscreen wiper, Lucas SW4-A109 with twin blades. Road springs, front 1+5 rebound leaves, rear 1+5 rebound leaves up to chassis 2170, later 1+8 rebound leaves. Wheels, 5-stud, Easiclean, 3in × 16in with Dunlop 500-16 tyres. 'Jackall' built-in jacks optional extra. Radiator shell design changed at chassis 44277. Wheelbase 94in. Track 50in.

Ten Utility car

Engine type XPJM/U. Footbrake, Lockheed hydraulic to four wheels. Handbrake, cable to rear wheels. 9in diameter drums. Petrol tank at rear, 9 gallons. SU electric petrol pump. Open propeller shaft with Hardy Spicer needle-roller bearing universal joints. Rear axle ratio 6.333:1 (6/38). Luvax hydraulic shock absorbers. Bishop Cam steering. 12-volt electrical system. Single Lucas LWD-H1 headlamp. Convoy lamp, Lucas LWD-AF. Stop/tail lamp, Lucas LWD-T1. Windscreen wiper, Lucas A109-SW4. Side lamps, Lucas LD109/A. Semaphore electric indicators up to chassis 27595, later models had mechanical device fitted. Road springs, front 8 leaves, rear 10 leaves. Wheels, 5-stud, Easiclean, 4in × 16in with Dunlop 600-16 tyres. Spare wheel on cab roof. Wheelbase 94in. Track, front $50\frac{1}{2}$in, rear 52in.

Series Y, 10cwt van

Engine type TK3. Footbrake, Lockheed hydraulic to four wheels. Handbrake, cable to rear wheels. 9in diameter drums. Petrol tank on left-hand side, $5\frac{3}{4}$ gallons. SU electric petrol pump. Open propeller shaft with Hardy Spicer needle-roller bearing universal joints. Rear axle, 5.428:1 (7/38). Armstrong shock absorbers. Bishop Cam steering. Semaphore type indicators. 12-volt electrical system. Lucas LG133P head/side lamps. Lucas ST51 stop/tail lamp. Windscreen wiper, Lucas A119LG-CW1. Road springs, front 9 leaves, rear 10 leaves (home and export). Wheels, 6-stud, Easiclean, 3in × 18in with Dunlop 500-18 tyres on early models, 525-18 on post-war models. Spare wheel on early models housed inside van, on later models this was accommodated under panel at rear. Wheelbase 90in. Track, front 50in, rear 54in.

Series Y, 14hp ambulance

Engine type. TSDMA. Footbrake, Lockheed hydraulic to four wheels. Handbrake cable to rear wheels. 9in diameter drums. Petrol tank on left-hand side, 8 gallons. SU electric petrol pump. Open propeller shaft with Hardy Spicer needle-roller bearing universal joints. Rear axle ratio, 5.428:1 (7/38). Armstrong hydraulic shock absorbers. Bishop Cam steering. 12-volt electrical system (2 × 6 volt). Lucas L133/3 headlamps. Lucas LD109 A/1 side lamps. Windscreen wiper, Lucas A119 single blade. Road wheels, 6-stud, Easiclean, 4.5in × 16in with 700-16 Dunlop Fort tyres. Wheelbase 100in. Track, front 50in, rear 54in.

Body Colours	
Series Y, 10cwt van	Shop grey or choice of blue or green at extra cost.
Series Y, 10cwt truck	Shop grey or brown with black wings at extra cost. Khaki canvas tilt extra.
Ten utility car	Military camouflage finish.
Series Y, 14hp ambulance	Finish not known.

Engine Data

Type USHM

Four-cylinder, side-valve. 57mm bore. 90mm stroke. 918.636cc. 8.057hp. 3-ring aluminium alloy pistons. Three-bearing counterbalanced crankshaft. Valves, inlet and exhaust, 25mm diameter. Valve lift, inlet and exhaust, 6.5mm. Compression ratio, 6.5/6.7:1. Carburetter, SU 1in diameter with AC air cleaner. Clutch, 6¼in diameter Borg & Beck single dry plate. Distributor, Lucas DKH4A/110-1. Sparking plugs, Champion L10. Lucas belt driven dynamo, constant voltage control, C45YV. Coil ignition. Cooling system, thermo syphon with fan. 29.6bhp at 4,400rpm. Four-speed gearbox, synchromesh in 2nd, 3rd, and top. Ratios: 1st, 3.95:1; 2nd, 2.3:1; 3rd, 1.54:1; top, 1:1; reverse, 3.95:1.

Type USHMV

Four-cylinder, side-valve. 57mm bore. 90mm stroke. 918.636cc. 8.057hp. 3-ring aluminium alloy pistons. Three-bearing crankshaft (not counterbalanced). Valves, inlet and exhaust, 25mm diameter. Valve lift, inlet and exhaust, 6.5mm. Compression ratio, 6.5/6.7:1. Carburetter SU 1in diameter horizontal. Clutch 6¼in diameter Borg & Beck single dry plate. Distributor, Lucas DKH4A. Sparking plugs, Champion L10. Lucas belt driven 3-brush type C35MV to engine 169825 (circa 1951); later models and export models, Lucas 2-brush CVC type C45YV. Coil ignition. Cooling system, thermo syphon with fan. 29.6bhp at 4,400rpm. Three-speed gearbox with synchromesh in top and 3rd. Ratios: 1st, 3.187:1; 2nd, 1.81:1; top, 1:1; reverse, 4.25:1.

Type XPJM and XPJM/U

Four-cylinder overhead-valve, push-rod. 63.5mm bore. 90mm stroke. 1,140cc. 9.99hp. 4-ring aluminium alloy pistons. Three-bearing crankshaft. Valves. inlet 30mm diameter, exhaust 26mm diameter. Valve lift, inlet and exhaust, 8mm. Compression ratio, 6.6/6.8:1. XPJM carburetter, SU 1¼in diameter with air cleaner. XPJM/U carburetter, Solex type HBFDO 30mm with AC air cleaner EAC148. Clutch, 7¼in diameter Borg & Beck single dry plate. XPJM distributor, Lucas DK4A-A/109/1. XPJM/U distributor, Lucas DKZ4A-A131. Sparking plugs, Champion L10. XPJM dynamo, constant voltage control, Lucas C45YV3. XPJM/U dynamo, constant voltage control, Lucas C45P/3. Coil ignition. Cooling, thermostatic pump in front of block and fan. 37.2bhp at 4,600rpm. Four-speed gearbox with synchromesh in top and 3rd. Ratios: 1st, 3.807:1; 2nd, 2.253:1; 3rd, 1.506:1; top, 1:1; reverse, 3.807:1.

Type TK3

Four-cylinder, side-valve. 69.5mm bore. 102mm stroke. 1,547.82cc. 11.978hp. 3-ring aluminium alloy pistons. Three-bearing crankshaft. Valves, inlet and exhaust, 33mm diameter. Valve lift, inlet 7.704mm, exhaust 7.817mm. Compression ratio, 5.91:1. Carburetter, SU horizontal 1⅛in diameter. Clutch, 8in diameter Borg & Beck single dry plate. Distributor, Lucas DKH4A. Sparking plugs, Champion L10. Dynamo, Lucas C45NVO, 3-brush. Coil ignition. Cooling, thermostatic pump and fan. 34bhp at 3,400rpm. Three-speed gearbox with synchromesh in top and 2nd. Ratios: 1st, 3.55:1; 2nd, 1.73:1; top, 1:1; reverse, 5.03:1. Oil filler on off-side of engine after engine number 53194.

Type TSDMA

Four-cylinder, side-valve. 75mm bore. 102mm stroke. 1,802.5cc. 13.94hp. 3-ring aluminium alloy pistons. Three-bearing crankshaft. Carburetter, SU 1⅛in diameter with AC oil bath air cleaner. Clutch, 8in diameter Borg & Beck single dry plate. Coil ignition. Cooling, thermo syphon and fan. Distributor, Lucas DKH4A. Sparking plugs, Champion L10. Four-speed gearbox with synchromesh in top, 2nd and 3rd. Ratios: 1st, 3.974:1; 2nd, 2.29:1; 3rd, 1.48:1; top, 1:1; reverse, 3.974:1.

Chassis Numbers

Series E Eight	Pre-war saloons and tourers: SE/E542 - SE/E54675 Post-war saloons: SE/E54677 - SE/E122191

Series Z, 5cwt	Vans: SZ/EV101 - 51358
	GPO vehicles initially carried the prefix 'SZ/PO . . .'. In 1952, although the chassis numbering sequence continued, the prefix letters were changed on both GPO and civilian models to indicate the type of body, colour, etc. For example a Series Z, 5cwt, GPO engineer's van, green, RHD, home model, finished in synthetic paint, was 'GJE11 . . .'. The following code was used for the prefix:
	
	1st letter indicates model
	  G = 5cwt
	2nd letter indicates type
	  E = van H = mail J = engineers K = chassis
	3rd letter indicates colour
	  A = black B = grey C = red D = blue E = green F = beige G = brown H = CKD finish
	1st number indicates market
	  1 = RHD home 2 = RHD export 3 = LHD 4 = North America 5 = CKD (RHD) 6 = CKD (LHD)
	2nd number indicates paint
	  1 = synthetic 2 = Synobel 3 = cellulose 4 = metallic 5 = primed
	'CKD' denotes a vehicle shipped in 'completely knocked down' condition, ie dismantled. This prefix code continued to be used when the Morris Minor MM was introduced in 1948.

Series M Ten	Pre-war saloons: SM/TN101 - SM/TN27120 Post-war saloons: SM/TN36001 - SM/TN89570
Ten utility car	Known in military terms as 'Motor Car, 2 seater, 4 × 2, Light Utility'. SM/TNU27121 - SM/TNU35791 Note that these numbers are part of the gap in the Series M pre-war to post-war numbering sequence. Numbers 35793-36000 were not used.
Series Y, 10cwt	Vans and trucks: SY/TWV101 - SY/TWV21607 Post-war production commenced at chassis SY/TWV5294
Series Y, 14hp	Ambulance: commenced FN/SY/A930 Final figure not known, but numbers appear to have been shared with the early Series Y 10cwt van and truck vehicles and lie between 930 and 5293.

Series E Eight Models. Body and Colours Upholstery

	Pre-war models	1945 to late 1947	1948, chassis 99967 onwards
Two-seater, and four-seater tourer	Blue or green cellulose with matching Karhyde. Black cellulose with brown Karhyde. Maroon cellulose with red Karhyde.		
Saloon, two-door and four door (Fixed-head versions upholstered in Karhyde, sliding-head versions in leather.)	Blue or green cellulose with matching upholstery. Black cellulose with brown upholstery. Maroon cellulose with red upholstery.	Black cellulose with brown upholstery.	Black cellulose with brown upholstery. Platinum grey or green cellulose. (Fixed-head version only on four-door saloon.)

Note: Some late models of the Series E Eight may have been finished in synthetic, Synobel, or cellulose finish.

Series M Ten Models. Body Colours and Upholstery

	Pre-war models	Post-war models
Saloon (Fixed-head versions upholstered in Karhyde, sliding-head versions in leather.)	Black cellulose with brown upholstery. Blue or green cellulose with matching upholstery. Maroon cellulose with red upholstery. Beige cellulose body with light brown wings and brown upholstery.	Black cellulose with brown upholstery. With the addition of the following from December 1947 onwards: Platinum grey or green cellulose.

Note: Some late models of the Series M Ten may have been finished in synthetic, Synobel, or cellulose finish.

Magazine Bibliography			
Model	Subject	Source & Date	Pages
Series E	Introducing the new Eight	*Practical Motorist 22/10/38*	p926-7
Series E	Morris develop new Eight	*The Autocar 14/10/38*	p708b, 708c
Series E	Eight Saloon, Road Test	*Practical Motorist 11/2/39*	p492-3
Series E	Eight Tourer, Road Test	*The Autocar 6/10/39*	p523-4
Sereis E	Eight Saloon, Road Test	*The Light Car 16/12/38*	p146-8
Series E	Eight Saloon, description	*The Light Car 14/8/38*	p . . .
Series E	The larger Eight	*The Morris Owner 11/38*	p862a, 862b, 895
Series E	Electrical Test Data	*Automobile Electricity 4/41*	p35-6
Series E	Post-War Eight	*The Autocar 21/9/45*	p673-4
Series E	Servicing Data Sheet 85	*The Motor Trader 28/2/40*	pi-viii
Series E	Servicing Eight steering	*Car Mechanics 4/58*	p51-2
Series E	Servicing Eight engine	*Car Mechanics 7/58*	p51-4
Series E	Servicing interior trim	*Car Mechanics 9/59*	p42-4
Series E	Servicing Series E Eight	*Car Mechanics 9/61*	p72-3, 82
Series E	Servicing Series E Eight	*Practical Motorist*	
		& Motor Cyclist 8/56	p132-4
Series E	Servicing Series E Eight	*Car Mechanics 9/59*	p45-7
Series E	Getting the best from		
	your Morris Eight	*Practical Motorist 6/5/39*	p13-14
Series E	Eight wiring diagram	*Motor Commerce 4/41*	p35
Series E	Eight service data	*The Motor Trader 28/2/40*	pi-viii
Series E	Eight Saloon, Road Test	*Motor Commerce 3/47*	p80, 82
Series E	Used Cars on the Road	*The Autocar 14/3/52*	p341
Series E	Used Cars on the Road	*The Autocar 5/9/52*	p1104
Series E	Eight, Data Sheet No 43	*Practical Motorist*	
		& Motor Cyclist 5/58	p936
Series M	Details of new model	*Practical Motorist 10/9/38*	p702-4
Series M	Ten Saloon, Road Test	*Practical Motorist 10/12/38*	p189-90
Series M	Ten Saloon, Road Test	*The Autocar 2/9/38*	p . . .
Series M	New features of Morris Ten	*The Motor Trader 21/8/46*	p525
Series M	Morris Ten on Test	*Motor Commerce 1/47*	p112, 114
Series M	Morris Ten	*The Light Car 2/9/38*	p . . .
Series M	Morris Ten	*The Light Car 11/11/38*	p . . .
Series M	The new Morris Ten	*The Morris Owner 9/38*	p658-9
Series M	Morris Ten	*Motor 13/11/46*	p . . .
Series M	Description, Morris Ten	*The Autocar 7/9/45*	p644-6
Series M	Description, Morris Ten	*The Autocar 10/1/47*	p31-2
Series M	Electrical Test Data, D459	*Motor Commerce 10/46*	p159-60
Series M	Morris Ten Improvements	*Motor Commerce 10/46*	p100
Series M	Ten Saloon, Road Test	*Motor Commerce 5/47*	p86, 89
Series M	Ten Service Data 75	*The Motor Trader 11/10/39*	pi-viii
Series M	Post-War Morris Ten	*Morris Owner &*	
		Nuffield Press 10/45	p39-40
Series M	Morris Ten, new radiator	*The New Outlook*	
		on Motoring 10/46	p48
Series M	Ten, Electrical Test Data	*Motor Commerce 7/43*	p79-80
Series E & M	Current Models	*Morris Owner 10/39*	p746-7
Utility Car	Description of model	*Morris Owner &*	
		Nuffield Press 2/44	p544-5, 554
Series Z & Y	Van description	*The Morris Owner 3/40*	p22
Series Z & Y	Van description	*The New Outlook*	
		on Motoring 7/46	p39
Series Z	Electrical Test Data D505		
	Export, D506 Home	*Motor Commerce 10/48*	p219-22
Series Y	Electrical Test Data, D463	*Motor Commerce 12/46*	p162. 163

The Post War Generations

When considering the immediate post-war activities of Morris Motors Ltd, one man and an outstanding Morris car figure prominently in motoring history. The man was Alexander Arnold Constantine Issigonis (later Sir Alec Issigonis, KBE) and the car the Morris Minor.

Alec Issigonis had joined Morris Motors Ltd at Cowley in 1936 as a development engineer, after gaining something of a reputation as a suspension expert at Humber. During the war he had, among other things, designed the independent suspension arrangement for the Mark I Morris light reconnaissance car; a design which came to fruition, unlike his early attempt to introduce this type of suspension on the Morris Series M Ten, which Lord Nuffield refused to sanction. It was during the hostilities that, with the encouragement of the late Lord Thomas of Remenham (then Miles Thomas), Issigonis began to jot down design ideas for a car of chassisless construction, capable of withstanding the reaction stresses that his suspension system would impose.

In 1943, with the end of the war in sight, he began to translate his ideas into a hand-built body shell, constructed in the experimental shop at Cowley. The styling of the 'Mosquito', as it was initially called, made it one of the first English car bodies to incorporate the front wing shape as part of the door (the

Some of his design ideas in sketch form by Alec Issigonis showing his speculations on the general shape of his 'Mosquito'. Taking into account the absence of the front bumper, the centre drawing closely resembles the final design.
(Morris Motors Ltd)

Jowett Javelin, then being developed, had a similar feature) and at a time when 16in diameter wheels were normal, the choice of small diameter wheels (to give the advantage of increased room within, in addition to the external suggestion of length) caused something of a problem. This was quickly solved when Dunlop co-operated and made available the then unusual 14in diameter wheels and tyres.

Another problem, and one which the motor industry as a whole had to contend with, was the system of taxation based on the RAC horsepower formula. Some experiments were made with an unconventional three-cylinder, double-piston, two-stroke engine which would attract low rates of taxation, but this proved unsuitable. Such an engine was built before a prototype Mosquito body was available, so a Series E saloon was used as a test-bed. Although developing about 16hp with a good low speed torque considering its intended use, it had the usual deficiencies of a two-stroke: smoking, four-stroking, and excessive fuel consumption. A horizontally-opposed flat-four engine (type YF) was substituted, driving through a three-speed gearbox which was basically the Morris Eight gearbox components in a compact casing with linkage to a steering column lever. The small pistons in this engine would have kept it down to a capacity of 800cc to qualify for a low tax rating in the United Kingdom yet could easily have been bored out to 1,100cc for a better performance and to attract sales in those markets abroad where taxation was not so penal as it was in Britain at that time. A considerable number of these flat-four engines were made at a factory in Canada Road, Byfleet, where, in more recent years, the vintage-style Panther car was constructed. It has been said that W.O. Bentley was involved as a consultant at this stage in the design. A similar engine had already been used in the unsuccessful Jeep-like four-wheel drive mono-construction military vehicle, to be called the Nuffield Gutty FV1800, that Issigonis had been working on. In the Mosquito body the flat-four engine, having a low height to width ratio, was to be mounted well to the

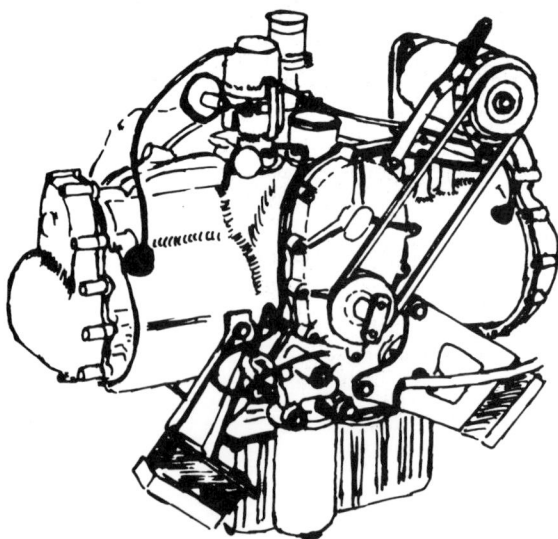

Author's drawing of the flat-four engine with rotary valves designed for the Mosquito, but dropped in favour of an existing Morris 918cc side-valve engine.

*Design mock-up of the Mosquito saloon. Note the position
of the headlamp.* (BMC Ltd)

front of the car with the radiator situated behind, and consequently a low
centre of gravity. Unfortunately, the louvres on top of the bonnet provided
insufficient cooling and the radiator had to be moved to a more conventional
forward position.

*Another theme on the Mosquito grille, with the headlamps
behind the vertical slats. Seats in this prototype were
simple tubular frames covered with canvas.* (BMC Ltd)

The prototype Mosquito bodies underwent various styling changes from the initial concept, from a flattened oval grille in 1944 to a long, low, slatted grille with built-in headlamps. By 1946, with the war over, a prototype with a 56in track was being tested up and down steep Cotswold gradients aroung the Midlands factories. The ride and handling of the car was impressive, but the quest for power and performance continued to pose problems. Added to this was the high cost of tooling and the over-optimistic idea that a very robust crankshaft would compensate for two closely spaced main bearings. Lord Nuffield's reaction on seeing the prototype did not help matters, he likened it to a 'poached egg', and was against going into production. (Considering Issigonis' contribution to the success of the Nuffield organisation in the immediate post-war years, it is curious that Lord Nuffield only ever met Issigonis twice — the second occasion was in 1960 when he finally thanked the designer of the Minor.) Nor was time on the side of the Cowley management, with other motor manufacturers having ambitions for the small car market.

The answer to some of the problems came from an unexpected quarter. In the 1947 Finance Act, the Chancellor of the Exchequer, Hugh Dalton, abolished the old RAC horsepower tax and instituted a £10pa flat rate duty for all new cars, operative from New Year's day 1948. A decision was reached to abandon the flat-four unit and utilise the 918cc side-valve engine with a four-speed gearbox, which was basically the same power unit used in the Morris Eights since their introduction in 1934. (History tends to repeat events. It is on record that when Austins were in the design stage of the original Austin Seven,

On the side-valve Series MM Minor accessibility was excellent, the USHM 918cc engine is almost lost in the wide compartment.

it was planned to use a flat-twin engine, a copy of the Rover Eight German designed air-cooled unit, but the prototype was so rough that Herbert Austin decided on a tiny four-cylinder 696cc side-valve unit.) The use of the Morris Eight power unit necessitated a few changes in the steering geometry, and no further consideration was given to the proposed independent suspension at the rear. Additional savings were also made by dropping the idea of a split rear axle. By late summer 1947 a pre-production Mosquito successfully completed a 10,000 mile test.

It was while the tooling of the body dies were underway, and some actually completed, that Alec Issigonis decided that for aesthetic reasons he would increase the width of the body by a few inches. To get the proportions right he had a pre-production model cut in half lengthwise and the two halves moved apart until he felt that the proportions were right. The increase decided upon was 4in, and as a result the track dimension increased from 56in to 60in. There remains a legacy of this change of mind on subsequent production cars, for the bumpers, designed for the original narrow body, had already been produced by Wilmot Breeden. Rather than scrap them they were cut in half and a steel spacing piece added. Another feature to be found on the eventual production bodies, bearing witness to this last minute change of mind, is the raised 4in wide centre section on the bonnet.

The name 'Mosquito' had been dropped by the time the first production Morris Minor saloon rolled off the line at Cowley on 8 October 1948; followed a week later by the first of the tourers. That such a car was on the way was generally known in the trade and indeed Sir Miles Thomas had himself voiced some very broad hints the previous year in his after dinner speech on the eve of the Geneva Motor Show, with his prophecy of a small-engined family saloon

*The prototype Mosquito was 4in narrower than the final
Minor production model. In this form the Mosquito was
driven by a flat-four engine through a three-speed gearbox
with a steering column shift.* (Morris Motors Ltd)

Morris Minor Type MM with the 918cc side-valve engine.
In this form the Minor was first produced in 1948.
(Morris Motors Ltd)

'capable of cruising at 60mph' with bodywork 'constructed on the stressed skin principle . . . producing a car of extremely low weight and high power/weight ratio.' The first the general public knew of the Morris Minor was through the pages of a national Sunday newspaper which had broken the press embargo by ten days and published details which were intended for release on 26 October.

At the first post-war International Motor Exhibition held at Earls Court, London, late in October 1948, the public were able to see for the first time the results of the previous years of concentrated development work at Cowley. There was not only the Series MM Minor in two-door saloon and tourer form, but a new Morris Oxford saloon with a four-cylinder 1,476cc side-valve engine and a six-cylinder overhead-valve Morris Six saloon. Nearby, on the Wolseley stand, it was not difficult to see the economical use of common pressings in the last-minute appearance of the four-cylinder Four-Fifty and the Wolseley Six-Eighty saloon.

The success of the Morris Minor exceeded the expectations of even the optimists at Cowley, who anticipated that production might run until 1952. When the Minor passed into motoring history twenty-three years later, over $1\frac{1}{2}$ million, in various forms, had been produced. However, before that landmark was reached there were to be many changes in specification, starting in June 1949 when twin tail lamps anticipated forthcoming regulations. The first new option came in October 1950 with the four-door saloon, introducing to the home market the new front end, with headlamps mounted in the wings. Starting the previous year, export models bound for America had been made with these raised wings to confirm with State laws, for example, California implemented regulations in 1949 requiring headlamps to be not less than 24in above the road surface. By early 1951 all models had this new headlamp location and what has become known — especially in the Antipodes — as the 'low-lite' ceased. As a consequence of the mere 15 per cent of Minors released

The first production Morris Minor competing in the Diamond Jubilee 12-hour Endurance Run at Silverstone, organised by the Morris Register in 1973.
(K. Martin)

A legacy of a change of mind, the steel spacing piece added to the bumpers on the early Series MM Minors.

for the home market in the early years, the low lamp model is comparatively rare in the United Kingdom. It was estimated in 1952 that only five out of every hundred people who had ordered new cars had any hope of taking delivery during the year.

Raw materials for Britain's manufacturing industry were severely restricted following the war, allocation being dependent on a high export total of the finished product. In 1946 the supply was conditional to 50 per cent of production going to overseas markets, by 1947 the figure was raised to 60 per cent, then 70 per cent, and by 1949 75 per cent. These shortages continued well into the 'fifties and resulted in the occasional use of aluminium where sheet steel had been normally used. The outbreak of the Korean war in 1950 put a further strain on supplies, so it is not surprising to find that some manufacturers replaced chromium plate with paint on traditional bright parts. This occurred on the Minor in March 1951 when such a finish was specified for the radiator grille, although chromium plating returned to the nave plates by September that same year. Earlier, in December 1949, painted door window frames were fitted instead of plated ones.

Up to June 1951 the Series MM Minor tourer had been equipped, at the rear, with detachable fabric-covered celluloid side-screens similar to its open model predecessors. Now referred to as the Minor convertible, the open version was constructed with fixed rear quarter-windows of framed glass. Interestingly,

*Morris Minor facia arrangement on the MM and early
Series II models.* (Morris Motors Ltd)

*Revised facia with a single centre instrument introduced at
the 1954 Motor Show.* (Morris Motors Ltd)

*The Morris Minor Type MM tourer in its initial form with
the removable rear side-screens.* (Morris Motors Ltd)

*The Minor front suspension (unchanged throughout the life
of the model) was copied by a number of car
manufacturers in Europe and the United States of
America.*

J.H. Keller, the Zurich agent for Nuffield products in Switzerland, had
produced their own conversion with fold-down glass quarter lights as early as
1949. Further changes in specification were to follow, such as the substitution
by 1957 of the canvas hood for one of plastic material, in addition to changes
common to the other body types. In June 1969 the very last Minor convertible
(registered YPK345H) was delivered by Stewart & Ardern.

An example of demand turning full circle is illustrated when back in the
'fifties the Northampton coachbuilding firm of Airflow Streamline were busy
converting soft-top models into saloons. This enterprise is contrasted, in recent
years, by a number of small motor engineers, finding a ready market for open
Minors, stripping and rebuilding two-door saloons as convertibles. One such
concern was Paul Holland Motors of Sussex.

In February 1952 the Nuffield Organisation merged with the Austin Motor

Company to form the British Motor Corporation, and one of the first fruits of
the amalgamation was the introduction of Austin components into the Morris
Minor. With the Minor in mind, work had been progressing on a Morris
version of an overhead-valve engine based on the 918cc unit, type UPHW, used
in the Wolseley Eight. Six Minors were fitted experimentally with such an
engine, in the summer of 1951, before this project was abandoned when the
merger took place. In its stead the 803cc ohv A-Series engine was used when a
Minor 'Series II' four-door saloon was introduced in the following July. Some
months prior to the merger of the two old rivals, Austin had brought out a
small car in competition with the Minor, called the Austin A30 (originally
hailed as the new Austin Seven), and it was the engine designed for this car,
mated to a gearbox originally used in the Austin Eight, that provided the new
Power unit. In October, the Austin transmission was used, resulting in an
alteration in the gear ratios, making intermediate ratios slightly 'wider', being
1.68:1 as against 1.545: in third, 2.59:1 compared with 2.3:1 on second, and
4.09 in place of 3.95 in first. In February 1953 the two-door saloon and
convertible, designated Series II, took on the 803cc engine also, and a new 'M'
motif appeared on the bonnet.

During the production of the Minor the older type of British Whitworth and
British Standard Fine fixings were replaced piecemeal as British industry
adopted Unified threads (similar to American National Fine) in an attempt at
standardisation. This resulted in many Morris cars and commercials having a
mixture of threads and they had to be identified by grooves or circles on the
hexagon flats.

In May 1953 came the 5cwt van and pick-up, known as the 'O type $\frac{1}{4}$ ton'.
This light commercial was never made in side-valve form (unless the one-off
prototype van, finished in blue and completed in February 1951, is considered.
Unlike most prototypes, this unique vehicle, carrying the chassis number
OEH/15/500, was sold.) The 803cc overhead-valve engine was used from its
inception until 1956 when, like the car, the capacity of the power unit was

Morris Minor pick-up as an articulated work-horse in
Lancashire. (Arthur Ingram)

Light commercial, the O type 6/8cwt van, with a 1,098cc engine.

increased to 948cc. Initially the vans and pick-ups were assembled at Cowley, but gradually more and more were made at Morris Commercial Cars at Adderley Park alongside the larger commercials. Interestingly, the MG lines at Abingdon also had a hand in their production, being responsible for something over 9,000 van units between 1960 and 1963, and about fifty pick-ups in 1962. Other than the Abingdon involvement, the entire output of the $\frac{1}{4}$-ton van was transferred to Birmingham in 1962 just as the plant was recovering from a disastrous fire, equalled only by the wartime air raids. This inferno occurred in May and caused damage estimated at £2-3 million, cut production by over 60 per cent, and destroyed irreplaceable historic records.

Unlike the saloon, which relied on its intregral body to provide rigidity, the light commercial was based on a true chassis with box-section framework continuing on from the mid-way cross member to the rear of the vehicle, where it was given a curve over the axle to accept the spring shackles. This arrangement of a simple box-section frame made it possible to supply a basic chassis-cab for the customer who wished to fit a body of their own design.

In its introductory form, the mechanical features and 803cc A-Series engine were almost identical to the Series II car. One area of difference was the arrangement of the rear leaf spring damping, which followed that of the last of the Series Z vans, having telescopic (instead of lever type) shock absorbers. The badge, too, had a precedent as the fitting of a motif separated from the bonnet styling strip was in the manner of the earlier side-valve MM Minor; on the later Series II Minor cars this badge was integral with the strip.

The late appearance of a van derivative is accounted for by the continuance of the 8hp Series Z van until 1953, and the considerable numbers of these used by the GPO may have had some bearing on the retention of the design to such a late date. Nevertheless the GPO were early users of the $\frac{1}{4}$-ton van, with the first (registered NLW583 and designated 'Morris Minor Type Q van') going into service in May 1953. Early examples of the vans supplied to the GPO had, like the Series Z vans, the headlamps mounted on top of the wings, which were of moulded rubber construction. Another curious feature was the opening windscreen on the driver's side of the split screen, necessitating a top mounting for the windscreen wiper arm. In the main, however, the vans used by the GPO were fairly conventional and supplied in large numbers; 12,000 Minor vans were included by the Post Office fleet in the first eight years, a figure that had totalled 52,745 by 1969. Abroad, other postal authorities also found the Minor van ideal for their purposes. These included Bahamas, Australia, Turkey, Iran (who ordered seventy vans, made to the GPO specification, in 1957), and the Norwegian postal services.

First design of the traveller in 1952 had the 803cc ohv engine and split windscreen. (Morris Motors Ltd)

Five months after the appearance of the van, in October 1953, a brake type vehicle called the Minor traveller was added to the range. This newcomer consisted of normal saloon car front and floor pan parts, mated with a traditionally built body of ash framework and aluminium. The station wagon or 'woodie' had been around as a popular utility vehicle in America from the early 'forties, but had largely lost its timber content to all-steel construction by 1953. Nevertheless on the traveller the front end of the steel saloon-type roof finished at a point just before the door pillar and mated, via a rubber fillet, to the aluminium rear roof section. The remainder of the body, and the van-like rear doors, comprised some fifty separate pieces of seasoned ash and aluminium panels.

At the same time a de luxe version of the Minor was introduced with a heater, leather seats, bumper over-riders, and a sun-visor for the passenger.

In its fifth year the front was revamped with a new chromium-plated grille

*Revamping of the Morris Minor front in late 1953. Note
the modified badge and repositioning of the side-lamp
from inside the surround to a position under the headlamps on the later model.*

with horizontal bars on all models. This grille had been shortened and the side
lamps moved from inside the surround to a position on the wings. At the rear,
larger lamps were now fitted as a direct result of The Road Transport Lighting
Act, 1953, which dictated the minimum size of reflectors and rear lamps, and
which became law on the 1 October 1954. The other major change at this time
was a revised facia with a single central instrument in lieu of the previous
separate speedometer, fuel and oil pressure gauges. This was probably done to
facilitate the assembly of the left-hand-drive models. The re-styled radiator
grille and facia panel was not seen on the ¼-ton vans until about February
1955.

*Morris Minor four-door
saloon of early 1956. Later
in the same year a single-
piece curved windscreen
replaced the split screen
shown on this model.*

*Export, left-hand-drive,
version of the two-door
Minor 1000.*

Apart from the use of stainless steel for the hub caps, the Minor design
remained virtually unaltered for the next two years. Then in October 1956 the
engine capacity was raised to 948cc, using a BMC A-Series overhead-valve
engine bored out to 63mm, increasing the power output by 23 per cent. The
name was changed the the 'Minor 1000'.

Other changes that came with the different name were deeper sectioned rear
wings, dished steering wheel, short 'stick shift' gear lever, a larger fuel tank
with a capacity of $6\frac{1}{2}$ instead of 5 gallons, and a repositioning of the direction
indicator switch from the facia to the steering column. More important, the
driving visibility was improved by combining thinner front-door pillars with a
single-piece curved windscreen and a larger rear window. At the same time the
$\frac{1}{4}$-ton van and pick-up were re-designated 'Series III' and given the larger
engine, together with the single-panel windscreen and safety type steering
wheel. This single-piece windscreen was not without its initial teething
troubles, caused by incompatible radii on the body shell and glass, which made
the rubber sealing extrusion reluctant to lie flat in the corners. An immediate
solution was found by the initiative of the line workers (no doubt mindful of
their piece-work rates) who unofficially introduced slivers of cane between the
glass and rubber. The result was a neat appearance and a good seal.

Some remarkable performances on the road can be claimed for the Morris
Minor, not only when the car was just introduced, but also in old age. As

recently as 1980 Lord Montagu of Beaulieu purchased a 1968 Morris Minor (registered UYU741F) from the Archbishop of Canterbury, Dr Robert Runcie, and entered it in the gruelling 3,000-mile Air India Himalayan Rally. Philip Young, editor of *Collector's Car* magazine, was accompanied by former rally driver, the Rev Rupert Jones, on an exciting event that saw the Minor finishing 15th, while only half the seventy-five starters completed the Rally. The car was sold to a New Zealand collector in 1982. There had, of course, been the factory testing and proving runs; for example the pre-production tests on the larger engined Minor 1000 involved 25,000 miles at an average 60mph on the German autobahnen in 1956. One of the earliest record runs was that done by rally driver Ian Appleyard in March 1949 when he took a side-valve Minor MM (registered MNW970) for a winter sport weekend in Switzerland. Leaving London after office hours on Friday night, a journey was made to Geneva where the party of three enjoyed some skiing before turning the car's bonnet homeward, to arrive back in London on Monday morning having covered 1,130 miles. Four years later Appleyard drove a Minor from Italy to England, crossing the Channel to Le Touquet from Gatwick. Leaving Rome at 9.30am in the morning, he and his passenger arrived at Westminster exactly 24 hours later having averaged 45mph on the 1,067-mile journey. The Ulster Automobile Club's Circuit of Ireland Trial in 1949, covering 1,000 miles, and which included three Minor saloons (consecutively registered MZ4701, 4702, and 4703), may have inspired Eric Fry, director of Morris distributors G.A. Brittain, to drive a Morris Minor saloon (ZJ2641) a similar distance round Ireland in just over 26 hours in February the same year — an achievement that he was to repeat in 1952, over the same circuit that started and finished at the Gough monument in Pheonix Park. The same car was used, which by then had travelled 50,000 miles. Another example was the 1,050 miles covered by John Rabson and his co-driver early in 1951. In 58 hours they covered a route that took them from London to Fort William in Scotland, and back. Average speed for the Morris Minor used (PWL761) was rather better than 34mph. More of a publicity stunt perhaps, but nevertheless an object lesson in sustained reliability, was a run round Goodwood by one of the first batch of the new 'export only' four-door 803cc cars in 1952, when a thousand miles a day was maintained for ten consecutive days. To ensure that the Minor (registered SJO624) never had to stop for re-fuelling, the Experimental Department at Cowley had built a perambulating 'dry-dock' which ran along with the car from time to time.

Among the record 225-car entry for the 1949 Monte Carlo Rally was a Minor (NWL858) with an all-woman crew of Mrs Elsie M. Wisdom, Miss B.M.C. Marshall and Miss Betty Haig; this car was placed third in the 750 to 1100cc category. Pat Moss, sister of Stirling Moss, took to the Morris Minor in the late 'fifties and was active in many rallies, including a creditable 23rd in the general classification in the 1957 Tulip Rally. The following year the blizzard-ridden Monte Carlo Rally was not such a success, as all Paris starters, including her Morris, were defeated by the weather conditions. But she was fourth out of 130 finishers, and took the Ladies Prize, when driving her Minor in the 1958 RAC International Rally. (Fifth in the same rally was W.H. Wadham, also in a

Morris Minor 1000.) Her Minor 1000, registered NJB277, lasted so long (in rally terms) and bore such a charmed life, that she nicknamed it 'Granny'. Another successful Tulip Rally entrant was P.D.C. Brooks who, with his co-driver R.E.W. Wellwest, came seventh with their Morris Minor NNF374 in the 1956 event.

A Morris not found on the roads of Britain was the Series I Morris Major, introduced in Australia in 1958 by Nuffield (Australia) Pty Ltd of New South Wales. An almost identical design, but with two doors, had been considered by BMC as the replacement for the Morris Minor, as the 'B-Series Minor', but this got no further than the prototype stage. Internally, the instruments were set in a central binnacle, in the manner of the contemporary Minor, but in the case of the Major the combined instrument was rectangular rather than round.

No further major changes were made for several years (the lids added to the glove boxes during 1957, and the courtesy light switches fitted to the front doors in October of the same year, hardly qualify as major) but an important milestone in the Minor story was reached in early 1961. The thirteenth year of production was a notable one, for the millionth Morris Minor was built. Of this staggering total over half were Minor 1000s and 48 per cent were exported. To celebrate the occasion the millionth vehicle, a saloon, was finished in a special shade of lilac with upholstery in 'white gold' leather. and in place of the usual 'Minor 1000' nameplate this saloon carried a 'Minor 1000000' motif. This special car was taken to Grosvenor House in London where, at a simple ceremony, Alec Issigonis, for the British Motor Corporation, presented it to the National Union of Journalists to use in any way they like to assist the finances of their Widow's and Orphan's Fund. It was offered as a prize in a raffle and subsequently won by a solicitor's wife living in Cardiganshire. This car, with a Bristol registration 1MHU, is at present owned by a Birmingham Morris Minor enthusiast.

349 replicas of the Minor 1000000 were also produced. Strangely, the replicas were produced first. These, carrying the chassis numbers 1000001 - 1000349, were built mostly on 13 and 14 December 1960. The actual millionth car was built last of all, on 22 December. Many of these special Minors still exist including car number 1000001, which is believed to have been exhibited at the New York Motor Show.The majority of the replicas were sent to main

Morris distributors in the UK, but thirty were assembled as left-hand-drive cars, so it is reasonable to assume that these were exported. The fact that these 350 Minor Million cars carried the chassis numbers 1000000 - 1000349, demands an explanation, as the first Minor carried the number 'SMM501'! The cars (including travellers and convertibles) had a separate number series from the light commercials, although both series started at 501. When the sum total had reached 999,999 (ie the private car series were up to approximately 886,161 and the commercials made up the remaining 114,338 units) the Minor Million car numbers were

In 1961, to celebrate the production of the millionth Morris Minor, it and 349 replicas were finished in a special shade of lilac and given a 'Minor 1000000' nameplate. (Morris Motors Ltd)

introduced. Later, when the Series V Minor car chassis numbers actually reached the figures 1000000 - 1000349, in late 1962, they were not re-allocated; instead numbers allocated to production cars jumped from 999999 - 1000350. As the commercial series never exceeded six figures, the problem did not arise with the van numbering.

When the production of the millionth Minor was announced, Morris Motors tried to find the oldest post-war Minor, and offered a new Minor 1000 in exchange for it providing the car had covered 100,000 miles. The response was staggering, with letters being received from all over the world. But the earliest of all turned out to be the first one off the production line numbered 'SMM501', and was found in Sheffield. It had started life as a works demonstrator, the owner was presented with one of the replica Minor 1000000 saloons at Kennings Group depot in Sheffield and his old car, registered NWL576, was subsequently restored by the manufacturers. This first Morris Minor is now part of the BL Heritage Collection at Syon Park.

The tendency to make changes around Motor Show time continued. In 1961 flashing indicators were fitted in place of the now outdated semaphore type (flashing type signals had been made legal in the United Kingdom in January 1954). There was provision for screen washers, although actual washers were only fitted to the de luxe models. The alteration in specification for the following year was more involved, for in November 1962 the engine size was increased once more, this time to 1,098cc. In addition, the Borg & Beck 'A' type clutch plate was enlarged from the original $6\frac{1}{4}$in diameter to one of $7\frac{1}{4}$in, and the 7in front brake drums replaced by 8in ones. Shortly afterwards the manufacturers reverted to the original practice of uncovered glove boxes. Lengthened windscreen wiper arms working in tandem and combined side/flasher units were introduced in October 1963, while in October 1964 there was combined starter/ignition switch, oil warning light, two ashtrays, a two-spoked steering wheel, and (once again) lids on the glove compartments.

In the 'sixties the police introduced the 'Panda Car' with the word 'Police' boldly marked along the side. The obvious choice of a light car for such a force, able to travel quickly to trouble spots, was the Morris Minor 1,098cc two-door saloon. A two-door vehicle was selected as it made difficult the escape of anyone in custody at the rear. These cars were in two colours with the main body in light Bermuda blue with a white finish was given to the doors and roof surfaces. They were one-man operated and equipped with radio communication sets. An instant identification feature for present-day owners of such cars, apart from the fixing holes in the roof for the illuminated sign and a radio aerial, is a long transverse zip-fastener built into the headlining to provide access to the roof sign.

The end of the road for the Minor 1000 came in late 1970 when the last body shell, numbered 1582302, left the Common Lane body plant. This was in September but naturally assembly at Cowley continued for a short period and the saloon model was still being listed a month later. As mentioned earlier, the last convertible was delivered in 1969, so the Minor 1000 range continued with just the traveller and the Series III 6 and 8cwt vans. A considerable number of these vans, from 1968, carried an Austin badge, and around April 1971 the final traveller (chassis 1294082) was produced. With the demise of the Morris Minor traveller went another piece of Morris history — the Coventry Body Plant, where the framed semi-coachbuilt bodies were made.

In fact the Coventry Body Plant pre-dated Morris history by a century. It was started in 1812, in a building on a hill alongside what became the London-Coventry railway line, and for almost ninety years the firm made coaches and carriages. In 1876 Edward Henry Hollick bought the coachbuilding business; in later years he was joined by a partner named Pratt and as 'Hollick & Pratt' they began making motor car bodies. Lancelot W. Pratt's involvement with William Morris and the purchase of Hollick & Pratt by Morris in 1923 has already been mentioned in an earlier chapter. In 1926 it was renamed Morris Motors body Branch and so continued until the formation of the British Motor Corporation who, in 1966, combined their four body-building subsidiaries (located at six centres) into a single organisation named Pressed Steel Fisher Ltd. The subsidiaries were Fisher & Ludlow at Castle Bromwich and Llanelli, Nuffield Metal Products Ltd at Washwood Heath in Birmingham, Pressed Steel Company at Cowley and Swindon, and Morris Motors Bodies Branch. The latter, as the Coventry Body Plant, concentrated on the requirements of special and open bodies using the high degree of skill available for the traditional ash-framed semi-coachbuilt vehicles such as the early T-type MGs, Morris Eight tourers, Post Office and other commercial van bodies, and taxicabs. During World War II, in addition to constructing bodies and canopies for military transport, the Bodies Branch made ailerons for the Horsa gliders, jerrricans, seat panels for utility buses, shell carriers, transport crates, compressor boxes, tool bags, etc. It was the most heavily 'blitzed' Nuffield plant during the war; more than 1,000 bombs fell on the factory during the Coventry air raids.

One Minor 1000 that is likely to confound future historians is a two-door saloon built in 1974! Edwin Law of Halifax bought the first of the ten he had owned in 1954, replacing them at suitable intervals, until a time came when he discovered that they had gone out of production. His son, Roderick, made enquiries of the cost of a new car assembled from new spares. When told that it would be little short of £4,000, he was not daunted, and told the firm to go ahead and build it. The parts division at British Leyland, although amazed at the request, provided the 595 essential parts within days, some from stock and others from Morris distributors throughout the country. Of the smaller parts, the only new part that could not be located was a tiny, but vital, locking lever for the driver's door — in the end one was obtained from a scrap yard. The new car, registered GHR800N was handed over on 31 October 1974, together with a telegram of greeting from Lord Stokes. Sadly, Edwin Law was too ill to be present and he never drove his new Minor 1000, for he died a few days later.

The last Morris to be graced with the traditional radiator was the Morris Six. The four-door saloon as introduced in 1948. (British Leyland)

The two other Morris cars to make their début at that first post-war Motor Show in 1948 were, in name, modern descendants of famous ancestors, the 'Morris Six' and 'Morris Oxford'. The integral steel constructed body design for both cars followed the general lines of the Minor, incorporating the tail end of the front wings in the door pressings, and a vertically split windscreen. In the case of the Morris Six the radiator shell followed the 'thirties pattern, contrasting with the Oxford's wide excess of chromium-plated decoration. This traditional radiator shell, reminiscent of the pre-war Series M Ten (complete with a shield shaped badge), was incorporated, so the story goes, on Lord Nuffield's insistence. (Similarly, when the Riley Pathfinder and the restyled 1½-litre Riley came on the scene in late 1953, Lord Nuffield boasted that it was entirely due to him that the familiar radiator filler cap had been retained on both these models, and on the new MGs. The designers' desire to conform with

A Wolseley Six-Eighty of 1951. The Four-Fifty and Six-Eighty utilised the same body pressings as the Morris Six.

modern radiator styling by dropping the orthodox filler cap was overruled by Nuffield, who felt that the retention of the cap was essential to the traditional radiator design. The cap was not merely ornamental however, as it served as a release catch for the bonnet.)

The Morris Six, Series MS, had a six-cylinder engine of the same bore and stroke as the side-valve four-cylinder Oxford, but with overhead-camshaft and valves. This engine was later to gain a reputation for burning exhaust valves, partly due to the fact that each valve had a built-in threaded adjuster preventing the valves from rotating in use, allowing the destructive effect of any slight gas leak to concentrate on one part of the valve. Only one body, the four-door saloon, was marketed, though it was subsequently made in de luxe form. For an extra £28 the owner of the de luxe Morris Six had leather covered seat squabs in place of the standard leathercloth upholstery, and the addition of a heater and centre arm rest to the back seat. A considerable number of Sixes were exported between 1948 and March 1953 when the model ceased; perhaps the relative few that found their way into hands of British motorists accounts for its undistinguished record, attracting little publicity after the initial unveiling in 1948. (A Morris Six, registered EJT282, was driven by E.H.

and P.H. Channon in the Monte Carlo Rally of 1951.) Something like 12,500 vehicles were produced (7,000 of which were exported) and during its currency few modifications were made to the specification. These were mostly minor changes, made around October 1949, such as improvements to the rear lights (twin stop/tail lamps to enlarge the arc of visibility to 180°), the use of leather as standard on the bench-type front seat and a reshaped rear squab to increase width, improved choke control so that the ratchet position could be more easily found, and inclined hydraulic dampers fitted to eliminate the separate anti-sway bar. The 1951 Motor Show saw the introduction of a pressurised system of cooling, modifications to the cam-gear steering to reduce the car's turning circle, adoption of a headlight warning light, and the fitting of a tinted rear view mirror inside the car. In 1952 standardised ventilators in the rear door windows and an oil-bath air cleaner on cars for the home market were introduced.

Frontal design changes on the Series MO Oxford. The wide latticed look on the early cars was changed, in October 1952, to a fabricated and chromium-plated grille.

By contrast, the Oxford underwent many changes both in body design and mechanical arrangement during the following twenty-three years. The first Oxford, the Series MO four-door saloon, had in its enlarged Minor-style body a bench-type front seat which, combined with a steering-column gear change and umbrella handle handbrake lever beneath the instrument panel, made possible the seating of three persons in the front compartment. Anyone who has driven an early Oxford will remember the feeling of being persuaded to the off-side of such a wide seat by the slight angle of the steering column. The option of a traveller version (initially described as a 'Station Wagon' in original Morris publicity) with a wood-framed rear came in during September 1952. The modifications to the Oxford by this date included rear telescopic shock absorbers (1949), painted radiator grille instead of chromium-plating (1949), deletion of rear blind (1951), battery moved from dash panel to floor of engine compartment (1951), and fabricated plated radiator grille (1952).

In January 1954 production of the flat-head Series MO Oxford ceased; replaced by a 'Series II' saloon and traveller, in May and October respectively. This new traveller body was essentially a normal Series II saloon front half and appeared similar to its predecessor, although in fact the wood did not form the

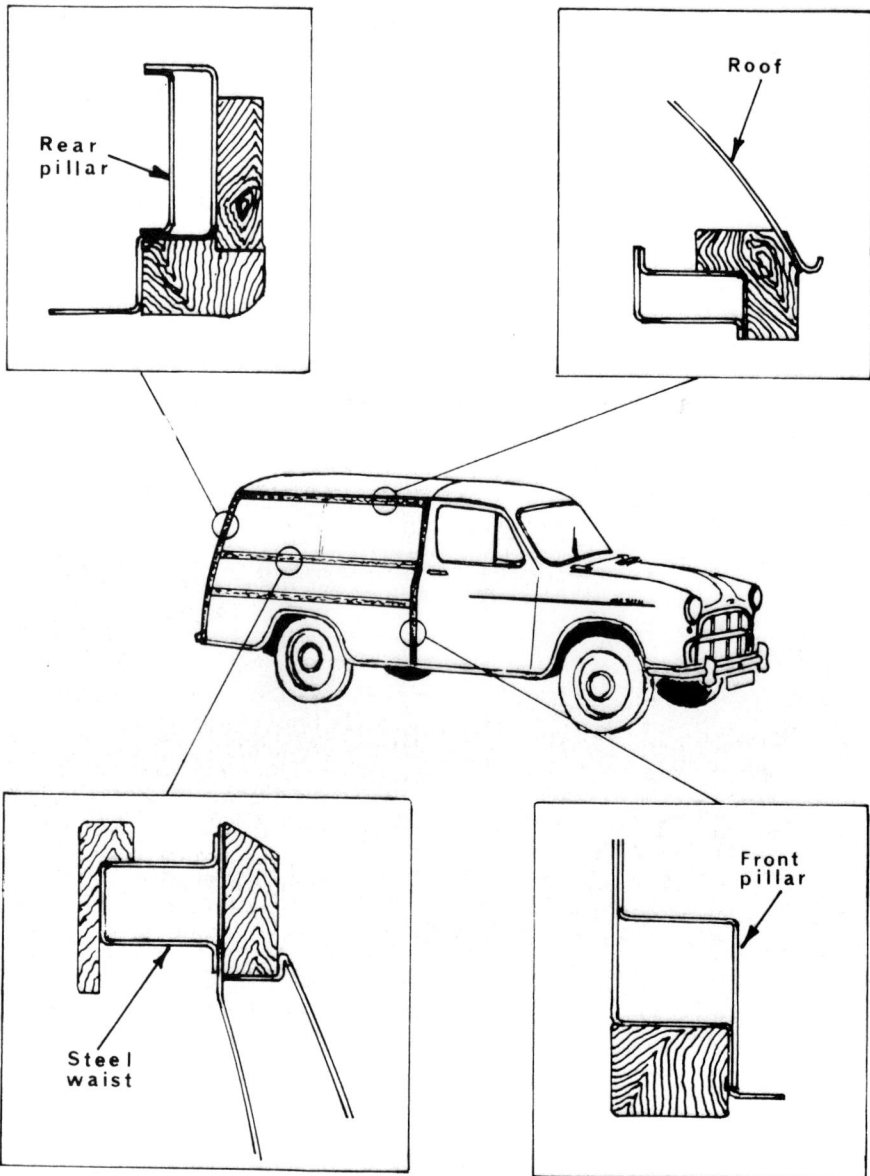

Rear pillar

Roof

Steel waist

Front pillar

*Although the Series II traveller had a conventional wood
frame and metal panel appearance, the steel box sections
actually formed the main structure of the body. These
detail sections show how they were camouflaged.*

main structure of the body, but was fixed to steel box section members — with
the single exception of a wood interface between roof and upper window frame
piece. The engine now used, replacing the original 1,476cc, was an Austin four-
cylinder overhead-valve unit of 1,489cc and features such as a one-piece
windscreen, handbrake lever at the side, and an air intake on the bonnet, were

to be found. These Series II Oxfords were subsequently replaced by Series III versions in 1956. Again the 1,489cc engine was utilised but minor styling changes, like the fluted sides to the bonnet, absence of the air intake, finned rear wings, and hooded headlamps, were immediately noticeable. Internally, a dished safety type steering wheel and lids on the glove boxes were added.

Of the Series III Oxfords, the first to disappear from the dealer's showrooms was the wood overlay/metal bodied traveller, to be replaced in June 1957 with an all-steel version with four side doors and tailgate, and two fuel filler caps, one on either side of the body. This carried the 'Series IV' label. The Series III four-door saloon continued on through 1957 (with new external body mouldings added in November) to March 1959.

A completely redesigned exterior belies the almost unaltered Morris Minor 'chassis' on the Wolseley 1500. Despite appearances the overall length is only 3¾in longer.

The 'Series V' Oxford four-door saloon which took its place was a completely new body styled by Pininfarina (only written as one word after 1952), with 14in diameter wheels and finned rear wings. The car came as either standard or de luxe and there were many mechanical changes compared with the preceding models. The original 12 gallon full tank was now reduced to 10 gallons, gearbox ratios were lower, floor change or column mounted lever for the gearbox was optional, the final drive of 4.875:1 was changed to 4.55:1, suspension was by means of coil springs instead of torsion bars, there were lever-type shock absorbers, and the rack and pinion steering was replaced with a worm and peg arrangement. What was really being offered was the Morris version of a basically common model, for apart from the radiator grille design, detail body fittings and interior layout (and, of course, the badge), the Series V Oxford was identical to the Mark II Austin A55 Cambridge. The Wolseley 15/60 was another BMC saloon to utilise the same Pininfarina designed body. Dimensionally, the Series V Oxford had a $2\frac{1}{4}$in longer wheelbase, giving an increase of 7in to the overall length, compared with the earlier car, but the track was narrower by 5in at the front and 3in at the rear.

In April 1960, the all-steel Series IV traveller ceased production and a traveller version of the Series V Oxford replaced it in the following September. Surprisingly, this newcomer was a reversion to the wood overlay/metal construction, but was only short lived, being withdrawn in March 1961. Meanwhile modifications to the Series V Oxford saloon were being made; in March 1960 a redesigned number plate incorporating two bulbs and two months later new ashtrays and alterations to the engine involving the camshaft. The Series V cars finished in 1961.

Austin's equivalent to the Morris Oxford VI, the Austin Cambridge A60.

Under bonnet arrangement of the Series VI Morris Oxford, with the four-cylinder 1,622cc overhead-valve push-rod engine.

The last of the Morris Oxfords came in October 1961, as the 'Series VI', and in standard or de luxe saloon and traveller form they were to continue until 1971. This final Oxford had a larger four-cylinder ohv engine of 1,622cc, a longer wheelbase of 100¼in compared with the previous 99⁷⁄₁₆in, and a wider track. The redesigned bumpers and stop/tail light cluster was a feature on both saloon and station-wagon type bodies, and there was the option of Borg-Warner '35' automatic transmission. A diesel engined Series VI Oxford was available from November 1962, powered by a BMC 1,489cc diesel derived from the B-Series petrol engine and which owed much to the Saurer diesel. A similar power unit was to be found in the J2 and J4 vans and, in later years, in the diesel version of the Morris Marina. As the Cambridge A60 was Austin's version of the Oxford VI, this too was sold in diesel form. BMC claimed that this was the first time a diesel engine smaller than 1½ litres capacity had been offered in a private car, and that after 9,000 miles the extra cost of the diesel engine was offset by the superior fuel consumption of 40mpg against 30mpg on petrol versions for normal use. Top speed was quoted as 65mph.

Hindustan Motors Ltd of Calcutta also built a version of the Morris Oxford. Their first adoption of a Morris design was the Hindustan Ten based on the Morris Series M Ten (as already briefly mentioned in an earlier chapter), superseded in due course by a version of the Series MO Oxford known as the

Wolseley's equivalent to the Morris Oxford VI, the
Wolseley 16/60.

Hindustan 14, and the Baby Hindustan looking, not surprisingly, like the Series II Morris Minor upon which it was based. India's motor industry had gradually expanded from the 'twenties when two leading American manufacturers established factories in India for the assembly of their cars and commercial vehicles for the local market. By 1945 two Indian companies had been formed for local production of cars, one in Calcutta (started by G.D. Birla and his nephew, B.M. Birla, with technical help from Nuffield) for the manufacture of the Hindustan Ten and also Studebaker Champion vehicles; and a second in Bombay with a manufacturing licence from Chrysler. By the end of 1949 there were five companies, two had technical backing from American firms and the other three support from British manufacturers. All five were soon alleging unfair competition in India's limited market, so a Tariff Commission was set up in 1953 which laid down rules that manufacturing programmes were to be limited to one each of small, light, medium and heavy cars, and that within three years the indigenous content should be 60 per cent of the value of CKD components.

Hindustan Motors adhered faithfully to the timetable so that by 1954, when the Morris Oxford Series II began production as the Hindustan Landmaster, dependence on imported components had been reduced substantially. The introduction of the Oxford Series III in the United Kingdom in late 1956 was reflected in the new Hindustan Ambassador in 1957. The following years, to the present day, have seen a succession of Ambassador models, gradually

The Hindustan Ambassador was based on the Morris Oxford. By 1978 the Mark IV was announced and this is the current model. (David Garrett)

differing in detail from the basic Oxford Series III with its fluted bonnet, but still recognisably an Oxford. The Ambassador Mark 4 of 1982 still has the B-Series 1,489cc four-cylinder overhead-valve pushrod engine and steering column gearchange in standard four-door saloon form. Optional extras include a 1,760cc engine, vacuum servo-assisted brakes, and floor mounted gear shift. For the Indian motorist the choice of a motor car is limited. Apart from the Ambassador there is a 1960-type Fiat 1100 built near Bombay and a modified Triumph Herald called the Standard Gazelle. Imported cars are extremely rare. At the current exchange rate the Ambassador's price of R78,635 equates to something like £4,626 sterling, but considering the average middle-management take-home pay is about £100 per month, the purchase of a car in India is a luxury few can afford.

The MCV Cowley van and pick-up introduced in May 1950 was based on the front and design of the Oxford MO car and utilised a similar 1,476cc side-valve engine.

In the 'fifties two other old Morris names were to reappear, the Cowley and the Isis. Replacing the Series Y 10cwt van, the new Cowley of May 1950, in both van and pick-up form, was based on the front-end design of the Oxford MO car and utilised a similar 1,476cc four-cylinder side-valve engine. Designated 'Series MCV', it had the car's independent torsion-bar suspension at the front and a steering column gear change — necessary in the case of the pick-up version as this had the car comfort of a full width bench seat so that, if desired, three persons could be accommodated. The van, on the other hand, was provided with a single bucket seat for the driver as standard on home models, allowing full use of access to the load from both front and rear doors; if required, a second bucket seat or full bench seating was available as an extra. When this light commercial gave way to a Series III in September, the name Cowley was dropped and it was simply listed as the Morris ½-ton. As before the

*The Austin A55 type commercial, available as van and
pick-up had counterparts in the Morris ¹/₂-ton commercial,
introduced in late 1962, types M/HV4 and M/HK4.*

available alternatives were van, pick-up, and basic chassis-cab. The engine of
the restyled vehicle was the B-Series 1½ litre overhead-valve unit, developing
50bhp compared with the 41bhp for the side-valve. The altered front
appearance brought it into line with the contemporary Oxford and Cowley
saloons with their single-piece windscreens and wide grille with the central
horizontal bar curved to extend to the side lamps. The Cowley four-door
saloon, which had appeared in July 1954, followed historic precedent by being
a less powerful version of the Oxford. By this time the Oxford had been fitted
with a floor mounted handbrake lever, an air intake on the bonnet and a one-
piece windscreen, so not surprisingly the Cowley had these features on its
introduction. The engine, from the Austin side of BMC, was an overhead-valve
unit of 1,200cc. In January 1955 the brake drums were increased in size from
8in to 9in diameter, then less than nineteen months later this short lived model
ceased production. Meanwhile a second, and more powerful, Cowley had been
announced. This one, called the 'Cowley 1500' started to appear in the motor
showrooms in January 1955 with (in the manner of the Series III Oxford) a
four-cylinder ohv 1,489cc engine, dished type steering wheel, finned rear wings,
hinged windows on the front doors, lids on the glove boxes and absence of the

air intake protrusion. The Cowley 1500 finished in 1959.

Allowing for the development time required before a new model is unveiled, it is evident that the Isis of 1955 was the last pre-BMC design to achieve production; and it also proved to be the last Cowley vehicle to carry an old pre-war Morris name. In this case the letters ISIS appeared on each front wing above a long chrome flash. The first Isis dated back to 1929 when it was displayed for the 1930 season. The modern Isis came as a four-door saloon monoconstructed, as was the more expensive eight-seater Isis traveller, although the non-structural timber addition gave the impression of a wood frame construction. The suspension followed the lines of the contemporary Oxford, with independent torsion bar suspension at the front and half-elliptic springs at the rear, and the body shell was structurally similar, although additional bracing was to be found at the front end to cater for the heavier and more powerful engine. The steering layout was quite different from that used on the Oxford, in which a rack and pinion system was placed behind the power unit. As a result the Isis rearrangement eliminated the offset in the Oxford steering wheel position already mentioned.

A C-type 2,639cc six-cylinder overhead-valve, push-rod engine driving through a four-speed synchromesh gearbox, operated by steering column control, provided the power to drive what was a large and heavy car, well over $1\frac{1}{4}$ tons with an overall length approaching 15ft. This combination of weight and size probably accounted for the increase in brake drum size early in 1956.

One of the optional extras offered by BMC on the Isis was an overdrive installation of the epicyclic gearing type. This was available late in 1955. An electrical control on the facia gave the driver a choice of overdrive or normal transmission. When selected, the overdrive cut in automatically at about 32mph when the driver lifted his foot slightly from the accelerator.

Both models were replaced at Motor Show time in 1956 by Series II versions. These had been restyled to include fins on the rear wings, revised rear lights, added brightwork (forming a convenient superstructure separation which facilitated the five duotone colour schemes offered as an extra), and fluted sides to the bonnet. Inevitably, the internal changes included the dished steering wheel and covers to the glove boxes, in addition to the floor-mounted gear change lever. By October 1957 production of the traveller ceased but the saloon continued and another 1,500 vehicles were produced during the next six months before that too became motoring history.

Just over 13,000 Isis cars were made in total, a not very impressive figure, so it is not surprising to find that the Isis attracted little publicity. One exception was in 1956 when an Isis in standard form, apart from long range fuel tanks, was driven 7,056 miles from Nairobi to Cape Town and back at an average speed of 60.7mph. The drivers, J. Manussis and P. Davies, took 116 hours 9 minutes, which included a $7\frac{1}{2}$-hour stop in Cape Town, this time being claimed as a new record.

Minor, Chassis Numbers

Early Morris Minors, up to April 1952, carried the prefix 'SMM'. After this date the Nuffield code was introduced which consisted of three letters followed by two numbers:

First letter, model
  F = Minor   O = light commercial
Second letter, type
  A = saloon 4-door   B = saloon 2-door   C = tourer   E = van   F = pick-up   G = chassis/cab   H = mail (GPO)   J = engineers (GPO)   K = chassis
Third letter, colour
  C = red   E = green (GPO vehicles)
First number market
  1 = RHD home   2 = RHD export   3 = LHD   4 = North America   5 = CKW (RHD)
Second number, paint
  1 = synthetic   2 = Synobel   3 = cellulose   4 = metallic   5 = primer
In 1958 the BMC prefixes were introduced:
First letter, make
  M = Morris   A = Austin
Second letter, engine type or size
Third letter, body type
  E = engineer's (GPO)   G = Mail (GPO)   S = saloon 4-door   2S = saloon 2-door   T = tourer   U = pick-up   V = van   W = traveller
  Q = chassis/cab
Fourth letter (or number), model Series

Model						
Series MM, 918cc, side-valve	501 - 159189	159202 - 160000	161026 - 170000	171001 - 178500	179801 - 179839	
Series II, 803cc, ohv	159190 - 159201	160001 - 161021	170001 - 171000	178501 - 179800	180001 - 438788	438704 - 438804
	439001 - 446000	448001 - 448041	448043 - 448430	448432 - 448550	448552 - 448666	448668 - 448686
	448688 - 448707	448712 - 448714				
Series III, 948cc, ohv	438789 - 438703	438805 - 439000	446001 - 448000	448042 - 448431	448551 - 448667	448687
	448708 - 448711	448801 - 990289				
Minor 1000000	1000000 - 1000349					
Series V, 1,098cc ohv	990290 - 999999	1000350 - 1294082				
Series II, 1/4-ton, 803cc	501 - 49767	GPO postal: OHC3554 - OHC67196; Engineer: OJE2999 - OJE85136				
Series III, 1/4-ton, 948cc	49801 - 149536	GPO postal: OHC67197 - OHC177640; Engineer: OJE85137 - OJE176000				
Series V, 6cwt, 1,098cc / 8cwt, 1,098cc	149537 - 327369. GPO postal: OHC177641 onward; Engineers: OJE176001 onward (Austin and Morris) 238597 - 327369 Series C Austin 6cwt van, pick-up and chassis/cab; 236504 onward					

Morris Minor

Series MM

Tourer. Production period, October 1948 - February 1953
Saloon, 2-door. Production period, September 1948 - February 1953
Saloon, 4-door. Production period, September 1950 - July 1952

Engine, side-valve, 918.6cc, USHM2. Four-speed gearbox with synchromesh on all forward gears, floor gear change lever. Clutch, single dry-plate Borg & Beck, $6\frac{1}{4}$in diameter. Brakes, Lockheed hydraulic, 7in diameter brake drums, two leading shoes at front. Handbrake by cable to rear wheels. 5-gallon petrol tank. SU electric fuel pump. 12-volt electrical system, positive earth. Steering, rack and pinion. Cooling system, fan and thermo siphon (impeller when heater fitted). Hypoid bevel rear axle, 4.55:1. Independent front suspension front suspension by torsion bar and wishbones. Armstrong hydraulic double-acting shock absorbers. Rear springs, semi-elliptic leaf springs with rubber bushes, 7 leaves. 500-14 Dunlop tyres on pressed steel disc wheels, four-stud. Wheelbase 7ft 2in. Track 4ft $2\frac{5}{8}$in front, 4ft $2\frac{5}{16}$in rear. Twin stop/tail lamps, in place of single lamps, added in January 1949. In January 1951 headlamps of two-door saloon raised above the grille and separate side lamps fitted. March 1951, painted radiator grille in place of chromium plate. June 1951 tourer now known as convertible and with fixed rear side windows. December 1951 ash tray fitted centre of top facia. March 1952 headlamp warning light on facia added to four-door saloon models, and added to other models in April 1952. This same period saw the glove box emblem changed from chrome/enamel to plastic.

Series II

Convertible. Production period, February 1953 - September 1956
Saloon, 2-door. Production period, February 1953 - September 1956
Saloon, 4-door. Production period, July 1952 - September 1956
Traveller. Production period, September 1953 - September 1956

Engine, overhead-valve, 803cc, APHM. Four-speed gearbox with synchromesh on all forward gears, floor gear change lever. Clutch, single dry-plate Borg & Beck, $6\frac{1}{4}$in diameter. Brakes, Lockheed hydraulic, 7in diameter brake drums, two leading shoes at front. Handbrake by cable to rear wheels. 5-gallon petrol tank. SU electric fuel pump. 12-volt electrical system, positive earth. Steering, rack and pinion. Cooling system, fan and impeller. Hypoid bevel rear axle, 5.375:1, 5.286:1. Independent front suspension by torsion bar and wishbones. Armstrong hydraulic double-acting shock absorbers. Rear springs, semi-elliptic leaf springs with rubber bushes, 7 leaves. 500-14 tyres on pressed steel disc wheels, four-stud. Wheelbase 7ft 2in. Track 4ft $2\frac{5}{8}$in front, 4ft $2\frac{5}{16}$in rear.
Front of Series II Minor changed October 1954 with new style radiator grille of horizontal slats painted off-white or body colour. Side lamps moved to position beneath headlamps. Revised dashboard with centrally mounted speedometer.
Traveller. Rear door fitted with automatic check arms to hold them open at a convenient loading angle. Spare wheel and tool compartment beneath floor at rear. Two bucket seats at front tip up to allow easy access to rear from either side. Rear folded down flush with the floor when desired to increase goods space. Large windows in side of body with sliding panels. Bumpers, full width at front and quarter bumpers with central number plate at rear.

Series III

Convertible. Production period, September 1956 - September 1962
Saloon, 2-door. Production period, September 1956 - September 1962
Saloon, 4-door. Production period, September 1956 - September 1962
Traveller. Production period, September 1956 - September 1962

Engine, overhead-valve, 948cc, APJM (9M). Four-speed gearbox with synchromesh on all forward gears, floor change gear lever. Clutch, single dry-plate Borg & Beck, $6\frac{1}{4}$in diameter. Brakes, Lockheed hydraulic, 7in diameter brake drums, two leading shoes at front. Handbrake by cable to rear wheels. 5-gallon petrol tank to 1957, then $6\frac{1}{2}$ gallons. SU electric petrol pump. 12-volt electrical system, positive earth. Steering, rack and pinion. Cooling system, fan and impeller. Hypoid bevel rear axle, 4.55:1. Independent front suspension by torsion bar and wishbones. Armstrong hydraulic double-acting shock absorbers. Rear springs semi-elliptic leaf springs with rubber bushes, 7 leaves, reduced to 5 leaves on cars in 1959. 500-14 tyres on pressed steel disc wheels, four-stud. Wheelbase 7ft 2in. Track 4ft $2\frac{5}{8}$in front, 4ft $2\frac{5}{16}$in rear.
Single-piece windscreen. Stalk type switch for indicators and horn, horn button repositioned on steering wheel boss from chassis 672268. Flashing indicators replaced semaphore type for 1962 models. Courtesy light switches added in front doors October 1958. Self-cancel indicator switch, wider doors on saloon and traveller, March 1959. De luxe specification included leather seat upholstery, passenger sun-visor, bumper over-riders, and fresh-air heater. Screen washers and seat belt anchorage standard on de luxe models after October 1961.

Series V

Convertible.   Production period, September 1962 - June 1969
Saloon, 2-door.   Production period, September 1962 - November 1970
Saloon, 4-door.   Production period, September 1962 - November 1970
Traveller.   Production period, September 1962 - April 1971.

Engine, overhead-valve, 1,098cc, 10MA. Four-speed gearbox with baulk-ring synchromesh on upper three forward gear ratios, floor mounted gear change lever. Clutch, single dry-plate Borg & Beck, 7$\frac{1}{4}$in diameter. Brakes, Lockheed hydraulic, 8in diameter brake drums, two leading shoes at front. Handbrake by cable to rear wheels. 6$\frac{1}{2}$-gallon petrol tank. SU electric petrol pump. 12-volt electrical system, positive earth. Steering, rack and pinion. Cooling system, pressurised, engine driven pump. Hypoid bevel rear axle, 4.3:1.
Independent front suspension by torsion bar and wishbones. Armstrong hydraulic double-acting shock absorbers. Rear springs semi-elliptic leaf springs with rubber bushes, 5 leaves. 520-14 tyres on pressed steel disc wheels, four-stud. Wheelbase 7ft 2in. Track 4ft 2$\frac{5}{8}$in front, 4ft 2$\frac{5}{16}$in rear.
Specification changes from previous models included, fresh-air heater, repositioned tandem wiper spindles giving a greater screen sweep, zone-toughened windscreen, self-propping boot lid support, combined side/amber flashing lights, amber flasher added to larger fluted rear light unit on cars (traveller had extra round amber light at rear), near-side door on two-door saloon had key lock. Changes after September 1964 include, two-spoke steering wheel, starter/ignition switch, oil filter warning light, individual swing-out ash trays under parcel shelf, new design sun-visors, plastic framed driving mirror, improved heating. October 1967, basic models of the saloon and traveller had windscreen washers as standard fitting. Some late 1971 traveller models fitted with steering lock.

O type 5cwt ($\frac{1}{4}$-ton) light commercial vehicle

Van and pick-up.   Production period May 1953 - September 1962

Engine, overhead-valve, 803cc, APHM. After October 1956, 948cc APJM. Four-speed gearbox with synchromesh on upper three forward gears, floor mounted gear change lever, modified to remote control after October 1956. Clutch, single dry-plate Borg & Beck, 6$\frac{1}{4}$in diameter. Brakes, Lockheed hydraulic, 7in diameter brake drums, two leading shoes at front. Handbrake by cable to rear wheels. 6$\frac{1}{2}$-gallon petrol tank. SU electric petrol pump. 12-volt electrical system, positive earth. Steering, rack and pinion. Cooling system, fan and impeller. Hypoid bevel rear axle, 4.55:1. Independent front suspension by torsion bar and wishbones, hydraulic piston type dampers. Rear springs semi-elliptic leaf springs, 7 leaves, controlled by telescopic hydraulic dampers with anti-sway mounting. 500-14 tyres on pressed steel disc wheels, four-stud. Wheelbase 7ft 2in. Track 4ft 2$\frac{5}{8}$in front, 4ft 2$\frac{5}{16}$in rear. Split windscreen to October 1956 then single panel, other changes at this time included a new dashboard with central speedometer. Hinged squab on passenger seat added 1959. Van body capacity, 78 cu ft, with additional 12cu ft beside the driver.
Pick-up with fold-down tail gate. Optional extra, canvas tilt supported by tubular hoops fitting into sockets let into top face of the fixed side panels.
Post Office vans had a special specification. Early versions up to 1954 had headlamps mounted on top of rubber wings. To about 1957, vans had split windscreen with driver's side hinged at top. Special fittings on Post Office vans included wire mesh on inside of rear door windows, wood framed partition, covered with wire mesh, behind driver. Rear doors had exterior locking bar on postal vans; on telephone vans a hasp and staple was fitted. Ladder rack fitted to telephone vans. Yale pattern lock on driver's door after 1958.

O type 6cwt light commercial vehicle

Van and pick-up.   Production period October 1962 - 1971

Engine, overhead-valve, 1,098cc 10MA. Specification generally as Series V cars, differences include telescopic hydraulic dampers front and rear, semi-elliptic rear springs 7 leaves. Larger rear doors from 1968.
Telephone and postal vans supplied without hub caps. 1964/5, publicity board clips on van sides.

O type 8cwt light commercial vehicle

Van and pick-up.   Production period 1968 - December 1971

Engine, overhead-valve, 1,098cc 10MA. Specification generally as Series V cars, differences include telescopic hydraulic dampers front and rear, semi-elliptic rear springs 8 leaves. Front kingpins, drag links, and trunnions strengthened to cater for new rating. Larger rear doors. Tyres 560-14 on 4in wide pressed steel disc wheels. Some vehicles supplied with Austin badge. These differed from Morris version by 'crinkled' radiator grille slats, Austin motif on steering wheel and plain hub caps.

Morris Minor, Body Colours and Upholstery

Research into the colour variations has posed many problems, not least of which are the contradictions found when comparing catalogues, manufacturers, shade cards and other sources of data. From the first half of the 'sixties, colour schemes ceased to be quoted in catalogues, and model descriptions to be found in the motoring press very seldom quote the options available. In the following list the years quoted are necessarily approximate. Upholstery colours are variously described as red or maroon in different publications. To avoid confusion the term red has been used in these tables. The figures quoted in brackets are the ICI shade reference numbers.

Series MM 1948 - about 1950

Romain green, with beige upholstery piped in green.
Platinum grey or black (122) with beige upholstery piped in brown.
Maroon (2473) with beige upholstery piped in maroon also available for a limited time during this period.

Series MM approximately 1950-2

Thames blue with beige Vynhide; except on four-door saloon with green upholstery, front seats in leather.
Gascoyne grey with beige Vynhide; except on four-door saloon with brown upholstery, front seats in leather.
Black (122) with beige Vynhide; except on four-door saloon with brown upholstery, front seats in leather. This trim colour appears to have changed to red on all other models after mid-1951.
Mist green with beige Vynhide; except on four-door saloon with green upholstery, front seats in leather. This body colour appears to have been added around mid-1951, for other than the four-door saloon.

Series II approximately 1953-5

Empire green (2653) with green upholstery.
Clarendon grey (2508) with brown upholstery, except on saloon models with red upholstery.
Black (122) with red upholstery.
Birch grey (2507) with red upholstery. This body colour appears to have been superseded about 1954 by smoke blue with red upholstery. In turn, sandy beige (6187) with red upholstery took its place in March 1955. Convertible hood from chassis 433571, mottled green for empire green models, others fitted with mottled red.

Minor 1000 approximately 1956-9

Dark green (5107) with grey upholstery.
Clarendon grey (2508) or birch grey with red upholstery.
Black (122) with red or green upholstery.
The hood material on the convertible model changed from double-duck to plastic material in February 1957, at chassis number 477961.

Minor 1000 approximately 1959-61

Porcelain green (5091) with beige upholstery.
Yukon grey (5505), black (122) or Old Engligh white (2379) with red upholstery.
Smoke grey (3301) or clipper blue (3300) with blue-grey upholstery.

Minor 1000 approximately 1962-8

Almond green (3483) with porcelain green upholstery, except on de luxe saloon with silver beige/porcelain green upholstery.
Rose taupe (6185), Old English white (2379), dove grey (3346) or black (122) with red upholstery, except on de luxe saloon with silver beige/tarton red upholstery.
Trafalgar blue (6189) with light blue upholstery.
Highway yellow with bluc-grey upholstery, except on de luxe saloon with silver beige/blue-grey upholstery.
Smoke grey with blue-grey upholstery on de luxe saloon.
The two-tone tartan upholstery on the de luxe saloon models appears to have been confined to the 1962-3 season.
Hood material on the convertible was pearl grey.

Minor 1000 approximately 1968-70

Almond green (3483) with green upholstery.
Black (122) snowberry white (3012), maroon (2473) or peat (6198) with red upholstery.
Trafalgar blue (6189) with light blue upholstery.
Smoke grey (3301).
Blue royale (5186) was probably used on late models.

Minor 1000000 1961

Lilac with white gold leather upholstery piped in black. Pale grey mottled roof lining. Cream wheels and radiator grille.

Minor traveller 1953-71

No data available on early models of the traveller, these may have had the same colours as the contemporary saloon models. Known colours and probable dates are:
Black (12) with red upholstery, 1953-71.
Smoke grey (3301) with blue-grey upholstery, 1959 onwards.
Old English white (2379) with red upholstery, 1959 onward.
Yukon grey (5505) with red upholstery, 1959-61.
Rose taupe (6185) with red upholstery, 1962 onward.
Almond green (3483) with green upholstery, 1962 onward.
Arianca beige (6190), 1966-7.
Limeflower (7968), 1971.
Aqua green (7932), 1970-1.
Glacier white (4309), 1971.
Bronze yellow (7681), 1971.
Harvest gold (BL number BLVC19), 1971.

$\frac{1}{4}$-ton 6/8cwt van and pick-up 1953-71

Initially available in beige, azure blue, dark green (5107) and platinum grey. Complete data not known. In addition to the usual practice of supplying commercials in primer for the customer's own livery, the following additional colours are known to have been available at some time:
Whitehall beige/Nevada beige (3305), probably 1960-7. Almond green (3483), probably 1962 onward.
White and Persian blue have also been recorded.
Vans supplied to the Post Office for telephone service were green up to 1968, then yellow. GPO postal vans red.

Morris Minor Engines	
Type USHM2	Four-cylinder, side-valve. 57mm bore. 90mm stroke. 918.6cc. 8.057hp. Compression ratio 6.5/6-7:1. SU 1⅛in diameter carburetter, type H1. 27.5bhp at 4,400rpm.
Type APHM	Four-cylinder, overhead-valve. 57.9 bore. 76.2mm stroke. 803cc. Compression ratio 7.2:1. SU 1⅛in diameter carburetter, type H1. 30bhp at 4,800rpm.
Type APJM or 9M	Four-cylinder, overhead-valve. 62.9mm bore. 76.2mm stroke. 948cc. Compression ratio 8.3/7.2:1. SU 1¼in diameter carburetter, type H2 on early models, H2S later. 37bhp at 4,750rpm.
Type 10M	Four-cylinder, overhead-valve. 64.5mm bore. 83.7mm stroke. 1,098cc. Compression ratio 8.5/7.5:1. SU 1¼in diameter carburetter, type HS2. 48bhp at 5,100rpm.

Morris Minor. Magazine Bibliography

Model	Subject	Source	Date	Pages
Minor	Model description	The Autocar	29/10/48	p1063, 1064
Minor	Model description	The Motor	27/10/48	p ....
Minor	Road Test	The Autocar	26/11/48	p1175-7
Minor	Road Test	The Motor	16/2/49	p58-60
Minor	Road Test	The New Outlook on Motoring	5/49	p36, 37, 39
Minor	Wiring diagram and data	Motor Industry	5/49	p160, 161
Minor	Detailed description	Automobile Engineer	8/49	p295-303
Minor	Model description	The Autocar	9/9/49	p926, 927
Minor saloon	Road Test	Motor Industry	10/49	p76, 77
Minor	Data Sheet 167	Motor Trader	8/3/50	Supplement
Minor saloon, 4-door	Model description	Motoring	12/50	p48, 49
Minor tourer	Road Test	The Autocar	30/6/50	p ....
Minor saloon, 4-door	Model description	Motoring	11/50	p36, 37
Minor saloon, 4-door	Model description	New Exchange	11/52	p14
Minor saloon, 4-door	Road Test	The Autocar	28/11/52	p1579-81
Minor	Wiring diagram & data	Motor Industry	4/53	p163, 164
Minor ¼-ton van & pick-up	Model description	News Exchange	8/53	p12, 13
Minor traveller	Model description	News Exchange	11/53	p8, 9
Minor	Wiring diagram & data	Motor Industry	2/54	p129, 130
Minor	Wiring diagram & data	Motor Industry	3/54	p161, 162
Minor traveller	Road Test	The Motor	5/54	p ....
Minor Series II	Data Sheet 220	Motor Trader	20/10/54	Supplement
Minor ¼-ton van & pick-up	Model description	News Exchange	5/55	p24
Minor MM	Overhaul data	The Autocar	18/11/55	p832-4
Minor MM	Supercharged saloon	The Autocar	22/7/55	p114-17
Minor, A40 engined	Road Test	The Autocar	16/12/55	p1001-3
Minor ¼-ton van & pick-up	Model description	News Exchange	12/56	p23
Minor 1000	Road Test	The Autocar	14/12/56	p897-900
Minor 1000	Wiring diagram & data	Motor Industry	Automobile Electricity	Data Sheet D643
Minor 1000	Road Test	The Autocar	24/6/60	p ....
Minor 1000	Road Test	Motoring	6/66	p23, 24
Minor	Servicing data	Motoring	8/66	p41-4

Morris Six. Specification

Series MS Saloon.   Production period October 1948 - March 1953.
Chassis numbers 501-12965.

Engine type VC22M, 6-cylinder, overhead-valve, overhead-camshaft, 73.5mm bore, 87mm stroke, 2,214.8cc. 4-bearing crankshaft. Pistons, 'Y' alloy wire-wound up to engine 12054, plain alloy from engine 12055, 3-ring. Compression ratio, 6.5/6.6:1. Cooling, thermo-syphon with impeller assistance, pressured system adopted late 1951. Carburetter, SU horizontal. Air cleaner and silencer fitted until chassis 11889, then AC oil bath type as fitted to all export models from introduction.

Petrol tank, rear mounted, 12 gallons. Electrics, 12-volt positive earth. SU electric petrol pump. Clutch, Borg & Beck, 9in diameter, dry-plate. Gearbox, 4-speed with synchromesh in second, third and top (ratios for middle gears differ after engine 2864), steering column gear-change lever. Steering, Bishop cam and lever, two-pin type, modified late 1951. Brakes, foot brakes Lockheed hydraulic, hand brakes by pistol grip and cable to rear wheels. 10in diameter brake drums. Independent front suspension with torsion bars, Armstrong hydraulic piston-type dampers up to chassis 7956, then Monroe or Girling telescopic hydraulic type. Semi-elliptic leaf springs at rear with Armstrong dampers up to chassis 4084, then Monroe or Girling type.
Commencing car number SMS9537 (home) and SMS6405 (export) a rear axle ratio of 4.555:1 (9/41) was fitted in place of the original 4:1 ratio (10/41).
Wheelbase 9ft 2in. Track, front 54in, rear 53in.

Morris Six. Body Colours and Upholstery

As introduced in 1948:	Romain green or platinum grey with beige upholstery and contrasting piping (upholstery listed as green in 1949). Maroon with beige upholstery and constrasting piping (upholstery listed as maroon in 1949). Black with beige upholstery and contrasting piping (upholstery listed as brown in 1949).
About 1950	Thames blue with breen trim, gascoyne grey with brown trim, mist green with green trim, or black with brown trim (this trim being listed as red in 1951).
About 1952	Birch grey with maroon trim, clarendon grey with maroon trim, empire green with green trim, or black with maroon trim.

Some paint finishes were cellulose, others synthetic or Synobel.

Morris Six. Magazine Bibliography

Subject	Source	Date	Pages
Model description	*The Autocar*	29/10/48	p1063, 1064
Model description	*The Motor*	27/10/48	p . . .
Model description	*The Autocar*	9/9/49	p927, 928
Wiring diagram and data	*Motor Industry*	1/50	p145, 146
Road Test	*The Motor*	16/8/50	p66-8
Road Test	*Motor Industry*	3/51	p92-4
Wiring diagram and data	*Motor Industry*	7/52	p181, 182
Data Sheet No 216	*Motor Trader*	30/6/54	

Morris Isis. Specification

Saloon   Production period, July 1955 - October 1956.
    Chassis numbers 501 - 7192.
Traveller.   Production period, July 1955 - October 1956.
    Chassis numbers 501 - 7192.
Series II saloon.   Production period, October 1956 - April 1958.
    Chassis numbers 7201 - 13614.
Series II traveller.   Production period, October 1956 - October 1957.
    Chassis numbers 7201 - 11963.

Engine, type C26M. Six-cylinder, 2,639.4cc overhead-valve. Four-speed gearbox with synchromesh in 2nd, 3rd and top. Steering column gearchange lever, until Series II models. Clutch, Borg & Beck, 9in diameter, single dry-plate, hydraulic operation. Brakes, Lockheed hydraulic, two leading shoes at front. Handbrake lever on right-hand side. 10in diameter drums until chassis 3588 on traveller, 4566 on RHD saloon, 4391 on LHD saloon, then 11in diameter drums. Fuel tank, 12 gallons on saloon, 10 gallons on traveller. SU electric petrol pump mounted at rear. 12-volt electrical system, positive earth. Steering, Bishop Cam (16:1), modified box introduced at chassis 3133 on RHD models and 4391 on LHD models. Independent front suspension, upper wishbones, lower links and torsion bars; system modified at chassis 5646. Half-elliptic leaf springs at rear. Hydraulic telescopic shock absorbers (reinforced bracket fitted at chassis 7201). Pressurised cooling system with fan, pump and thermostat. New type radiator grille and the addition of centre arm rests to front seats, circa July 1956. Tyre size, 6.00-15 on 5K × 15in four-stud steel disc wheels. Wheelbase 8ft 11½in. Track, front 53⅝in, rear 53in. Rear axle ratio (10/41) 4.1:1. Modified engine steady rod introduced at chassis 3618, modified front engine mounting at chassis 5446, modified axle casing and carriers at chassis 1457.

Series II vehicles had bonnet with fluted sides and a mascot added on front of bonnet. Rear wings were finned, and other changes included dished safety type steering wheel, lids on glove boxes, side position for gear change lever. Screen washer, wheel rimbellishers and courtesy light made standard equipment. Vehicles exported to North America fitted with sealed-beam headlamps and flashing indicators.

De luxe version of saloon fitted leather upholstery and heater as standard.

Optional extras on Isis included Borg-Warner overdrive from 1955. Duotone colour finish on Series II cars.

Morris Isis. Engine	
Type: C26M Low compression C26M/L Low compression C26M/H High compression	2,639.4cc. 6-cylinder overhead-valve, push-rod. 79.375mm bore, 88.9mm stroke. Aluminium alloy four-ring pistons. Compression ratio, low compression 7.25:1, high compression 8.3:1. Carburetter, SU semi-downdraught. Oil-bath type air cleaner. External oil filter. 86bhp at 4,250rmp. Four-bearing crankshaft. Modified camshaft at engine 2341, modified distributor housing at engine 3789, modified timing case at engine 4513, modified oil pump with three studs at engine 7122, and modified cylinder head at engine 7683.

Morris Isis. Body Colours and Upholstery

A complete list of colours used on the Isis is not available. Early models are known to have been finished in black, clarendon grey, empire green, and smoke blue. Sand beige with red upholstery substituted for smoke blue in early 1955.

From about May 1957 the Isis was available in duotone colour schemes. The two-colour upholstery listed was part nylon-cloth and part leathercloth. Allowing for the existing single colour range available at that date, these new colour schemes provided thirty-eight colour combinations:

Black top body colour with lower part swiss grey with red, grey, or green upholstery, or two-tone upholstery which combined pale beige with black, red or green. With red (or red two-tone) upholstery the carpets were crimson; grey or black upholstery had black carpets, and green had matching green carpets.

Turquoise top body colour with lower part Swiss grey with grey upholstery, or two-tone upholstery of black/pale beige. Black carpets.

Dark green top body with lower part island green with grey upholstery, or two-tone upholstery of black/pale beige. Black carpets.

Sage green top body colour with lower part twilight grey with grey or green upholstery and green carpets, or two-tone upholstery which combined green or black with pale beige. Carpets were green or black respectively.

Birch grey top body colour with lower part red with red or grey upholstery, or two-tone upholstery which combined pale beige with either red or black. Carpets were crimson in all cases.

Alternatively the Isis was available with the entire body painted as the top colour, or an all cream, or all clarendon grey finish. Upholstery colours were two-tone.

Morris Isis. Magazine Bibliography

Model	Subject	Source	Date	Pages
Saloon and traveller	Model description	*The Autocar*	15/7/55	p89-92
Saloon and traveller	Model description	*Motor Industry*	8/55	p107, 108
Saloon	Model description	*News Exchange*	9/55	p4, 5
Traveller	Model description	*News Exchange*	9/55	p6
Saloon	Road Test	*The Autocar*	27/4/56	p440-3
Saloon and traveller	Model description	*News Exchange*	11/56	Supplement
Saloon and traveller	Wiring and Test data	*Motor Industry*	Automobile Electricity	Data Sheet D656
Saloon	Model colour range	*News Exchange*	5/57	p16, 17

Oxford Series MO

Saloon, four-door. Production period October 1948 - January 1954. Chassis numbers between 501 and 160460.
Traveller 'Station Wagon', composite wood and metal body. Production period September 1952 - January 1954. Chassis numbers between 150594 and 176522.

Engine, 1,476cc, four-cylinder VS15M1001. Four-speed gearbox, steering column gear change, gear ratios modified about October 1949. Clutch, Borg & Beck single dry-plate. Brakes, Lockheed hydraulic with two leading shoes at front, 8in diameter drums. $9\frac{1}{2}$-gallon petrol tank. SU electric petrol pump. 12-volt electrical system, positive earth. Steering, rack & pinion. Armstrong double-acting, piston-type, shock absorbers at front. Telescopic double-acting hydraulic shock absorbers at rear after chassis 14832, previous models had Panhard-type anti-sway bar. Tyre size 5.25-15 up to June 1949, later models (home chassis 14944) 5.50-15. 5-stud steel disc wheels. Independent front suspension and half-elliptic leaf springs at rear. Wheelbase 8ft. Track 4ft 5in. Rear axle ratio, 4.55:1. Vertically split two-piece windscreen.

Painted, instead of chromium-plated radiator grille, at chassis 24354. A new fabricated chrome plated radiator grille was introduced on the saloon at chassis 117566 and on the traveller at chassis 120859. Twin stop/tail lamps at chassis 14832. Rear blind deleted from the specification at chassis 71373. Other changes about October 1951 include addition of tinted driving mirror, new type headlamps, Dunlopillo overlays on seat cushions, and stainless steel window channels. Arm rests added to rear doors at chassis 25100. Pressurised cooling system at chassis 18325.
Optional extras included HMV radio built into dash and Smiths interior heater.

Oxford Series II

Saloon, four-door. Production period January 1954 - October 1956. Chassis numbers between 161001 and 248341.
Traveller, metal body with wood finishing strips. Production period October 1954 - November 1956. Chassis numbers between 175523 and 248339.

Engine, 1,489cc, four-cylinder BP15M. Four-speed gearbox with synchromesh in all gears except first. Steering column gearchange lever. Clutch, Borg & Beck single dry-plate, 8in diameter, hydraulic operation. Brakes,Lockheed with two leading shoes at front 9in diameter drums. 12-gallon petrol tank on saloon, 10 gallon on traveller. SU electric petrol pump, HP, mounted at rear. 12-volt electrical system, positive earth. Steering, rack & pinion. Telescopic shock absorbers front and rear, with cooling fins. Tyre size, 5.50-15, to chassiss 232115 Dunlop Fort, 5.60-15 from chassis 232116 on 4in × 15in four-stud steel disc wheels. Independent front suspension and half-elliptic rubber mounted leaf springs at rear. Wheelbase 8ft 1in. Track, front 4ft $5\frac{1}{2}$in, rear 4ft 5in. Rear axle ratio, 4.875:1 (8/39).

Standard fittings included semaphore indicators for home models, flashing units for left-hand-drive export models; twin sun-visors single-piece windscreen, air intake on top of bonnet for heating and ventilating unit; two rear doors, vertically hinged, and timber finishing strips on traveller body.
Optional extras include Radiomobile car radio.

Oxford Series III

Saloon, four-door. Production period October 1956 - March 1959. Chassis numbers between 248401 and 300636.
Traveller, metal body with wood finishing strips. Production period October 1956 - June 1957. Chassis numbers between 248445 and 262129.

Engine, 1,489cc, four-cylinder BP15M. Four-speed gearbox with synchromesh in all gears except first. Steering column gearbox change initially, later floor mounted change optional. Clutch, Borg & Beck single dry-plate, 8in diameter hydraulic operation. Brakes, Lockheed hydraulic with two leading shoes at front, 9in diameter drums. 12-gallon petrol tank. 10 gallon on traveller. SU electric petrol pump, HP 12-volt electrical system, positive earth. Steering, rack & pinion. Tyre size 5.60-15 on steel disc wheels. Independent torsion bar front suspension and half-elliptic leaf springs at rear. Wheelbase 8ft 1in. Track, front 4ft $5\frac{1}{2}$in, rear 4ft 5in. Rear axle ratio, 4.875:1 (8/39), special 5.125:1 (8/41).

Fluted bonnet, finned rear wings, hooded headlamps, dished safety-type steering wheel. Additional body trim added to saloon at chassis 270007 to facilitate optional extra Duotone finishes in five combinations.
Optional extras included selective automatic two-pedal control.

Oxford Series IV

Traveller, all-steel body. Production period June 1957 - April 1960. Chassis numbers 265001 - 306517

Engine, 1,489cc, four-cylinder BP15M. Four-speed gearbox, steering column gear change. Clutch, Borg & Beck single dry-plate, hydraulic operation. Brakes, Lockheed hydraulic, 9in diameter drums. 11-gallon petrol tank with filler caps on both sides of body. Tyre size 5.90-15 on steel disc wheels. Independent front suspension, torsion bar, and half-elliptic leaf springs at rear. Telescopic dampers front and rear. Pressurised cooling system. Wheelbase 8ft 1in. Track, front 4ft 5½in, rear 4ft 5in. Rear axle ratio 4.88:1.

All-steel body with single large tailgate hinged at top. Bench type seats front and rear with centre arm rests. Rear seat folding flush with floor. One-piece curved windscreen. Optional extras included ratio, windscreen washers, rimbellishers, Duotone body colour schemes.

Oxford Series V

Saloon, four-door.   Production period January 1959 - October 1961. Chassis numbers between 101 and 79205.
Traveller.   Production period May 1960 - October 1961. Chassis numbers between 101 and 8427.

Engine, 1,489cc, four-cylinder BP15M. Four-speed gearbox with synchromesh in all gears except first. Steering column or floor mounted gear change optional. Clutch, Borg & Beck single dry-plate, 8in diameter, hydraulic operation. Brakes, Girling hydraulic, 9in diameter drums. 10-gallon petrol tank. SU electric petrol pump, type PD, at rear. 12-volt electrical system. Flashing indicators. Steering, cam & lever. Tyres, Dunlop 5.90-14 tubeless on saloon, 6.40-14 or 5.90-14HD on traveller. 14 × 4½ four-stud disc wheels. Independent coil spring front suspension and semi-elliptic leaf springs at rear. Armstrong piston-type lever arm shock absorbers. Wheelbase 8ft 3³/₁₆in. Track 4ft ⁹/₁₆in on saloon, 4ft ⅛in on traveller. Rear axle ratio, 4.55:1 (9/41) on saloon, 4.875:1 on traveller.

Pininfarina styled body. Traveller, all-steel body, with upper and lower hinged tailgate. Upholstery in leather with leathercloth on non-wearing surfaces. Traveller with full bench seats with facing of leather or woven fabric. Additional items on the de luxe saloon model included Smiths interior heater, screen washer, twin sun-visors, twin horns, bumper over-riders, clock, and leather covered seats.
Optional extras included Smith's Radiomobile car radio, wheel rimbellishers, and five Duotone finishes.

Oxford Series VI

Saloon, four-door.   Production period September 1961 - April 1971. Chassis numbers 79301 - 255327.
Traveller.   Production period September 1961 - February 1971. Chassis number 8501 - 41296.

Engine, 1,622cc, four-cylinder 16AMW, 16KAMW, 16AA, 16CA; on the saloon, BMC 1.5 litre diesel, available from January 1962. Four-speed gearbox, steering column or floor mounted gear change optional. Clutch, single dry-plate, 8in diameter, Girling hydraulic operation. Brakes, hydraulic, 9in diameter drums. 10-gallon petrol tank. SU electric petrol pump, type SP or AUF204. 12-volt electrical system, positive earth. Steering, cam & roller. Tyres 5.90-14, except on traveller up to car number 8739 which had 165-14SP. Wheel size 4J × 14. Independent coil spring front suspension and semi-elliptic rubber mounted leaf springs at rear. Armstrong hydraulic dampers. Wheelbase 8ft 4¼in. Track, front 4ft 2⅝in, rear 4ft 3⅜in. Rear axle ratio 4.3:1 (9/39) except on traveller up to chasis 8739 and diesel saloon which had 4.55:1 (11/41).

Heater optional extra on standard saloon. Three-speed Borg-Warner model 35 automatic transmission, optional.
Traveller, all-steel body, with upper and lower hinged tailgate. Windscreen wash and heater standard. Saloon, de luxe model fitted with heater, demister, and windscreen washer as standard.

Morris Oxford Engines
Type VS15M1001
1,476.5cc. Four-cylinder, side-valve. 73.5mm bore, 87mm stroke. Compression ratio 6.8:1. SU carburetter. Tecalemit oil filter fitted after about October 1949. 40.5bhp at 4,200rpm.
Type BP15M
1,489cc. Four-cylinder, overhead-valve, push-rod. 73.025mm bore. 88.9mm stroke. Compression ratio 7.43:1 to engine BP15M21644, later alternative BP15M & BP15ML 7.15:1, or BP15MH 8.3:1. Three-bearing crankshaft. Aluminium alloy concave-top, split-skirt, four-ring pistons. SU carburetter semi-downdraught M2. Early models AC by-pass oil filter, later models full-flow Tecalemit or Purolator oil filter. Series II Oxford, 50bhp at 4,800rpm. Series III Oxford, 50bhp at 4,200rpm. Series IV Oxford, 55bhp at 4,000rpm. Series V Oxford, 55bhp at 4,350rpm.
Type 16AMW, 16KAMW, 16AA, 16AC
1,622cc. Four-cylinder, overhead-valve, push-rod. 76.2mm bore. 88.9mm stroke. Compression ratio, high 8.3:1, low 7.2:1. SU carburetter 1¼in semi-downdraught, type HS2. 61bhp at 4,500rpm.
BMC Diesel 1.5 litre
1,489cc. Four-cylinder, overhead-valve. 73mm bore. 88.9mm stroke. Compression ratio 23:1. CAV type DPA fuel-injection pump.

Morris Oxford. Body Colours and Upholstery

After the formation of the British Motor Corporation in 1952, the once detailed colour description ceased to be quoted in catalogues, making a comprehensive list of finishes impossible. The early Series MO Oxfords are fairly well recorded, although variations appear to have been fairly frequent. Initially, the finish choice was romain green with beige upholstery piped in green (green carpets), and platinum grey or black with beige upholstery piped in brown (brown carpets). Shortly afterwards a maroon version was being listed with maroon upholstery. By 1950 all but black had been superseded and colours then listed were Thames blue or mist green with green upholstery, and gascoyne grey with brown upholstery — all the roof interior trimmed with leathercloth to match the exterior. The wings on Oxford cars were the same colour as the body. In 1951 black was being listed with maroon upholstery (this change of trim colour may have been introduced earlier) and in place of the earlier colours the Oxford was now offered as clarendon grey with maroon upholstery, birch grey sharing the same maroon trimming, and empire green with a green interior.

Standard colour schemes on the Series II Oxfords appear to have continued with black or clarendon grey with maroon upholstery, birch grey again with maroon interior, and empire green with green upholstery. Another possible shade was smoke blue but cars finished thus would not have been produced for any length of time for early in 1955 this blue was superseded on all Morris cars by a new finish, sandy beige. Maroon upholstery normally complemented this new one. Research suggests that Series II Oxfords were also available in dark green and sage green.

When the Series III Oxfords were introduced in 1956 the sales literature gave little information other than a brief mention of a 'choice of seven standard colours'. A clue to what these may have been is to be found in the lists of touch-up paints made for the Oxford by firms such as Dupli-Colour and Unipart, these include black, empire green, clarendon grey, birch grey, sandy beige, dark green (otherwise known as Connaught green), sage green, and island green. The later introduction of additional bright trim on the sides of the Series III Oxfords allowed for optional-extra five duotone colour schemes from about June 1957, these are well documented and account for more than the seven shades quoted above. Duotone came as:

Black top body colour with Swiss grey on the lower part. Upholstery was red with crimson carpets, grey with black carpets, or green with green carpets.
Turquoise top body colour with Swiss grey on the lower part. Upholstery was grey on the lower part. Upholstery was grey with black carpets.
Dark green top body colour with island green on lower part. Upholstery was grey with black carpets.
Sage green top body colour with twilight grey on the lower part. Upholstery was grey or green with green carpets.
Birch grey top body colour with red on the lower part. Upholstery was red or grey with crimson carpets.

Less information is recorded when considering subsequent models. The Series IV traveller is known to have included a green/grey shade as well as optional duotone finishes. A choice of six standard colours or three optional duotone schemes were offered on the Series V cars. During the 'sixties the more popular shades used on Morris cars were maroon (including a slightly different shade called 'Maroon B'), dark or Connaught green, trafalgar blue, almond green, and dove grey. Glacier white, damask red, cumulus grey, blue royale, and El Paso beige figured among the newer shades in use on Morris vehicles in the early 'seventies.

Morris Oxford. Magazine Bibliography

Model	Subject	Source	Date	Pages
Series MO Saloon	Model description	The Motor	27/10/48	p....,
Series MO Saloon	Model description	The Autocar	29/10/48	p1063, 1064
Series MO Saloon	Road Test	The Motor	29/11/48	p480-2
Series MO Saloon	Road Test	The Autocar	24/12/48	p1271-3
Series MO Saloon	Road Test	Motor Industry	4/49	79,80
Series MO Saloon	Road Test	Motoring	5/49	p32-3
Series MO	Equipment & data	Motor Industry	5/49	p163, 164
Series MO Saloon	Model description	The Autocar	9/9/49	p927
Series MO	Service data sheet, No 162	The Motor Trader	19/10/49	Supplement
Series MO Traveller	Model description	News Exchange	10/52	p4
Series MO Traveller	Road Test	The Autocar	6/2/53	p169-71
Series MO Saloon	Model description	The Autocar	24/12/54	988-91
Series II Saloon	Equipment & data	Motor Industry	7/53	p173, 174
Series II Saloon	Model description	The Autocar	21/5/54	p727-31
Series II Saloon	Model description	Motor Industry	6/54	p101-3
Series II Saloon	Model description	News Exchange	7/54	p12, 13
Series II Saloon	Road Test	The Motor	29/9/54	p292-5
Series II Saloon	Model description	The Autocar	16/7/54	p727-31
Series II Saloon	Road Test	The Autocar	23/7/54	p117-20
Series II Traveller	Model description	The Autocar	15/10/54	p564
Series II Traveller	Model description	Motor Industry	11/54	p73, 74
Series II Traveller	Model description	News Exchange	11/54	p6
Series II Traveller	Service data sheet, No 239	The Motor Trader	21/9/55	Supplement
Series III	Servicing data	Motoring	9/66	p41-4
Series III Saloon	Model description	Motor Industry (Motor Show Issue) 1956		p166, 168-71
Series III Saloon	Model description	News Exchange	11/56	p12
Series III Saloon	Duotone Colour range	News exchange	5/57	p16
Series IV Traveller	Model description	Motor Industry	9/57	p77, 78
Series IV Traveller	Model description	Motor Industry (Motor Show Issue) 1957		p126
Series V Saloon	Model description	Motor Industry	4/59	p72
Series V	Servicing data	Motoring		53-5
Series V Traveller	Model description	Motor Industry (Motor Show Issue) 1960	11/66	p53
Series V Saloon	Road Test	The Motor	4/5/60	p511-14
Series V Saloon	Equipment & data	Motor Industry	Automobile Electricity	Data Sheet D653
Series VI Saloon	Road Test	Motoring	7/64	p34, 35
Series VI Diesel Saloon	Model description	Motor Industry	12/62	98-100
Series VI Saloon	Road Test	Motoring	7/64	p47
Series VI Diesel Saloon	Model description	Motoring	4/65	p30, 31

Cowley & Commercial. Specifications

Cowley

Saloon, four-door.   Production period July 1954 - September 1956. Chassis numbers 501 - 17903.

Engine type 12M. Four-cylinder overhead-valve, push-rod. 65.48mm bore, 88.9mm stroke. 1,200cc. 42bhp at 4,500rpm. Compression ratio 7.2:1. Aluminium alloy, concave crown pistons, 4-ring, split skirt. 3-bearing crankshaft. Inlet valve 1$\frac{3}{8}$in diameter. Exhaust valve 1$\frac{3}{16}$in diameter. SU carburetter, H2, semi-downdraught. AC oil bath air cleaner.

Four-speed gearbox with synchromesh in all gears except first. Steering column mounted gear-change lever. Clutch, Borg & Beck, single dry plate, 7$\frac{1}{4}$in diameter, hydraulic operation. Footbrakes, Lockheed hydraulic with two leading shoes at front. Handbrakes by cable to rear wheels, 9in diameter drums from chassis 6349, 8in diameter drums up to chassis 6348. Petrol tank, 12 gallons. 12-volt electrics, positive earth. SU electric petrol pump type HP. Steering, rack & pinion. Independent front suspension with torsion bar and hydraulic telescopic dampers, half-elliptic leaf springs at rear. Tyres, 560-15 Dunlop on 4in × 15in steel disc wheels, 5-stud. Wheelbase 8ft 1in. Track, front 4ft 5$\frac{1}{2}$in, rear 4ft 5in. Rear axle ratio 5.125:1 (8/41).

Equipment included a driver's sun-visor and twin self-parking windscreen wipers, upholstery in leathercloth, 49in wide front bench seat with foam rubber interior, single piece front windscreen, combined arm rests and door pulls at rear. Heater extra, but the necessary ducts were fitted.

Cowley 1500

Saloon, four-door.   Production period September 1956 - March 1959. Chassis numbers 18001 - 22633.

Engine type BP15M. Four-cylinder overhead-valve, push-rod. 73.025mm bore. 88.9mm stroke. 1,489cc. Aluminium alloy, concave crown pistons, 4-ring, split skirt. 3-bearing crankshaft. SU carburetter, semi-downdraught.

Four-speed gearbox with synchromesh in all gears except first. Steering column mounted gear-change lever. Clutch, Borg & Beck, single dry-plate, 8in diameter, hydraulic operation. Footbrakes, Lockheed hydraulic with two leading shoes at front. Handbrake by cable to rear wheels. 9in diameter drums. Petrol tank 12 gallons. 12-volt electrics, positive earth. SU electric petrol pump type HP. Steering, rack & pinion. Independent front suspension with torsion bar and hydraulic telescopic dampers, half-elliptic leaf springs at rear. Tyres, 560-15 Dunlop on 4in × 15in steel disc wheels, 5-stud. Wheelbase 8ft 1in. Track, front 4ft 5$\frac{1}{2}$in, rear 4ft 5in. Rear axle ratio 4.875:1 (8/39).

Equipment included safety type steering wheel, finned rear wings, hinged windows in front doors, lids on glove boxes. The Cowley was similar to the Oxford, but over-riders and trim flashes on the front wing crowns were not fitted to the Cowley models.

Cowley SMCV, 10cwt

Van and pick-up truck.   Production period May 1950 - September 1956. Chassis numbers between 501 - 44098.

Engine type VS15MV. Four-cylinder side-valve. 73.5mm bore. 87mm stroke. 1,476.53cc. 3-bearing counterbalanced crankshaft. Aluminium alloy pistons, 3-ring. External oil filter. 40.5bhp at 4,000rpm. Cooling, centrifugal water impeller with thermostat control. SU carburetter, 1$\frac{1}{4}$in diameter, type H2. Air silencer on home models, oil-bath type air cleaner on export models.

Four-speed gearbox, synchromesh on all gears except first. Steering column mounted gear-change lever. Clutch, Borg & Beck single dry-plate, 8in diameter. Footbrakes, Lockheed hydraulic. Handbrake pistol-grip type by cable to rear wheels. 9in diameter drums. Petrol tank 12$\frac{1}{2}$ gallons. Electrics, 12-volt, positive earth. SU electric petrol pump, type L. Steering, rack & pinion. Independent front torsion bar suspension with Armstrong, Woodhead-Monroe or Girling hydraulic telescopic dampers. Tyres, 600-15 Dunlop, on disc-type steel wheels, 5-stud. Wheelbase 8ft 1in.

On van a single bucket seat as standard. Additional bucket seat or a bench type seat extra. Standard to this model was a split windscreen, and rubber floor covering driver's cab floor. Van body capacity 120cu ft.

Morris $\frac{1}{2}$-ton, Series III

Van.   Production period October 1956 - October 1960. Chassis numbers 44201 - 60430.
Pick-up.   Production period October 1956 - October 1960. Chassis numbers 44201 - 60377.

Engine type BP15ML, B Series. Four-cylinder overhead-valve, push-rod. 73.025mm bore. 88.9mm stroke. 3-bearing crankshaft. Aluminium alloy pistons. 1,489cc. 50bhp at 4,200rpm. SU carburetter.

Four-speed gearbox with synchromesh on all gears except first. Steering column mounted gear-change lever on early models, later floor mounted lever. Clutch, Borg & Beck single dry-plate, 8in diameter, hydraulic operation. Footbrake, Lockheed hydraulic with two leading shoes at front. Handbrake by lever on right-hand side, cable to rear wheels. 10in diameter drums. Petrol tank 9⅞ gallons. Electrics, 12-volt, positive earth. SU electric petrol pump type HP. Steering, rack & pinion. Independent front suspension, torsion bar, hydraulic telescopic dampers. Half-elliptic leaf springs at rear. Tyre size 600-15 on 5in × 15in ventilated disc wheels. Spare wheel mounted under number plate panel. Wheelbase 8ft 1in. Track, front 4ft 6⅜in, rear 4ft 6½in. Rear axle ratio 5.125:1 (8/41).

Central horizontal grille bar curved to extend to side lights. One piece windscreen. Tool roll strapped to right-hand door pillar. Capacity of van 120cu ft plus another 18cu ft if passenger seat not fitted. Optional extras included windscreen washer, heater, canvas tilt on pick-up.

Morris ½-ton (Morris version of Austin A55 type, A/HV6, A/HK6 & A/HQ6, ½-ton commercial)

Van M/HV4 and pick-up M/HK4.   Production period October 1962 - circa 1971. Chassis numbers 179301 - . . .

Engine type 15AC, B Series, 73.025mm bore. 87mm stroke. 3-bearing crankshaft. 1,489cc. Compression ratio 7.2:1 (alternative 8.3:1). 47bhp at 4,100rpm. Carburetter, single Zenith downdraught.
Later engine (from 1964) type 16AC, B Series. 1,622cc. Carburetter SU. Oil bath air cleaner on export vehicles. Four-speed gearbox, steering column gearchange lever. Clutch, single dry-plate, hydraulic operation. Steering cam and peg. Independant front suspension by coil spring and wishbones. Rear suspension, semi-elliptic leaf springs with Armstrong hydraulic lever shock absorbers and stabilising bar. Drum brakes: foot, hydraulic to rear wheels; handbrake pistol-grip by cable to rear wheels. 12-volt electrics. Steel disc wheels, spare wheel housed in wind-down tray beneath rear floor. Single-piece windscreen. Wheelbase 8ft 7½in. Van capacity 110cu ft. All-steel mono-construction, with sides double skinned up to waist level with corrugated interior panels.

Morris Cowley. Magazine Bibliography				
Model	Subject	Source	Date	Pages
Cowley van & pick-up	Model description	*The Commercial Motor*	19/5/50	p458, 459
Cowley van	Model description	*Motor Industry*	6/50	74, 76
Cowley van	Road Test	*Motoring*	7/50	p35
Cowley van & pick-up	Model description	*News Exchange*	7/50	p16, 17
Cowley van & pick-up	Wiring diagram & data	*Motor Industry*	4/52	p159-60
Cowley pick-up	Road Test	*The Commercial Motor*	16/4/54	p302-4
Cowley saloon	Model description	*The Autocar*	16/7/54	p85-9
Cowley saloon	Model description	*News Exchange*	8/54	p12, 13
Cowley saloon	Model description	*Motor Industry*	8/54	p91, 92
Cowley saloon	Wiring diagram & data	*Motor Industry*	1/55	p141, 142
Cowley	Data Sheet No 239	*Motor Trader*	21/9/55	Supplement
Cowley saloon	Road Test	*The Autocar*	29/4/55	p587-90

The Mini Era

Alexander Issigonis' return to the British Motor Corporation in 1956, after his spell with Alvis, coincided with the Suez crisis. Colonel Gamal Abdel Nasser, the Egyptian president, nationalised the Suez Canal and the following events interrupted the flow of oil to Western Europe and compelled Great Britain, amongst others, to introduce petrol rationing — restricting the private motorist to something like ten gallons a month. Against this political background Leonard Lord, BMC chairman, gave Alec Issigonis a brief to produce a design of a small car with big-car performance, a miniature of the Minor with the maximum of space.

Issigonis gathered around him a close-knit team of some nine people including project engineer Jack Daniels who shared Issigonis' enthusiasm for suspension systems, Charles Griffin from Cowley, and body-jig-shop men John Sheppard and Vick Everton. Ron Davey was to build the experimental bodies from Dick Gallimore's layouts, checked by George Cooper. By March 1957 Issigonis' ideas had taken shape and Leonard Lord gave the go-ahead for the new vehicle which the team were unofficially calling the 'Austin Newmarket' — a reference to the new markets that they saw the car aimed at.

Before leaving BMC in 1952, Alec Issigonis had been working on a front-wheel-drive Morris Minor with a transverse mounted engine and gearbox which Jack Daniels had subsequently tested while Issigonis was at Alvis. The results were extremely satisfactory, but nothing further was done and the prototype (registered TFC717) was destroyed. It has also been recorded that in 1955 a Morris Minor fitted with all-round rubber suspension underwent a 1,000-mile test at the MIRA test track; presumably this was the same vehicle. This successful configuration obviously figured in the thinking behind the design of the Mini, which went through various experimental stages including an attempt to use half an 'A' Series engine, which in two-cylinder form proved too rough and was quickly abandoned. Problem areas were the constant-velocity joints which had to be capable of wide angles of change when steering, and the tiny 10in diameter tyres that Issigonis demanded. One of the major drawbacks with the experimental front-wheel-drive Minor had been the unacceptably large size of the constant-velocity joints and hub bearings used, but now the answer was found in a thirty-year-old invention, by Czechoslovak Hans Rozeppa, being made in Yorkshire for submarine conning-tower control gear by the British patent owners, Unipower of Shipley. Dunlop had originally come to the rescue when the (then) small 14in diameter tyres were required for

1961 Morris Mini Minor saloon. (British Leyland)

the Morris Minor. Alec Issigonis now required not only small tyres to withstand high rotational speeds, but tyre life that was not obtained at the expense of grip. Security of the tyre, on a rim with a beading length 12in shorter than on the Minor tyre, was just one further problem with which the Dunlop engineers had to contend. Eventually, using a much modified German two-cylinder Goggomobil as a test bed, they evolved a suitable cross-ply tyre.

By October 1957 the design team had two prototypes on the road. They were powered by the 'A' Series 948cc four-cylinder overhead-valve engine as used in the contemporary Minor 1000, mounted transversely with the radiator on the off-side. During the next nine months these two cars (camouflaged with existing Austin grilles and with the rear made to look like an Austin A30) were subjected to hard evaluation tests which included covering 30,000 miles in 500 running hours over appalling surfaces provided by a disused airfield at Chalgrove and on a long circuit in the Cotswolds. The 948cc car proved amazingly fast, reaching more than 90mph, compared with the contemporary Fiat with a 58mph maximum. This speed, combined with uneven weight distribution, meant that under very hard braking the rear wheels locked easily. Modifications increased the body width by 2in, repositioned the battery to the boot and introduced a rear-brake limiter valve. The most radical modification during this period of development was to reduce the engine capacity to 848cc and turn it through 180° which put the radiator on the nearside and gave a more sheltered position behind the engine block for the SU carburetter which previously had been prone to icing-up. As a result of turning the engine on its axis the Mini now had four reverse gears and one forward! This unacceptable arrangement was corrected by the introduction of three gears in the transmission chain.

In July 1958 Leonard Lord took a five-minute drive around the Longbridge site in a prototype, nicknamed the 'Orange Box' after its colour, and immediately gave Issigonis his decision; 'Have it in production within twelve months.' Almost twelve months later, on 26 August 1959, project 'ADO15', the Mini, was introduced.

The Morris Mini Minor Saloon and Austin Se7en Saloon caused a major sensation in the motor trade (although readers of the European magazine *L'Auto Journal* had already seen a scoop protograph published earlier in 1959). Its revolutionary features included conical rubber suspension elements giving a highly desirable variable-rate springing, designed by Alex Moulton of Moulton Developments Ltd. This was a company formed in 1956, at the instigation of Sir Leonard Lord, to develop such ideas for exclusive use by BMC. It was independent of any other firm and comprised a small development unit at Bradford-on-Avon under the direction of Alex Moulton, who was later to become a household name, when he introduced an equally revolutionary small-wheel bicycle using the same springing technique.

Whoever coined the nickname 'Orange Box' for the prototype Mini could not have realised that Outspan, the orange distributors, would use the basic Mini engine and running gear, some fourteen years later, to provide a fleet of six advertising vehicles made in the shape of giant oranges. These fibreglass oranges designed and built by Brian Waite Engineering of Flimwell, East

The 1965 'Drivable Orange', one of six such vehicles, based on Mini running gear, used by Outspan for promotional purposes. (The Outspan Organisation)

Sussex, were fitted to special 48in wheelbase space-frame chassis. The sub-frame at the front contained the engine, automatic gearbox, and suspension, while at the back, in addition to the trailing arm suspension, were supports for the 200lb of ballast found necessary to maintain stability — a stability that passed the 39° limit of London Transport's test rig. Steering involved the transmission of the turning movement through three universal joints and 114° to a rack and pinion mechanism giving the 'Drivable Orange' an impressive turning circle of just 16½ft. The Mini was also used by Duckhams who added a body in the shape of a two gallon can of oil. These special shapes revived an advertising trend from the earliest days of motoring when shoes, beer, bottles, radios, tyre sections, and even vacuum cleaners were to be seen in towns and cities.

Of course the Mini was not the first vehicle to introduce front-wheel-drive from a transverse mounted engine, and rubber suspension. At the turn of the century a contraption known as 'The Victoria Combination', or alternatively 'The Eureka', had a cross mounted air-cooled 2¼hp De Dion engine which drove the front wheels through a two-speed gearbox. Rubber suspension was first used for shock absorption on transport vehicles as early as 1852 when Rothwell & Co of Bolton, constructed a broad-gauge railway locomotive (which later recorded a speed of 81.8mph) where the engine was entirely suspended on rubber. A good example of the application of rubber as a suspension medium was the Harris-Leon-Laisne. This car was in production for ten years to 1937, and the French company turned out cars using various proprietary engines in order to demonstrate the merits of their rubber suspension system. One such vehicle was the 1932 Hotchkiss engined front-wheel-drive car built on a frame of heavy-gauge tubing with independent suspension all round. Each wheel was mounted on a radius arm so that the vertical movements of the wheels under road shocks were converted into horizontal thrusts up by discs working in rubber filled cylinders. In America, in 1904, Walter Cristie's lumbering front-wheel-drive monster, which ran in the Vanderbilt Cup races, boasted a transverse engine and independent front suspension. In Germany in the early 'thirties, the DKW F1 490cc and F2 584cc cars were designed by Rudolph Slaby. These small cars had a parallel-twin two-stroke unit fitted east-to-east with the gearbox and driveshafts mounted in front of the engine, allowing power to be transmitted via a duplex chain. Perhaps the nearest anyone got to the Mini, as we know it, was a concept described by Laurence E.W. Pomeroy, technical editor of *The Motor,* in an article he wrote for that magazine in February 1939. By coincidence he named his unconventional design the 'Minicar', and envisaged a transverse mounted four-cylinder 600cc watercooled engine with the front wheels driven by a chain from a gearbox tucked in behind the power unit.

'Mini' gave a new meaning to the abbreviation in the English language and a new word in other languages. A car that had originally been aimed at the housewife for shopping and taking the children to school, or the district nurse on her rounds, suddenly made nonsense of class distinction and became equally familiar in the forecourts of royal palaces or parked outside terraced houses.

*Van versions of the Mini Minor were introduced in June
1960, the month this example was registered in
Buckinghamshire.*

As a direct result of the fine summer of 1959, during which time the pre-production prototypes underwent their final tests, one major fault was overlooked. Thousands of proud owners of the new Minis found that in the following wet winter the carpets began to rot. This was caused by the ingress of water where a stiffener had been lapped the wrong way over a join in the floorpan. As a temporary solution, before modified metalwork could be introduced, each car as it came off the line was diverted to a special bay where the sills were drilled and injected with expanding plastic foam. Confirmation of the success of this stop-gap remedy came fifteen years later when Longbridge engineers recovered some of the early cars and found the metal inside the treated sections still bright.

At Longbridge the specially built assembly line in 'Car Assembly Building 1' turned out the Austin Se7en saloon (a name that the Austin version carried until January of 1962 when it became the Austin Mini) while the Morris Mini-Minor was assembled in the quarter-mile long 'E' block at Cowley. Inevitably, a $\frac{1}{4}$-ton van derivative was introduced, in June 1960, and as no purchase tax was payable on these commercials the basic price was a mere £360. There followed, shortly afterwards, brake versions known as the 'Countryman' when carrying the Austin label, and 'Traveller' when sold as a Morris. Colour choice depended on which badge the estate car carried. The Austin was speedwell blue, tartan red, or farina grey; while the mini traveller was white instead of grey, and shades of red and blue which differed to those of the Austin. Unlike the Morris Minor traveller, the ash frame had no structural importance, being simply wood battens attached to the body pressings; and even these disappeared on the all-steel variation dispatched to export markets from April 1961, and eighteen months later were sold in the United Kingdom. The next variation was the pick-up, which was one of the surprises at the International Geneva Automobile Exhibition in March 1961, before being seen in Britain.

The Mini was the ultimate in 'badge engineering'. Even the pretence at difference between Morris and Austin versions (exemplified by the Pininfarina Oxford where the Austin version had bucket seats at the front, and smaller details like the button for the horn, while the Morris differed by having a full width bench seat and a horn ring on the steering wheel) had gone, and now it was simply a case of changing the badge. Indeed there are recorded examples of cars carrying the Morris name at one end and Austin at the other! In addition there were the short-lived attempts at aiming for a more sophisticated market by restyling the front and incorporating a traditional radiator grille. In October 1961 the Riley Elf Mark I and Wolseley Hornet Mark I appeared in Duotone colour schemes, a lengthened boot, and wood veneer facia. Both models were replaced by the Mark II, 998cc engined versions in November 1962; then the Mark III in October 1966. By August 1969 both variations were withdrawn after a total of 30,912 Rileys and 28,455 Wolseleys had been produced.

Modifications and styling changes were made to the Mini range as time went on. One was the padded parcel shelf and the improved window catches of October 1960. The basic instrumentation (consisting of a speedometer incorporating a fuel gauge, and warning lights for ignition, main beam and oil pressure) had additions on the super and de luxe cars in 1961 when oil pressure and water temperature gauges appeared, together with a key-operated ignition/starter switch in place of the floor-mounted starter button (an idea, together with the Mini's sliding windows, borrowed from the 1929 Morris Minor) and a fresh air heater was offered as an optional extra. In the same year the painted front grille on the basic Mini was replaced with a plated one, behind which was a sixteen blade fan, introduced to reduce engine noise.

The first of the major changes came in September 1964 when all saloon models were fitted with the Hydrolastic suspension system, which had proved its worth in the contemporary 1100 models, where rubber springs combined with a water/alcohol-filled system linked front and rear suspension. Alex Moulton the inventor had years before worked with Alec Issigonis to incorporate this suspension innovation in a prototype 3-litre V8 Alvis of integral construction — a project that was abandoned because the high cost of capital equipment was prohibitive. During the experimental period with Hydrolastic suspension for the 1100, about half-a-dozen Morris Minor 1000s with independent suspension all-round were built for testing; they looked normal enough and one was driven as far as Monte Carlo. Mini vans and estate cars retained the rubber-cone springs and subsequently, due to the high cost of Hydrolastic, all models reverted to the rubber-cone arrangement in 1969. Other changes to the saloons in September 1964 included an improved gearbox with a diaphragm spring clutch, an increase in braking efficiency, safety mirror and sun-visors, oil-filter warning light, and a courtesy light. All saloons by then had the key-type starter switch.

A second major change came around about Motor Show time in October

*The new style grille and the increase in size of the rear
window were two features of the Mark II Morris Mini that
appeared in October 1967. The super de luxe model shown
here had the 998cc engine and automatic transmission was
an optional extra.* (BMC Ltd)

1967 when the new Mark II Austin and Morris, de luxe 998cc and standard
848cc saloons had a complete face-lift to stimulate interest. Most changes
affected the appearance, for example the new style grille, larger rear lamps, an
increase in the size of the rear window, and improved facia and upholstery
trim. Wider use was made of the remote control gear change (only the standard
848cc car now had the long gear-change lever) and innovations such as self-
parking wipers and a new audio flashing unit were introduced. There had been
a tragic freak accident in Solihull the previous year when a boy had been
injured by the forward facing door handle and the factory had immediately put
safety bosses under the leading edge of the lever action door handles; the Elf
and Hornet variations fitted push button handles later. Rauno Aaltonen, the
'Flying Finn', demonstrated the problem in a less dangerous way in the RAC
Rally of 1966. The door handle of his Mini neatly hooked the loop on the
timing clock and he unknowingly drove away from the control with the
instrument. Although the remaining competitors were timed with ordinary
stop-watches, there were so many complaints of discrepancies that the
organisers were forced to cancel the stage. The longer wheelbase countryman
and traveller still continued to be offered in all-steel or wood trimmed form,
but now with the 998cc engine only; the Mini van and pick-up had the larger
engine as an option.

Well over two million Minis (the millionth had been reached in the early
weeks of 1965) had left the production lines by October of 1969 when it was
announced that the named Austin and Morris variations were to be dropped

and henceforth all cars would simply be called the 'Mini' and carry the BL symbol. This was an attempt to tidy up the untidy dealer network inherited by British Leyland. Separate franchises for the six companies that had been British Motor Holdings still existed, together with their independant distribution chains and separate departments at head office. In addition to the Mini 850 and Mini 1000, the range now included the Mini Clubman two-door saloon, Mini Clubman Estate with the 998cc engine and drum brakes, and a Mini 1275GT two-door saloon powered by the 12H 1,275cc engine. The original conception of the Mini as a simple utilitarian body shell with a wheel at each corner and simple instrumentation and fittings, had now evolved with changes and optional extras that gave wind-up windows, a larger bonnet, bigger doors fully trimmed, face level fresh-air ventilation ducts, concealed door hinges, more comfortable seating, automatic transmission, adjustable rake front seats, electrically heated rear windows and, in the case of the 1275GT, Rostyle 4$\frac{1}{2}$in wide wheels, and disc brakes at the front.

Eventually an alternator, inertia-reel seat belts, car heater, laminated windscreen, passenger sun-visor, chrome door mirrors, Denova safety tyres, cloth seat trim, twin stalk controls, hazard lights, and larger pedals were among the items to become standard or optional extras on some or all Minis. What had been the Morris Mini-Minor priced at just under £497 in 1959 had become the Mini 1000 City, 1000HL, or the £3,761 1000HL Estate in 1981. By 1983 the Mini Mayfair was being offered with automatic transmission, wide-rim alloy road wheels and arch extensions, reclining seats, push-button radio, Raschelle knitted fabric seat trim, and tinted glass — a far cry from the original concept.

Before the Mini was many months old its potential in motor sport had been demonstrated. One of the first of the racing fraternity to realise this potential

Paddy Hopkirk's modified Mini Cooper S showing its paces in competition. (British Leyland)

was John Cooper who, together with Alec Issigonis, persuaded BMC's director George Harriman to let him have a Mini with which to experiment. Cooper contacted the Lockheed Hydraulic Brake Co Ltd, at Leamington Spa, where, as an engineering exercise the firm had been working on a disc brake system for Mini size wheels, and they arranged for Cooper's Mini to be fitted with 7in diameter disc brakes at the front. By careful tuning the maximum speed of the car was raised by 20mph and without much more effort the time taken to accelerate to 70mph was halved. The Mini, with twin SU carburetters and many other Cooper modifications, was returned to George Harriman within a fortnight and it was agreed that some of these specials would be made. Despite Harriman's opinion that 1,000 such models would never sell, John Cooper (who had suggested this figure to make the 'special' acceptable for competitions) had his way and the Mark I Austin or Morris Mini Cooper with a 997cc engine was available from July 1961.

The Cooper engine '9F' gave 55bhp at 6,000rpm when altered to include twin SU HS2 carburetters, a modified cylinder head with larger diameter valves, and a 9:1 compression ratio. A redesigned exhaust system was fitted and valve overlap at the top of the stroke had been increased from 15° to 37°, in conjunction with double valve springs. So as to avoid torsional vibrations (the major cause of trouble when attempting to tune the standard engine) the crankshaft diameter was increased adjacent to the out-rigged flywheel and the crank webs thickened. Attention to the gearbox involved the fitting of a remote control lever and the gearbox ratios were altered slightly, although the top gear remained at 3.765:1, giving a speed of 88.9mph at peak revs. Many detailed points were given Cooper's attention, such as a modified distributor and coil caps to prevent ingress of water, re-designed seats and interior trim, replacement of the cord by proper door handles, sound deadening material in the engine compartment, and carpets which were carried through to the boot which had been fitted with a separate floor. Changes in instrumentation added a watertemperature gauge and oil pressure gauge, but, surprisingly, no tachometer. To withstand the increased speed, Dunlop nylon ply 'Gold Seal' tyres were specified.

At the Montlhery Track one of the early production versions was put through its paces, running for 1,010 miles at an average speed of 84.27mph, when driven by BMC's chief experimental engineer G. Jones and carrying three members of the Experimental Department.

An additional development called the Mini Cooper S, carrying either the Austin or Morris name and a special motif on the front grille, was announced in March 1963. This newcomer was designed primarily for competition work with its highly-tuned over-square '10F' engine of 1,071cc, but because it did not fit into the 1964 racing classification it was discontinued. In the same year (1964) the Cooper S appeared with the option of the '12FA' 1,275cc or '9FC' 970cc engine, although by 1965 the smaller engine was discontinued as competition use showed a preference for the larger engine. In 1966 the specification listed twin petrol tanks and an oil cooler, then, in October 1967, the Cooper models anticipated a feature of later Minis by having a gearbox with synchromesh on first gear. The Mini Cooper S evolved: a Mark II was introduced in 1967, a Mark III with wind-up windows and concealed hinges in March 1970, and in July 1971 the Mini Cooper S was discontinued in the United Kingdom.

Meanwhile the Mark I Austin and Morris Mini Cooper continued to be made during those years following its début in 1961, albeit with a more reliable shorter stroke '9FA' 998cc power unit and radial SP41 tyres as standard by early 1964. The Mark II saloon replaced the Mark I in 1967, and by 1969 they had been superseded by the Mini 1275GT.

A merger of BMC and Leyland in February 1967 brought into being British Leyland with Donald (later Lord) Stokes as chief executive of the whole operation. Although British Leyland extended the agreement with the Cooper Car Company for three years in July 1968, a change in policy eventually ended the Mini Cooper variations in Britain, by which time a total of something like 150,000 Minis in various forms carrying John Cooper's name had been made.. It has been said that Stokes considered it absurd that British Leyland should pay considerable sums in royalties to use someone else's name and so the Mini 1275GT was put on the market. Cooper, on the other hand, is recorded as saying that: 'Stokes was not really interested in competitions and I don't think he liked paying me £2 a car. He thought there was difficulty in getting insurance for the Cooper because of its high performance and he decided to replace it with the 1275GT; but he should have called it the "1275 Executive" or something as they [the insurance companies] treated it the same as the Cooper S — and it wasn't anything like the same motor car.'

A detailed study was made of a proposed MG Mini to fill the gap left when the Cooper contract came to an end, but nothing came of this project and the MG octagon never appeared on the Mini.

Cooper specials, however, continued to be made for a further three years in Italy, during which period 583 Coopers came off the Innocenti production lines. The Innocenti family, who had made their fortune as makers of Lambretta scooters and three-wheelers, built a factory in Milan to assemble Minis and other BMC cars within the Continental tariff area. Later, as a subsidiary of British Leyland, Innocenti built an upgraded version called the Innocenti Mini with a hatchback (a design once thought to be the natural successor to the original Mini) and other variations of BL cars such as the 'IM3' based on the Morris 1100. By the late 'sixties, Minis were being assembled from CKD parts or being actually manufactured in twenty countries.

Variations on the Mini theme were numerous and some proved to be very curious. One was a 'Multi-Mini', 20ft 3in long, made by BMC's Research Department and displayed in the company's London showrooms over Christmas 1964. The late Peter Sellers, comedian and film star, was one of the early Mini owners. He had a series of Minis, each customised to incorporate his own ideas, such as the first Mini with a hatchback, and it was he who began a fashion for mock wickerwork on the side panels. The idea of a hatchback Mini was taken up by Alec Issigonis in 1967 when he designed a car, code named the '9X', which was the same size as the Mini but where the number of parts required in production would have been reduced by 42 per cent. The project

*Customer's own Minis were
altered to convertibles by an
Essex motor engineer. This
'Couplette' was originally a
1975 saloon.*
(Essex Chronicle Series)

was eventually abandoned so that the development money could be available
for research and development of the Marina. Yet another project that never got
any further than the experimental stage was the 'ADO88', a wider version of
the Mini with a boot at the rear. There have been bored-out Minis, Minis with
fuel injection, Mini beachcars, Mini-Mokes, Mini Marcos, a racing Mini called
the 'Landar', Chilean built Minis with glass fibre bodies, 'Unipower GT',
'Broadspeed GT' (a fastback conversion), 'GTM' (lightweight glass fibre two-
seater), a soft top conversion of the Mini called the 'Clayton Convertible' (six
of these were commissioned by the breakfast cereal firm Kelloggs in 1964 for a
promotional competition), 'Biota' (two-seater in kit form using a tubular
chassis), the glass fibre bodied 'Ogle Minis' and 'Mini Statys', the 'Mini-Power'
on the lines of the Lotus Seven, and another one from the same company with
a Rolls-Royce shaped radiator which went into short production until the
designer, Brian Luff, was threatened with legal action. There was even an
electric Mini, sponsored by the British Electricity Council using an Austin
traveller (DAE137C). It was converted to electric drive in 1965 by Associated
Electrical Industries Ltd; the nominal 96-volt power supply came from forty-
eight cells in the rear luggage area, leaving little space for anything else.
Performance was acceptable, with an acceleration from rest to 20mph in 7
seconds, a maximum speed of 40mph and a level-ground speed of 30mph. A
built-in charger permitted the battery system to be plugged in to a standard
13amp socket for re-charging.

   The 'Radford Mini de Ville' was a conversion with a sunshine roof, rear lift-
up door and fold-down rear seat in the manner of an estate car. Stewart &
Ardern's bodybuilding subsidiary in Wembley, which usually concentrated on
commercial vehicles, used the Mini Cooper S as a basis for the metallic finished
'Mini-Sprint' announced early in 1967. This was over 3in lower than the
standard Mini and was achieved by removing 1½in between the floor and

waistline and a further $1\frac{1}{2}$in between waist and roof. All the spot-welded flanges which stood proud were removed and the seams were then re-welded as butt or lapped joints to give an uncharacteristic smooth external finish, combined with reshaped front wings to accept rectangular Cibie headlamp units. The low roof necessitated reclining and lowering the front seat frames. They even lowered a Mini-Sprint by an extra $1\frac{1}{2}$in, and the standard guttering was replaced by an unobtrusive channel around the doors. Earlier, in 1953, Crayford Engineering Co, of Kent, had used a similar name for their open version. To make the 'Crayford Mini Sprint' conversion, the roof of a standard saloon was removed, together with the rear side windows, and the back window. Extra stiffening was added to the body shell to maintain rigidity. The finished conversion had a hood which folded right down and was concealed in a neat cover.

The dearth of open cars in the early 'eighties may have inspired two other conversions from Essex. The first of these, the 'EG', retained the original side body line of the saloon, together with the weather strips. The hood, supported and run on small rollers, could, it was claimed, be positioned up or down quite effortlessly by one person unaided. Whether 'EG' stood for Eastern Garages of Corringham, the firm who did the conversion, or for Eric Grant Motors, the exclusive suppliers, is not clear. The second such conversion available about the same time was the idea of a Widford motor engineer, Mike Cruise. He offered what in earlier days would have been called a coupélette, for only the rear half of the saloon superstructure was cut away to be covered in inclement weather by a conventional folding hood of plastic material; the roof of the forward portion was fitted with a hinged tinted sun port.

The Mini-Moke was orginally developed by the BMC Projects Department at Longbridge to meet military requirements for a vehicle which could be produced to be sold at a low cost commensurate with Army standards of reliability and therefore 'expendable'. In addition the vehicle had to be 'stackable', meaning that to simplify storage and transportation they could be stacked one on top of the other to a height of two or three. Overall length was specified at under 100in in order to permit rapid and efficient transportation by air in sizeable quantities. Development work on the Moke started while the Mini car was still in prototype form and the unusual nature of this variant warranted a separate proving programme at Longbridge. The first prototypes, with 950cc engines, were too long and a reduction to the 100in required by the Army meant virtually a complete redesign of the body construction to permit the same carrying capacity of four persons. Weight saving (the Army stipulated a limit of half a ton) was a problem when starting with a particularly light saloon model of only 12cwt, but the engineers managed to get the body weight down to 235lb.

By 1962 a new shorter model was ready, designated in military nomenclature 'Truck $\frac{1}{4}$ ton 4×2 BMC (Mini-Moke)'. Only two pressings were common to the saloon — part of the bulkhead and the toe-board — the rest was a completely new roofless unitary construction allowing for a shorter wheelbase of $72\frac{1}{2}$in. This new bodyshell was thoroughly tested at Chobham, on the notorious test tracks of the Fighting Vehicle Research and Development

Establishment. Ground clearance was improved by inserting packing between the suspension units and bodyshell to make the vehicle more adaptable to rough going and boggy terrain. Early tests indicated a capability of around 55mpg at a constant 30mph, dropping to 30-5mpg when used on wheel-spinning cross-country work.

As the majority of the Moke's weight (about 65 per cent) was situated immediately above the two front driving wheels, there was no problem with traction or hill climbing ability. However, as an experiment in 1962, Alec Issigonis and his colleague Charles Griffin made a twin-engined version with one engine driving the front wheels and the other the rear. Although the clutches were interconnected there were two gear levers, and it was possible to drive the car with one transmission in top and the other in third, or third and second (top and first would have been inadvisable!). It was also possible to switch off one engine and with the transmission in neutral run as a single-engined vehicle. The result was amazing to say the least. With a total power output of around 72bhp and twice the pulling power of the production Mini, this eight-cyliunder 1,696cc 'Twini-Mini' (as it was nicknamed) had tremendous acceleration and hill-climbing abilities, the ascent of a 1 in 2 gradient being feasible. The model was demonstrated to the British Press during an unusually severe snow-bound January in 1963. Several Twini-Minis were produced, one being driven by Jim Clark during the BBC TV-Autopoint in Hampshire mud, between the London Motor Club and the British Army Motoring Association, in 1963. The following year's event saw a reappearance of the 4 × 4 driven by John Surtees. About the same time, a left-hand-drive model, fitted with two 1,100cc engines (giving a total power output of 110bhp), was sent to the United States Army Tank-Automotive Centre in Warren, near Detroit, along with a normal 850cc Mini-Moke, for evaluation.

In the end the military interest came to nothing, but civilian interest grew. In 1964 the 850cc Mini-Moke (with the same wheelbase as the saloon) went into full production, despite a ruling by the Customs and Excise that insufficient space for parcels or goods disqualified it from the purchase-tax-free commercial classification. When introduced it cost £25 more than the Mini van. Soon variations appeared, with Barton Motor Co Ltd of Plymouth, marketing a glass-fibre hard-top runabout, and Bradbury & Wedge Ltd of Wolverhampton, a wooden-sided pick-up conversion which somehow achieved a purchase-tax-free price of £348 in 1966. Two years later, in May 1968, Mini-Moke production in the United Kingdom ceased. Was it the successful overland tour of Australia by Mini-Moke in 1966 that ensured a ready market for the utility vehicle in that continent, where they were still being produced until 1980? The Australian versions had a longer wheelbase and 13in diameter wheels. The gap left by the withdrawal of the Mini-Moke in the UK inspired the Connaught Garage to offer a similar vehicle called the 'Scamp', and another imitator of the Mini-Moke, called the 'Jimini' was built at Weybridge until 1977.

Three years after the revolutionary Mini appeared, came the Morris 1100. Generally considered as the big brother to the successful Mini, the comparison between the two was justified, for its designer had incorporated all the good

*Morris 1100 in 1962 with the BOAC Vickers VC10
airliner in the background.* (The Nuffield Organisation)

features of the Mini as well as adding some entirely new ones, such as the
Hydrolastic suspension. 'A new model is never a real success when the engine
and every single other part are new,' said Alex Issigonis when discussing the
new vehicle. 'Some big part of a new model must be inherited from previous
models if a good car is to be made. If everything is completely new, the
development problem is too big and the result is not good. This applies to the
engine in particular. The ADO16 [1100] power unit is within the general
framework of the ADO15 [Mini] but in making it bigger we have added a
stiffer crankshaft with vibration damper. Front suspension architecture is the
same too, it has the same ball-joints, but with bigger and more dished wheels
we now have centre-point steering.'

It was as a Morris that the 1100, with four-door and two-door saloon
bodywork, made its appearance in August 1962. A novel trade preview was
arranged by Stewart & Ardern Ltd who virtually took over Croydon Airport to
entertain 281 dealers, each of whom was allocated one of the new cars to be
driven away for subsequent showroom display on the release date.

Development of the 1100 had actually begun early in 1959, months before
the public saw the Mini. The bigger scale of everything made solving problems
easier for the designer, in particular the larger size avoided stark economy
features such as the out-turned joints between body panels. The placing of all
occupants within the wheelbase obviated rear wheel-arch intrusion and the
curvature of all door panels and windows ensured maximum width where it
was most needed, at elbow and shoulder height; across the rear seats, for
instance, the interior width was almost the same as the rear wheel track. Once
again Alex Issigonis had gone out on a limb and created a new standard in tyre

size by specifying 12in diameter wheels and tyres, a size not then generally available. Early trials of the 1100 took place in Germany, Italy and Switzerland; while Southern Italy and Norway provided ideal terrain to test the suspension system.

The first alternative badge came in September 1962 when the two- and four-door saloon MG1100 was announced. The twin carburetters and improved cylinder head design increased the 1,098cc engine's output by 7bhp, and there were many refinements to trim and body, including the option of leather upholstery instead of the standard leathercloth. Externally, the most noticeable difference was the MG octagon mounted on a facsimile earlier Abingdon-style radiator grille, but this MG never saw the now defunct Berkshire works and enthusiasts for the marque regard it disdainfully. Inside the body was a facia of imitation wood veneer material (later superseded by real wood veneer) while the speedometer was of a new type displaying a red ribbon moving horizontally across the dial.

Inevitably an Austin version was added to the range, and this came in September 1963. Identical in all mechanical aspects to the Morris 1100, the major identifiable differences were the Austin badge, small trim changes on the boot panel, and that singularly Austin feature: a rippled radiator grille, in this instance of anodised aluminium. The ribbon-type speedometer and a full width facia distinguished the interior of the car from its Morris counterpart. During

*Rebirth of an old name, the Riley Kestral version of the
1100 was introduced in 1965. The Mark II 1300, shown
here, differed little externally from the earlier model, the
major identification features being the ventilated wheels
and repeater flashers on the front wings.*

the following month the up-market Vanden Plas Princess was offered with its luxurious appointments such as walnut wood facia, extra gauges and much extra equipment. Mechanically, the model was as the MG1100.

Another year passed before further newcomers came on the market. In September 1965 BMC brought out the Riley Kestrel and the Wolseley 1100, both models having their own distinguishing traditional radiator shell design and a slightly more exclusive note was added by the use of walnut veneer and seats covered in real leather. One feature confined to the Riley version was the addition of an engine revolution counter and rimbellishers to the road wheels. The introduction of these new variations to the theme coincided with the availability of automatic transmission as an optional extra on 1100s.

By early 1966 the prospective 1100 purchaser had a choice of Morris, Austin, MG, Riley or Wolseley versions, but only in saloon form. Then in March of that year the long awaited estate version of the 1100 was announced. These were the Morris 1100 traveller and the Austin 1100 countryman — mechanically identical, with an all-steel body and single upward-opening door, giving access to a load carrying platform when the rear seat assembly was folded down. To cater for the loads that this utilitarian transport was expected to carry, the rear Hydrolastic units were changed for a slightly stiffer type.

Soon after the appearance of the estate models a larger 1,275cc engine became available on certain models, followed by a minor option, in May 1966, of reclining front seats on all the BMC 1100 and 1300 cars. Although the giant British Leyland Motor Corporation was soon to be created, it was still under the BMC flag in March 1967 that the one millionth 1100/1300 car was made. By the October 1967 Motor Show the permutations of 1100 and 1300 models were seemingly endless, for in addition to new Mark II 1100s under all the brand names, there were the Mark I 1300s which could be had in four-door or two-door, de luxe or super de luxe, as Morris or Austin; two-door only for the MG, but with the option of automatic transmission. Automatic transmission

Exclusively Australian, the Morris Nomad based on the Morris 1100. Immediate differences are the extra quarter lights and the lift-up tailgate. (Ian Gliddon)

was also offered on the Riley, Wolseley, and Vanden Plas versions, with the twin-carburetter engine as a standard fitting. To justify the Mark II designation on the 1100 range the face-lift involved new radiator grille designs, better seating and interior trim, revised facia layouts with combination switches mounted on the steering column, repeater flashers on the front wings, new rear light clusters set at a shallower angle (thus altering the profile of the car), and road wheels with slotted cooling ducts. For the estate versions, imitation wood now decorated the body waist.

Such a multitude of models seemed unwieldy to the new BLMC management (added to the knowledge that the Maxi was in the pipeline) for by January 1968 a run-down of 1100 types was underway. First to go was the Mark II Riley 1100, followed a month later by the Mark II Wolseley 1100. In March production of the Morris 1100 traveller and the Austin 1100 countryman ended, joining the Mark II two-door de luxe and two- and four-door super de luxe 1100s, the Mark II MG 1100, and the Mark II Vanden Plas 1100. 1300 cars to be discontinued at the same time were the Mark II Austin and Morris de luxe saloons, in both two- and four-door types. Later that same year, in October, the remaining Mark I MG1300, a two-door model, ceased production. This coincided with the 1968 Motor Show when its replacement, the Mark II MG1300 (again a two-door only model) was announced, together with the Mark II version of the Riley, Wolseley, and Vanden Plas 1300s which were destined to have extremely short production runs.

Only Austin and Morris labels were carried on the 1300GT announced in October 1969. Good performance was expected from these four-door saloons with an A-Series 1,275cc engine using twin SU carburetters, rated 70bhp at 6,000rpm. The image was assisted by the three colour options of flame red, glacier white, and bronze yellow; each had a vinyl-covered roof, chrome gutter surround, and black interior trim. Added embellishments were black side flashes, red 'GT' badges on both quarter panels, chromium-plated tipped exhaust pipe, and a matt black grille with a chrome surround incorporating a '1300GT' scroll. Mechanically the specification followed that of the MG1300 but optional extras included steering column lock, laminated windscreen, heated rear window, and servo-assisted brakes. Behind the launch of this ostentatious 1300 was an energetic, cigar smoking, American with the most improbable name of Filmer Paradise, who had moved from Ford's Italian Company to become successful in looking after BMC's European marketing. European buyers required more performance, he insisted, and the 1300GT was the result.

By the autumn of 1971 the Mark III 1300 range was introduced, with the usual new grille styling (matt black with a thin bright metal surround and a similar plated strip running horizontally across the centre), improvement in interior trim, thicker carpets, redesigned dash layout, and such styling gimmicks as a smaller padded steering wheel and a different choice of colours. Internal competition from the now established Austin Maxi and the new Morris Marina, which came onto the roads some months earlier, had resulted in only one 1300 model by then to carry the Morris name, the traveller, and this finally disappeared from the dealer's showrooms in 1972.

Largest of the transverse engined cars in the 'sixties was the 1800. In some quarters it soon gained the nickname 'Land Crab', and if for no other reason it made history as the first car introduced with a firm assurance from the manufacturer that it would continue in production, basically unchanged, for at least a decade. It was in October 1964 when the Austin 1800 came on the scene, and the promise was kept, for the redesigned 18/22 Series with its wedge-shaped body did not appear until 1975.

However, this is jumping ahead of the story. The first Morris 1800 was offered in March 1966, after some teething troubles with the gearbox, clutch, and cylinder head had been eradicated. There is a story told that early Morris 1800 cars were restricted to one colour only because chairman and managing director, Sir George Harriman, did not like the Old English White version. During a visit to the Cowley plant, Harriman, without consulting his managerial colleagues, ordered the paint shop superintendent not to finsh any more bodies in this colour. The sales director at Cowley subsequently heard of this unilateral decision when he found that orders for 1800s in white were not being delivered. The late appearance of the Morris version also avoided other snags with the early Austin 1800, such as the awkward centre-positioned handbrake lever (which was repositioned nearer the driver within the first year). Other alterations made to the Austin specification at the same time were the substitution of a lever and cam system for the original cable link for the gear change lever, and a final drive alteration from 4.188:1 to 3.88:1. That the car was a Morris was indicated by the Morris badge in the centre of a horizontal-bar grille. Another feature was a new wrap-around rear light unit.

Interestingly, the Morris version of the 1800 was not released in Australia. There, the 1800 carried the Austin badge and in addition to the sedan there was also a version in utility or 'Ute' form. It would seem that the 1800 car was popular in Australia, for in 1968 it was voted the car-of-the-year and had the best used-car valuation in that country. The same cannot be said for the 1800 utility as it fell within the same price range as the Holden, Ford Falcon and Chrysler which had the advantage of a six-cylinder power unit. The replacement for the Austin 1800 in Australia was the Austin Kimberley and the Austin Tasman, with a larger 2,200cc overhead-camshaft engine. The latter was the standard version with a single carburetter, and both models had Australian-built bodies with rear styling that eliminated the small rear quarterlights. The front grille also differed significantly from Cowley-made 1800s; the Kimberley, for example, having quad headlights.

In the United Kingdom the use of four headlamps was considered when a study was carried out with the object of producing a Riley 1800 with twin carburetters. Another still-born project was that of an 1800 van to replace the Austin A55; this was abandoned because it offered 20 per cent less space for goods and would have been more expensive than the older Austin.

For the prestige end of the 1800 range, from March 1967, there was the Wolseley 18/85 with a more luxuriously appointed interior and power-assisted steering as standard.

In addition to the specification changes already mentioned, the steering lock-to-lock ratio had been changed (September 1966) from 4.4:1 to 3.8:1 before the

Mark II version of the Morris 1800 saloon. (BMC Ltd)

introduction of the Wolseley 18/85. Later, in June 1967, all 1800 bodies had a modified black crackle facia to avoid the glare of the previous burnished aluminium finish. Other changes and additions were made at the same time such as the combined arm rests and door pulls, while the heater controls had been modified. Four months later it was announced that the standard Wolseley power-assisted steering was available as an optional extra on the other 1800 vehicles.

By May 1968, after some 143,000 1800s had been made (48.6 per cent of which had gone abroad), the Mark II Austin and Morris models were in the showrooms, attention having been given to the power unit, now with a raised compression ratio of 9:1 and with much detail rethinking. Alterations to the driving compartment included polished wood, a pull-out type handbrake to the left of the steering column, anti-burst locks and flexible window winding knobs on doors, facia changes to incorporate rocker switches, two-speed windscreen wipers, and different reclining seats. Externally, the front grille, incorporating the side lamps and flashers, had been made narrower, while at the rear the light clusters were restyled and fitted vertically. This enabled the designers to do away with the 'hump' on the rear wings which, together with an increase in wheel size from 13in to 14in diameter, made a discernible difference to the appearance. The optional extras at this time included automatic transmission and power steering.

Around Motor Show time in 1968 came the inevitable special, the Morris 1800S. The 1,798cc engine was made more powerful by increasing the compression ratio to 9.5:1, adding twin SU HS6 carburetters, and a new Downton induction manifold. This gave 95.5bhp at 5,700rpm, raised the top speed to 100mph compared with the standard 1800's top speed of 92mph, and necessitated an increase in diameter of the front wheel brake discs and the addition of power assisted brakes. 1800S models with the Wolseley and Austin badges were not available until a year later, coinciding with the introduction of the Mark II Wolseley 18/85.

The more powerful 2200 was introduced in 1972, the larger engine being an E Series, six-cylinder, 2,227cc unit. Similar vehicles carried either Morris or Austin badges, the Austin version is shown here.
(British Leyland)

With minor refinements and a re-designed grille in matt black and a new identification badge, the Mark III 1800 replaced the earlier model in March 1972, but it was put into the shade by the new 2200 that came out at the same time, powered by an E-Series six-cylinder overhead-valve, single overhead-camshaft, engine of 2,227cc. Both six-cylinder and four-cylinder cars continued to be listed for a further three years with little change, except the addition of extra warning lights, heated rear windows, reclining front seats and a cigar lighter, in mid-1974. Then in March 1975 the Austin/Morris 1800/2200 range underwent a metamorphosis into the Austin/Morris 18/22 Series with a completely new wedge-shaped body concealing revised suspension as well as many parts continuing on from earlier models, including the engines.

British Leyland policy later that year dictated a change in name with the disappearance of both Austin and Morris badges, so that all of the 18/22 Series became Leyland Princess cars. With the Leyland label came other minor changes, chiefly to the shape of the front grille which followed the bonnet line and grille style of the former Austin version. Circular headlamps, previously found on Morris cars, differentiated the 1800 from the 2200, which was fitted with trapezoidal halogen lamps. In 1982, along with a revamp, the Princess changed its name for one already used for years on the basic Morris Oxford design made in India: the Ambassador.

This book started with a single model, the Oxford, carrying the Morris label. It will end in the same way with only the Ital, successor to the Marina, keeping the name alive in the 'eighties. By April 1971, when the new Morris Marina was announced, the new marketing policy by British Leyland was well under way. Almost immediately, after the merger of British Motor Holdings and Leyland, commercials carrying Austin and Morris names had been given a common BMC badge together with the BL symbol. The following year the Morris name disappeared from the Mini. As the production of the Series VI Morris Oxford, Morris Minor 1000, and Morris versions of the 1300 saloon were discontinued in 1971, the only other cars to carry the name were the 1300 traveller (which only continued for a short period of time) and the Morris 1800/2200.

Introduced in April 1971, the Marina was deliberately designed to be

Morris Marina 1300 saloon super de luxe in the form
originally introduced in April 1971. (British Leyland)

relatively cheap to manufacture, and careful value-engineering ensured that only features that could justify a higher selling price would be included. To this end, existing A-Series 1,275cc and B-Series 1,798cc engines were incorporated, together with a front suspension system designed originally for the Morris Minor some twenty-five years earlier, and semi-elliptic leaf springs with telescopic shock absorbers at the rear. British Leyland made no secret of this, their publicity spoke of 'positive rack and pinion steering like that of Jaguars, the reliable race toughened engine of our MGB, rugged transmission and disc brakes like our Triumph, and the same parentage as our Land Rover.'

Nevertheless, many millions of pounds went into the development of the Marina to which could be added the cost of rebuilding the Cowley plant for a gate-line production system. This incorporated a new overhead conveyor system linking the body and final assembly buildings by crossing the Oxford bypass, thus ending a long established arrangement of transporting assembled body shells from Longbridge to Cowley. The roof of the plant was raised some 20ft to accommodate the new lines and conveyors. Sensitive labour relations were also involved, for Leyland refused to start production until the workers at Cowley would accept the American system of measured day-work in place of the inefficient piece-rates. Once in production, the Marina was planned to have a production rate of 300,000 per annum (a figure that was to prove optimistic as only $1\frac{1}{4}$ million had been made when it took its final bow in 1980) with the fleet market as one of the main objectives.

In the manner of the successful Ford Cortina, the Marina permutated the two engine sizes with coupé and four-door saloon bodies to produce the 1.3, 1.8, and 1.8TC versions. BL publicity people were no doubt sensitive to this comparison: 'The precedent set by our American-owned competitors is to make a single body shape do for both two and four-door versions. It is a engineering compromise we weren't prepared to make. In our book, four into two won't go. That's why we designed a different four-door car. It not only looks better, it works better.' Something of a back-handed compliment came from the Automobile Association: 'The Marina is, in fact, one of the most blatant current examples of the family car as a domestic utensil. It is devoid of any real flair or enthusiast appeal but eminently practical, safely and effectively designed for its job, and offered at a keen price.'

By clever mixing of plated brightwork, paint finish, wheel trim, badges, over-riders, and so on, the stylists managed to make the diversifications look identifiably different. In particular, the front grille was an area that could be drastically changed without any other alteration to the common body pressings. On the basic model, the de luxe (a term that by the 'seventies had long since been devalued to the point of being meaningless) the painted front pressing had two separate horizontal grilled areas. The special de luxe qualified for an inner and outer plated escutcheon with a painted horizontal insert, matching the body colour. For the top of the range, the 1.8TC, the centre piece was decorated with a 'TC' motif, while the bumpers had over-riders and additional plated trim was added to the wheel arches, sill, etc.

Mechanical specification, naturally, differed also. The larger engined cars had the option of manual or Borg Warner automatic gear box and, in the case

An estate version of the Marina, powered by the B-Series 1,789cc engine, was added to the range in September 1972. (British Leyland)

of the twin-carburetter 1.8TC, servo assisted brakes with disc at the front and drums at the rear; tyres were radial 165/70-13. The basic 1.8 model also had Girling disc brakes but with 145-13 tyres and the servo as an optional extra. Owners of the 1.3 models had to be content with 5.20-13 cross-ply tyres and drum brakes.

When the first 180 or so Marinas were assembled in Australia, they were to be called the 'Cavania' and to cater for the different nameplate suitable holes were drilled in the bodywork. Meanwhile a message was received from Lord Stokes in the UK instructing the BL management, at Zetland in NSW, to retain the 'Marina' name. Unfortunately, because of the slightly longer name Cavania, the new badge already produced had an additional stud and after fitting the Marina decoration the Australian BL people had an unused hole. The problem of the extra hole was solved and today an occasional early Morris can be seen with a deliberate full-stop fitted after the word 'Marina'!

Before the year was out the range comprised a two-door coupé and four-door saloon versions of the 1.3DL, 1.3SDL, 1.8DL, 1.8SDL, and 1.8TC, with only a £250 price difference between the cheapest and most expensive car. The following year, in April 1972, a Marina van appeared, then in September came the Marina Estate with the 1,789cc power unit.

Criticism from the motoring press was mainly concerned with the handling of the new car, although one magazine said that the Marina coupé was not so much a coupé, more a two-door version of the saloon with different, and less practical, hindquarters. Road test journalists were not at all happy with the pronounced understeer characteristics, even after the lower suspension link had been modified to raise the front roll-centre. The fault lay in the use of the unmodified Morris Minor front suspension arrangement, which worked well on that model with its relatively stiff rear springs, keeping the car upright and providing some understeer compensating weight transfer at the rear. On the softly sprung Marina, the rear wheels stayed upright on the live axle while the front wheels assumed alarming positive camber angles. Leyland responded to the complaints by making alterations, and existing 1.8TC cars (something like 1,600 had been produced by October 1971) were recalled for modification. Early Marinas also suffered from annoying brake squeal and even hot running front-wheel bearings. (In the 'twenties William Morris was once asked why Morris car brakes squealed. He replied that the squeal was the outward and oral manifestation of efficient braking!) These faults were overcome by manufacturing changes, as was the tendency for water ingress past the bulkhead caulking sealers into the passenger compartment and past the rear wheel arch into the boot. The results of early electrocoat paints were diappointing, but changes in paint formulation and better control of phosphating metal pre-treatment brought about an improvement.

The estate, in Mark I form, was marketed with the 1798cc engine only. Although slightly longer overall, it was based on the 1.8SDL specification, with stiffer springing at the rear to cater for the 900lb payload that could be carried within the 58cu ft goods area, accessible from the rear top-hinged door. The three versions of the Marina van ranged from a 7cwt standard van with a small

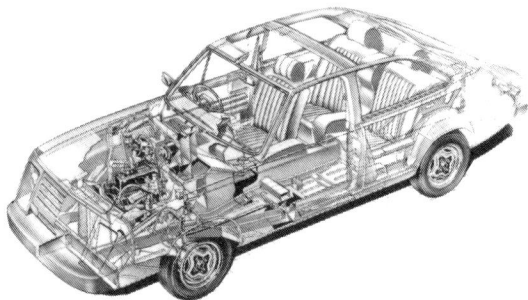

Ones that never were. The ESV Marina Phase I and ESV Maina Phase II were to be introduced in Australia in 1974, using an existing E-Series six-cylinder overhead-camshaft engine, but were never produced.
(British Leyland)

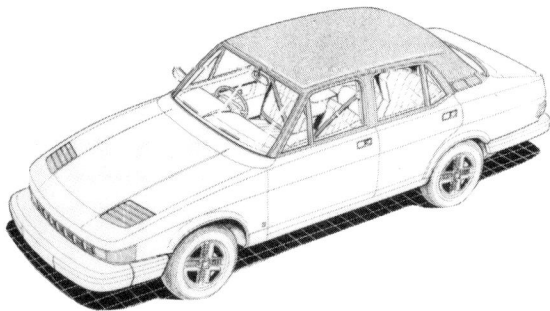

1,098cc four-cylinder ohv engine (the 1,275cc could be supplied at extra cost), a
7cwt de luxe, and the 10cwt version. These last two light commercials were
powered by the 1,275cc A-Series engine. Certain equipment on the 7cwt
standard van was of a utility nature such as the single seat for the driver, and
the use of a paint finish on door panels, front bumper and external driving
mirror. By contrast, the other two versions had chromium-plated bumpers
front and rear, vinyl trim on the door panels, driver and passenger seats, fresh
air heater, twin sun-visors, cab headlining, and plated external mirrors.

Some detail changes were made to the Marina in mid-1974 such as at
provision on all models of an exterior mirror, hazard warning lights and heated
rear windows. Other extras were added in proportion to the price of the model,
for example a cigar lighter and reversing lamps from the 1.3SDL upwards. All
1.8 cars had brake servo and twin horns in addition to those items mentioned,
until the top of the range 1.8TC specification included a vinyl roof, tinted glass,
nylon seat facings, and head restraints. A feature of the estate was the reclining
front seats. Another feature, and one not intended yet never rectified, was the
rear air vent which never fitted correctly.

If a future generation decides that the Morris Marina is a worthy old motor
to restore, it may be difficult to bring it back to 'original' specification, if only
because of the number of optional extras listed, to which could be added the
many factory supplied special tuning parts. A few examples of these
components from the British Leyland Special Tuning Department, at
Abingdon-on-Thames, were oil coolers, stiffer rear shock absorbers,
lightweight and figure-hugging seats, rally reclining seats, alternative axle
ratios, Weber twin-choke carburetters and manifolds, competition clutch-
cover assemblies and driven plates, lightweight steel flywheels, alternative
profile camshafts, special tappet adjusting screws, reduced-speed dynamo
pulleys, supplementary water radiator kits, special pistons, competition valve
springs, limited-slip differentials, alternative front torsion bars and rear road
springs, competition front and rear bump stops, alloy road wheels, wire wheel
conversions, mountings for auxiliary driving lamps, special sump guards,
lightweight body panels, extra instrument panels, petrol tank shields, fly-off
handbrake parts, competition heads, perspex window sets, and even leather
bonnet straps. One North London motor trader gave true meaning to the name
'doctor's coupé' in 1974 by modifying Marina 1.8 coupés specially for doctors.
The cars, said to have been equipped after recommendations from a team of
medical practitioners, were fitted with two-way radio, medical and rescue
equipment, double locking boot with alarm, spot lamp and cassette recorder.

When the Mark II Marina was launched in the UK, in October 1975, a face-
lift had given all variations a common re-designed radiator grille with round
sealed-beam headlights. The more expensive cars had additional rectangular
halogen driving lamps at the ends of a neat louvred centre strip. Other obvious
changes included a new facia layout, wrap-round bumpers, and new badge
symbol. Not so immediately obvious was the provision of disc brakes on the
front wheels of the 1.3 models and a new suspension system, with anti-roll bars
front and rear to improve the steering. These anti-roll bars were not fitted on
the estate. Top of the range 1.8TC coupé and saloon were renamed GT and HL

Mark I and Mark II Marinas. The use of a redesigned front grille to up-date the model is clearly illustrated.

respectively, and with the new identity came special equipment such as tinted glass, twin door mirrors, black body mouldings, twin coachlining, an underbonnet lamp, and front floor console. A vinyl covered roof was one of the extra refinements to be found on the new single-carburetter special two-and four-door models which replaced the discontinued 1.8SDL.

During the remaining years of the 'seventies there were to be the usual changes in the specification, mostly minor ones such as the standardisation of inertia reel seat belts in 1976 (in which year a 1.3 litre estate version appeared), and the adoption in 1977 of seats similar to those fitted in the contemporary Allegro. In January 1978, a 1.3 four-door special version was added to the range, with trim similar to the 1.8 special and a specification which included a clock, trip speedometer, head restraints, vinyl roof, twin halogen driving lamps, additional convenience lights fitted in the glove box and boot, two-tone horn, and opening front quarter lights. This was followed in April of the same year by a limited edition of a Marina 1.3 coupé type LE — easily identified by the striped upholstery and oyster metallic paint finish.

Changes of a more thorough nature coincided with the 1978 Motor Show when all the 1.8 models were discontinued and replaced by a new Marina 1700 range with four-cylinder, 1,695cc, overhead-camshaft O-Series engines. The smaller rated 1.3 cars, although still powered by the same ohv push-rod engine, were re-designated Marina 1300s. The revised specification for all models included new badges, a front spoiler, grill mounted driving lamps, larger rolled section bumpers with black end caps and recessed front indicator lamps, a rear fog light incorporated in new style rear lamp clusters, dual-circuit brakes with failure warning light, dipping mirror, and trip recorder. Other extras were allocated to the different variations, such as the bright wheelarch trim, chrome bumpers with rubber facing, radio aerial and built-in speaker, etc. The 1300L coupé, for example, had a vinyl roof, while the HL versions sported full moulded wheel trim, triple coachlining, intermittent screen wipers, and velour seat facings. Tinted glass and a radio receiver were two standard features on the up-market 1700HL. The Marina was to continue in production until July 1980, but before its successor was launched on the home market, one last variation was added in September 1979. This was the Marina 1700HL estate, mechanically similar to the 1700L estate but with HL equipment and trim.

Italian styling makes the Marina the Ital for the 'eighties.

One and a quarter million cars were made carrying the Marina badge before the name was dropped in mid-1980. Its replacement was still marketed as a Morris car but now known as the Ital, taking its new name from the Italian company, Ital Design, who were responsible for its styling. In an attempt to keep the Marina type of range selling well, particularly in the fleet market, BL Cars spent £5 million on the Ital revamp to take them through to its eventual replacement, the LC10 Austin Maestro, the LC11, and the end of the name Morris on cars — although, even then, the Ital itself may continue under another name, for it has been suggested that the design, because of its conventional engineering, would lend itself as an ideal replacement for the Indian built Hindustan Ambassador.

The Ital variations followed the lines of the previous range, with the 1.3 using the uprated 1,275cc engine (now referred to as the A-Plus Series), and the 1.7 litre retaining the O-Series engine. Equipment and trim varied as before, through a combination of L, HL, and HLS as coupé, saloon, and estate types. It was not until July 1982 that a pick-up derivative joined an increasing demand for such vehicles in the United Kingdom. In other countries the pick-up has been a long time favourite for its versatility, particularly in the commercial sector and in the leisure market. The Morris Ital 575 pick-up utilised the front end of the saloon and the 1,275cc engine. The rear end, which had a double skin to prevent damage to external panels, had a capacity of 36cu ft with a tailgate that folded to lay flush with the load platform. This feature, combined with a hinged number plate complete with lights, allowed the owner to carry extra-long items. In addition to the built-in mounting points for the tilt cover, extra lashing hooks were provided inside the cargo area for load-restraining ropes. The more powerful Ital made its début in October 1980 as the 2.0HLS four-door saloon and estate. The larger O-Series, four-cylinder, 1,993cc, single overhead-camshaft engine drives through a Borg-Warner Model 65 automatic gearbox as standard, giving a top speed in excess of 100mph — a family Morris car capable of a speed almost as fast as the world's land speed record when that first production Morris Oxford drove out of the old Military Training College at Temple Cowley less than seventy years earlier.

Mini Engines. 'A' Series. Four-cylinder, overhead valve by push rod

Type	Bore (mm)	Stroke (mm)	Capacity (cc)	Compression Ratio	Carburetter	Used in
8AM	62.94	68.26	848	8.3:1	single SU	Mini saloon, Mark I & II. Mini-Moke
8MB	62-94	68.26	848	8.3:1	single SU	Mini 850. Riley Elf Mark I
8SH	62.94	68.26	848	8.3:1	single SU	Wolseley Hornet Mark I
8AH	62.94	68.26	848	9:1	single SU	Mini 850 Automatic
9FC	70.6	61.91	970	10:1	twin SU	Mini Cooper S 1000
9F	62.43	81.28	997	high 9:1, low 8.3:1	twin SU	Mini Cooper
9WR	64.58	76.2	998	8.3:1	single SU	Mini Clubman, Riley Elf Mark II & III
99H*	64.58	76.2	998	8.3:1	single SU	Wolseley Hornet Mark II & III
9AG*	64.58	76.2	998	9.1:1	single SU	Mini Mark II 1000
9FA	64.58	76.2	998	high 9:1, low 7.8:1	twin SU	Mini Cooper
10F	70.6	68.26	1,071	9:1	twin SU	Mini Cooper S
10H	64.58	83.72	1,098	8.5:1		Mini Clubman 1100
12FA	70.6	81.28	1,274.8	9.75:1	twin SU	Mini Cooper S, Mark I & II
12H	70.6	81.28	1,274.8	high 8.8:1, low 8.3:1	twin SU	Mini 1275GT, Mark III

* Note: 99H and 9AG engines were fitted to Minis with automatic transmission.

Morris Mini-Minor. Specification

Mark I 1959 - 1964 Spec A	Saloon, chassis 101 onward. Traveller, chassis 19101 onward.
Mark I 1959 - 1964 Spec B	Mini-Cooper, chassis 13311 onward. Van $\frac{1}{4}$-ton, chassis 12601 onward.
Mark I 1964 - 1967	Pick-up, chassis 87551 onward (LHD models 88056 onward).
Mark II 1967 - October 1969	Mini-Moke March 1963 - May 1968.

Four-speed gearbox, synchromesh in all but first gear up to 1968. Combined transmission and oil sump below crankcase. Floor mounted gear-change lever. Automatic transmission optional extra from October 1965. Clutch, single dry-plate, $7\frac{1}{8}$ in diameter, diaphragm spring type from early 1964. Hydraulic operation. Brakes, Lockheed hydraulic foot brakes, front leading and trailing shoes up to September 1964, later models (chassis 296257 for saloon, 638879 for traveller) had twin leading shoes. 7 in diameter brake drums front and rear. Disc brakes at front on all Cooper models, with servo assistance on the Cooper S. Handbrake by cable to rear wheels. SU carburetter, single type HS2 on all Morris models with manual transmission. Automatic transmission models fitted type HS4. Twin SU HS2 units used on Cooper cars. Suspension, rubber cone springs with hydraulic telescopic shock absorbers on all travellers, vans, pick-ups, Mokes, and on saloon models before 1964. Saloons between September 1964 and 1969 (chassis 370004 RHD, and 370197 LHD, onward) fitted Hydrolastic. Steering, rack and pinion. Electrics, 12-volt, positive earth. SU electric petrol pump. Petrol tank, saloon $5\frac{1}{2}$ gallons, van and pick-up 6 gallons, traveller $6\frac{1}{2}$ gallons. Twin petrol tanks on Cooper S after January 1966.

Wheels, pressed steel ventilated disc, 5-stud, 3.508×10 with 145-10 tubeless radial tyres, or 5.20-10 tubeless crossply tyres. Cooper models, 3.508×10 standard wheel with 145-10 tubed radial or 5.20-10 tubed crossply tyres. Optional extra, $4.50J \times 10$ wheels with 500L-10 tubed tyres.

Wheelbase for saloon and Moke, 6ft 8in. Traveller, van and pick-up, 7ft $0\frac{5}{32}$ in.

Van capacity $46\frac{1}{2}$ cu ft with additional 12 cu ft if optional passenger seat not fitted.

Morris Mini-Minor. Body Colours and Upholstery

As originally introduced in 1959 the Morris Mini saloon was available in red, white, or blue, all with grey upholstery. The de luxe models were:

White bodywork with grey fleck and black two-tone upholstery, fitted dark grey carpets.

Red bodywork with grey fleck and black two-tone upholstery, fitted red carpets.

Blue bodywork with grey fleck and black two-tone upholstery, fitted blue carpets.

Road wheels on all these original saloon cars were finished in white.

Duotone colour schemes came in September 1961 for the super de luxe Morris Mini.

The Morris Mini van and pick-up, introduced in 1960, came originally in Whitehall beige or smoke grey, both with tan upholstery and pale grey headlining.

The Morris Mini Traveller, introduced in 1960, was first offered in the following colours:

White or blue bodywork with grey fleck and blue two-tone upholstery, blue carpets.

White bodywork with grey fleck and grey two-tone upholstery, red carpets.

Morris 1800 and 2200. Specification

Model	Specification
Morris 1800 Mark I March 1966 - May 1968	4-cylinder, overhead-valve, push-rod engine. 1,798cc type 18AMW, 18C, 18WB. 80.26mm bore, 88.9mm stroke. Aluminium solid skirt pistons, four-ring. Compression ratio: type 18AMW high compression 8.2:1 to 18AMW/U/H101631 then 8.4:1 low compression 6.8:1. Type 18C high compression 8.2:1, low compression 6.8:1. Type 18WB high compression 8.4:1, low compression 6.8:1. Carburetter, SU type HS6, $1\frac{3}{4}$in diameter. Petrol tank, $10\frac{1}{2}$ gallons. SU electric petrol pump. Brakes, Girling hydraulic, servo assisted, disc at front, 9in diameter drums at rear. Electrics, 12-volt, positive earth. Coil ignition. Clutch, diaphragm-spring, single dry-plate, Borg & Beck 8in diameter. Gearbox, four-speed synchromesh. Front wheel drive. Steering, rack & pinion; steering ratio changed from 4.4:1 to 3.8:1 in September 1966. Power assisted steering optional after October 1967. Hydrolastic suspension; anti-roll bar fitted up to A-HS10-32802. Wheels, ventilated disc, 4.5J-13. Tyres, 175-13 (eg Dunlop SP41 radial). Track, front 4ft 8in, rear 4ft $7\frac{1}{2}$in. Wheelbase, 8ft 10in.
Morris 1800 Mark II May 1968 - March 1972	4-cylinder, overhead-valve, push-rod engine. 1,798cc type 18H. 80.26mm bore, 88.8mm stroke. Compression ratio: high compression 9:1, low compression 8:1; export low compression 6.9:1. Carburetter, SU type HS6 on 'Special', $1\frac{3}{4}$in diameter. Petrol tank, $10\frac{1}{2}$ gallons. Mechanical fuel pump. Brakes. Girling hydraulic, servo assisted, disc at front, 9in diameter drums at rear. Electrics, 12-volt, positive earth. Coil ignition. Clutch, diaphragm-spring, single dry-plate, Borg & Beck 8in diameter. Gearbox, four-speed synchromesh; alternative ratios on standard models. Optional automatic transmission, Borg-Warner type 35TA. Front wheel drive. Steering, rack & pinion; power assisted steering optional. Hydrolastic suspension. Wheels, ventilated disc, 4.5J-14. Tyres, radial ply, 165-14. Track, front 4ft 8in, rear 4ft $7\frac{1}{2}$in. Wheelbase, 8ft 10in.
Morris 1800 Mark III March 1972 - March 1975	4-cylinder, overhead-valve, push-rod engine. 1,798cc. 80.26mm bore, 88.9mm stroke. Carburetter, SU type HS6, $1\frac{3}{4}$in diameter. Brakes, Girling hydraulic, servo assisted, 9.28in diameter disc at front, 9in diameter drum at rear. Electrics, 12-volt, negative earth. Coil ignition. Dynamo standard; alternator optional when not fitted with power assisted steering. Clutch, diaphragm-spring, single dry-plate, Borg & Beck 8in diameter, hydraulically actuated. Gearbox, four-speed synchromesh, rod operated gearchange mechanism; automatic transmission optional. Front wheel drive. Steering, rack & pinion; power steering optional. Hydrolastic suspension. Wheels, ventilated disc, 4.5J-14. Tyres, radial ply, 165-14. Track, front 4ft 8in, rear 4ft $7\frac{1}{2}$in. After July 1974, exterior mirror, hazard warning lights, heated rear window, and cigar lighter standard. Wheelbase, 8ft 10in.
Morris 1800 18/22 Series March 1975 - September 1975	4-cylinder, overhead-valve, push-rod engine. 1,798cc. 80.26mm bore, 88.9mm stroke. Brakes, hydraulic, dual system, disc at front, drum at rear. Electrics, 12-volt, negative earth. Clutch, diaphragm-spring, single dry-plate, hydraulically actuated. Gearbox, four-speed synchromesh. Front wheel drive. Steering, rack & pinion. Hydrogas suspension.
Morris 2200 March 1972 - March 1975	6-cylinder, single overhead-camshaft, overhead-valve, engine. 2,227cc, type 23H. 76.24mm bore, 81.28mm stroke. Compression ratio 9:1. 110bhp at 5,250rpm. Carburetter, SU type HS6 twin, $1\frac{1}{4}$in diameter. Petrol tank, $12\frac{1}{2}$ gallons. Brakes, Girling hydraulic, servo assisted, disc at front, drom at rear. Electrics, 12-volt, negative earth. Alternator. Coil ignition. Clutch, diaphragm-spring, single dry-plate, Borg & Beck $8\frac{1}{8}$in diameter, hydraulically actuated. Gearbox, four-speed synchromesh, floor mounted gear change lever. Optional Borg-Warner type 35 automatic transmission. Front wheel drive. Steering, rack & pinion; power assisted steering optional. Hydrolastic suspension. Wheels, ventilated disc, five-stud, 4.5J-14. Tyres, radial ply, 165.SR-14. Rostyle wheels optional. Track, front 4ft 8in, rear 4ft 7.5in. Wheelbase, 8ft 10in.
Morris 2200 18/22 Series March 1975 - September 1975	6-cylinder, single overhead-camshaft, overhead-valve, engine. 2,227cc. 76.24mm bore, 81.28mm stroke. Carburetter, SU type HS6 twin. Brakes, hydraulic, dual system, disc at front, drum at rear. Electrics, 12-volt, negative earth. Clutch, diaphragm-spring, single dry-plate, hydraulically actuated. Gearbox four-speed, synchromesh. Front wheel drive. Steering, rack & pinion; power assisted standard on 2200HL model. Hydrogas suspension.

Morris 1100 & 1300. Specification

Mark I 1100 saloon 4-door	August 1962 - October 1967
Mark I 1100 saloon 2-door	August 1962 - October 1967
Mark I 1100 traveller	March 1966 - October 1967
March II 1100 saloon 4-door	October 1967 - September 1971 for de luxe, March 1968 for super de luxe
Mark II 1100 saloon 2-door	October 1967 - March 1968 for de luxe, March 1968 for super de luxe
Mark II 1100 traveller	October 1967 - March 1968
Mark I 1300 saloon 4-door	October 1967 - March 1968 for de luxe, spring 1971 for super de luxe
Mark I 1300 saloon 2-door	October 1967 - March 1968 for de luxe, spring 1971 for super de luxe
Mark I 1300 traveller	October 1967 - 1972
1300GT	October 1969 - spring 1971

Engine, four-cylinder, overhead-valve, push-rod. Transversely mounted.

1100: 64.58mm bore. 83.72mm stroke. 1,098cc. 48bhp at 5,100rpm. Compression ratio 8.5:1/8.9:1.

1300: 70.61mm bore. 81.28mm stroke. 1,275cc. 58bhp at 5,250rpm. Compression ratio 8.8:1.

Carburetter, SU type HS (twin HS2 $1\frac{1}{4}$in diameter on 1300GT). Clutch, single dry-plate, diaphragm-spring, hydraulic operated. Petrol tank, 8 gallons. Electrics, 12-volt, positive earth (negative earth after August 1971). SU electric petrol pump type PD. Gearbox, 4-speed, synchromesh in all but first gear on early models, later, from October 1967, synchromesh in all gears. Floor mounted gearchange lever. Optional automatic transmission on Mark II. Steering, rack and pinion. Brakes, foot to all wheels, Lockheed hydraulic. Handbrake, cable to rear wheels. 8in diameter disc at front ($8\frac{1}{2}$in diameter disc on 1300GT) and 8in diameter drums at rear. Wheels, Mark I, $4J \times 12$ with 5.50-12 cross-ply tyres; Mark II and 1300, wheels $4C \times 12$ with 145-12 radial ply tyres. Suspension, Hydrolastic. Track, front 4ft $3\frac{1}{2}$in, rear 4ft $2\frac{7}{8}$in. Wheelbase 7ft $9\frac{1}{2}$in.

All standard Morris models finished in single colours. GT models in red, white, or bronze, with distinctive side flash. PVC covered roof panel and black upholstery. Traveller with imitation wood strip along sides, single lift-up rear door, maximum load 725lb.

Marina & Ital, Engines		
All engines four-cylinder, overhead valve, water cooled.		
A Series, Type 10	1,098cc, push-rod. 64.58mm bore. 83.72mm stroke. Aluminium four-ring pistons. Valve diameter, inlet 1.15in, exhaust 1in.	7cwt van
A Series, Type 12V	1,274.86cc, push-rod. 70.61mm bore. 81.28mm stroke. Aluminium solid skirt pistons. Valve diameter, inlet 1.3in (later engines 1.56in), exhaust 1.15in. Three-bearing crankshaft. Compression ratio 8.8:1 (low compression 8:1).	Marina & Ital 1.3 7cwt van de luxe 10cwt van & pick-up 440 van, 575 van & pick-up
B Series, Type 18V	1,798cc, push-rod. 80.26mm bore. 88.9mm stroke. Aluminium solid skirt pistons. Valve diameter, inlet 1.62in, exhaust 1.34in. Five-bearing crankshaft. Compression ratio 9:1 (low compression 8:1).	Marina 1.8
O Series, Type 17V	1,698cc, overhead-camshaft. 84.45mm bore. 75.8mm stroke. Duotherm solid skirt three-ring pistons with concave crown. Valve diameter, inlet 1.57in, exhaust 1.34in. Five-bearing crankshaft.	Ital 1.7
O Series, Type 20H	1,993cc, overhead-camshaft. 84.45mm bore. 89mm stroke. Duotherm solid skirt three-ring pistons with concave crown. Five-bearing crankshaft.	Ital 2.0

Marina & Ital, Specification
Clutch, Borg & Beck or Laycock diaphragm spring. Hydraulic operation, $6\frac{1}{2}$in diameter on 1.3 Marina up to chassis 185356. 8in diameter on 1.3 Marina after chassis 185357, also 1.8 Marina, and 7cwt van. $8\frac{1}{2}$in diameter on 1.7 and 2.0 Ital. Brakes, hydraulic, 8in diameter drums front and rear, on 1.3 Marina and 7cwt van. 9in diameter drums front and rear, with servo on 10cwt van and pick-up. Servo assisted with disc at front and 8in diameter drums at rear, on 1.8 Marina (actually the servo was optional on this model), the 440 van, 1.7 Ital, and 2.0 Ital. Servo assisted disc at front and 9in diameter drums at rear, on 575 van and pick-up.
Carburetter, SU (some models witht twin carburetters). Suspension, independent by torsion bars at front with lever type shock absorbers. Semi-elliptic leaf springs and telescopic shock absorbers at rear. Anti-roll bars front and rear. Gearbox, manual four-speed synchromesh, or Borg Warner type 35 automatic transmission optional or standard on some models. Tubular propeller shaft with Hardy-Spicer roller bearing universal joints. Rear axle hypoid. Steering, rack and pinion. Petrol tank, 11.5 gallons. SU mechanical pump. Electrics 12 volt, negative earth. Lucas alternator. Pressed steel wheels, $4\frac{1}{2}$in × 13in. Tyres, most models 155.13 radial ply (with reinforced type on 575 van and pick-up and 440 van. Early 1.3 models, 520-13 crossply. 145-13 radials on 1.8 and late 1.3. 165/70R-13 radial tyres on 1.8TC models. Wheelbase 8ft. Track 4ft 4in.

Addenda

Page 57
1927 & 1928 models. Tourer 4-door, four/five-seater. Colours should read:
   Claret with claret H.B. hide.
   Blue with blue H.B. hide.
   Brown with brown H.B. hide.
   Grey with grey H.B. hide.

1927 & 1928 models. Saloon 4-door. For 'maroon' read 'claret'. Both years upholstered in brown.

Page 61 – Line 4
For 'Approximately 700 MG cars' read '900'.

Page 70
1927 models two-seater. Colours should read:
   Blue with blue leather.
   Grey with grey leather.
   Maroon with red leather.
   Brown with brown leather.

1928 models two-seater. Colours should read:
   Blue with blue antique H.B. hides.
   Claret with antique H.B. hides.
   Brown with brown antique H.B. hides.
   Beige with beige semi-furniture hides.

1929 models two-seater. Colours should read:
   Stone/brown duotone cellulose with beige antique leather.
   Wine/maroon duotone cellulose with red antique leather.
   Deep maroon/bronze duotone cellulose with beige antique leather.
   Blue/black duotone cellulose with blue antique leather.

For all 1927 and 1928 models, delete the word 'cellulose' and replace with 'oil based paint'.

Page 71
1928 models Saloon, 13.9hp. Colours should read:
   Blue with blue semi-furniture hides or self-patterned moquette to choice.
   Claret with claret semi-furniture hides or self-patterned moquette to choice.
   Brown with brown semi-furniture hides or self-patterned moquette to choice.
   Beige with beige semi-furniture hides or self-patterned moquette to choice.

1929 models Saloon dome-back. For colour of monochrome cloth upholstery read 'fawn'.

For all 1927 and 1928 models, delete the word 'cellulose' and replace with 'oil based paint'.

Page 74 – Line 30
For 'the 750 cars later build by MG' read '736'.

Line 31
For 'between 1928 and 1932' read '1928 and 1933'.

Page 79
1928 models Saloon four-door and Coupé two-door add 'brown' to colour of furniture hide.

1929 models add 'Niagra' to blue.

Page 102
1927 Two-seater four-wheel brake model. Colours should read:
    Blue with beige Rexine.
    Grey with grey Rexine.

1928 Two-seater four-wheel brake model. Colours should read:
    Blue with blue antique Rexine.
    Beige with brown antique Rexine.

1927 Four-seater Tourer. Colours should read:
    Grey with grey Rexine.
    Blue with blue Rexine.

1928 Four-seat Tourer. Colours should read:
    Blue with blue antique Rexine.
    Beige with brown antique Rexine.

1927 Coupé. Colours should read:
    Blue with blue Rexine leathercloth.

1928 Coupé. Colours should read:
    Blue with blue antique Rexine.
    Beige with brown antique Rexine.

1927 Saloon. Colours should read:
    Blue with blue Rexine leathercloth.

1928 Saloon, four-door model. Colours should read:
    Blue with blue antique Rexine.
    Beige with brown antique Rexine.

For all 1927 and 1928 models, delete the word 'cellulose' and replace with 'oil based paint'.

For all 1929 models finished in 'stone/maroon', it should be noted that Morris literature of the period sometimes quoted this as 'brown/maroon'.

Page 103
All 1930 models where 'Morris brown' is listed. This was sometimes quoted in Morris literature of the period as 'stone/brown'.

1930 Four-seater Tourer. Add words 'lined white' to Niagra blue cellulose.

1931 Two-seater. Colours should read:
    Niagra blue cellulose, lined white, blue Karhyde upholstery.
    Morris maroon cellulose, lined white, red Karhyde upholstery.

1931 Coupé. Colours should read:
  Morris maroon cellulose, lined white, red Karhyde upholstery, putty Karhyde head.
  Niagra blue cellulose, lined white, blue Karhyde upholstery, fawn Karhyde head.

1931 Saloon. Colours should read:
  Fixed-head:
  Niagra blue cellulose, lined white, blue Karhyde upholstery, fawn Karhyde head.
  Morris maroon cellulose, lined white, red Karhyde upholstery, putty Karhyde head.
  Folding-head:
  Dark maroon cellulose with brown Karhyde upholstery.

1931 Commercial traveller's Saloon. Colours should read:
  Blue cellulose with blue Karhyde upholstery.

1932 Two-seater, Coupé, and Saloon. Colours should read:
  Atlantis blue, lined Azure blue, brown Karhyde upholstery.
  Autumn brown, lined cream, brown Karhyde upholstery.
  (Other Morris literature of the period quoted black with brown Karhyde
  upholstery).

Page 104
1933 Two-seater, Coupé, and Saloon (fixed and sliding-head). Colours should read:
  Hobar blue, lined cream, brown Karhyde upholstery.
  Autumn brown, lined white, brown Karhyde upholstery.
  Black, lined white, green Karhyde upholstery.

1933 Special Coupé. Colours should read:
  Green duotone cellulose with green leather upholstery.
  Red duotone cellulose with red leather upholstery.
  Grey duotone cellulose with blue leather upholstery.
  Black duotone cellulose with brown leather upholstery.

1934 Saloon, fixed and sliding-head. Colours should read:
  Blue cellulose with blue leather upholstery.
  Green cellulose with green leather upholstery.
  Black cellulose with brown leather upholstery.

Page 108 – Line 8
For 'ceased some 3,200 cars later' read '3,235'.

Page 124
1929 overhead camshaft chassis numbers 'M.101-M.2738' should read
'M.101-M.12738'.

Page 125
1929 Four-seater Tourer. Colours should read:
  Blue cellulose with blue antique Rexine.
  Brown cellulose with brown antique Rexine.

1929 Fabric Saloon, two-door. Colours should read:
  Pebble grain blue fabric, beaded blue and white, blue antique Rexine upholstery.
  Pebble grain brown fabric, beaded brown and orange, brown antique Rexine
  upholstery.

1930 Fabric Saloon, two-door. Colours should read:
  Blue fabric body, beaded blue and white, blue Karhyde upholstery.

1930 Coachbuilt Saloon. Other Morris literature of the period quoted 'stone/brown'
in place of 'dark brown & brown'. Morris sources also listed deep maroon
cellulose, lined orange, brown Karhyde upholstery for 'late 1930 models'.

1932 Family Eight Saloon. Colours should read:
  Atlantis blue cellulose, lined azure blue, brown Karhyde upholstery.
  Black cellulose, lined cream with green moulding, green Karhyde upholstery.

Page 126
1932 Two-seater and Coachbuilt Saloon. Colours should read:
  Willow green, lined cream, green Karhyde upholstery.
  Atlantis blue, lined azure blue, brown Karhyde upholstery.
(Morris literature of the period also quoted black with green Karhyde upholstery.)

1932 Four-seat Tourer. Colours should read:
  Atlantis blue cellulose, lined azure blue, brown Karhyde upholstery.

1933 3-speed Two-seater models. Colours should read:
  Hobar blue, lined cream, brown Karhyde upholstery.

1933 4-speed two-seater, Coachbuilt Saloon, and Family four-door Saloon.
Colours should read:
  Hobar blue cellulose, lined cream, brown Karhyde upholstery.
  Willow green cellulose, lined cream, green Karhyde upholstery.
  Black cellulose, lined white, green Karhyde upholstery.

1933 Four-seat Tourer. Colours should read:
  Hobar blue cellulose, lined cream, brown Karhyde upholstery.

1934 Two-seater and Coachbuilt Saloon. Colours should read:
  Blue cellulose with blue leather upholstery.
  Green cellulose with green leather upholstery.
  Black cellulose with brown leather upholstery.

1934 Family four-door Saloon. Colours the same as the 1933 version but with
leather upholstery.

Page 145
Oxford Six 1934 model chassis numbers '34/032832 - 34/035637' should read
'34/0.32383 - 34/0.35637'.

Page 147
1930 4/5 seater Tourer. Colours should read:
  Deep maroon cellulose, lined orange, brown C.B. hide upholstery.
  Niagra blue cellulose, lined white, blue C.B. hide upholstery.

1930 Saloon and Coupé (sliding-head). Colours should read:
  Deep maroon cellulose, lined orange, brown Vaumol leather.
  Niagra blue cellulose, lined orange, blue Vaumol leather.

1930 Fabric saloon. Colours should read:
  Black fabric, beaded red & white, red Vaumol leather upholstery.
  Red fabric, beaded black & white, red Vaumol leather upholstery.

1931 Saloon, sliding and fixed-head. Colours should read:
  Black cellulose, lined orange, brown Celstra hide upholstery.
  Niagra blue cellulose, lined white, blue Celstra hide upholstery.

1932 Saloon, sliding-head. Colours should read:
  Autumn brown/black, lined cream, putty colour leather upholstery.
  Atlantis blue/black, lined azure blue, putty colour leather upholstery.
  Wine/black, lined cream, putty colour leather upholstery.
  Black cellulose, lined green, putty colour leather upholstery.
  (Other Morris literature quoted 'brown' instead of 'putty'.)
  Black superstructure applies to all models.

Page 148
1933 Saloon (sliding-head) and Coupé (sliding-head). Colours should read:
    Wine/black cellulose, lined white, putty colour leather.
    Willow green/black, lined orange, putty colour leather.
    Hobar blue/black, lined cream, putty colour leather.
    All black, white lined, putty colour leather.
    (Some Morris literature of the period quoted 'brown' instead of 'putty' for the
    leather colour.)
    For all the colour combinations, black is the top colour.

Page 161
Isis 1930-35 specification. Add 'Radiator shutters on export models start at chassis
number IS1579'.

Page 162
1930 Tourer. Colours should read:
    Blue/black cellulose, blue M.B. hide upholstery.
    Maroon/wine cellulose, red M.B. hide upholstery.
    Hood material in black.
    (Some Morris literature of the period uses the description 'Lake' instead of
    'Maroon'.)

1931 Tourer. Colours should read:
    Niagra blue cellulose, lined white, blue M.B. hide upholstery.
    Lake cellulose, lined white, red M.B. hide upholstery.
    Hood material in buff.

1932 Coachbuilt Saloon with sliding-head. Colours should read:
    Autumn brown/black, lined cream, brown Celstra leather upholstery.
    Atlantis blue/black, lined azure blue, brown Celstra leather upholstery.
    Wine/black, lined cream, brown Celstra leather upholstery.
    Black, lined cream, brown Celstra leather upholstery.

Page 163
1933 Tourer. Colours should read:
    Blue or brown cellulose with leather upholstery.

1933 Saloon, sliding-head. Add note to colours:
    'Elsewhere Morris sources list the colours the same as the 1932 Isis saloon
    sliding-head models'.

1933. Add word 'Coupé' to 'Special'.

Page 177
1931 Saloon and Coupé with folding-head. Colours should read:
    Wolseley Lake cellulose, lined white, red Karhyde upholstery, putty Karhyde head.
    Black cellulose, lined green, green Karhyde upholstery, putty Karhyde head.

1931 Salonette fabric body. Colours should read:
    Black fabric body, beaded red and white, red Karhyde upholstery, putty
    Karhyde head.

1932 Saloon and Coupé. Colours should read:
    Atlantis blue, lined azure blue, brown Karhyde upholstery.
    Black, lined cream, green moulding, green Karhyde upholstery.
    (Other Morris literature of the period also list the additional colour of green
    cellulose with green Karhyde upholstery.)

1932 Sports Coupé. Add 'Sliding-head and fixed-head' to listed colours.

1933 Saloon and Coupé. Colours should read:
  Hobar blue cellulose, lined cream, brown Karhyde upholstery.
  Autumn brown cellulose, lined cream, brown Karhyde upholstery.
  Black cellulose, lined white, green Karhyde upholstery.

1933 Special Coupé. Colours should read:
  Green duotone with green leather upholstery.
  Red duotone with red leather upholstery.
  Grey duotone with blue leather upholstery.
  Black with brown leather upholstery.

1933 Tourer four-seater. Add note to colours:
  'Elsewhere Morris literature also lists a model in brown cellulose with brown
  Karhyde uphostery'.

Page 191
1933 Tourer four-seater. Colours should read:
  Black cellulose with brown Karhyde upholstery.
  Brown cellulose with brown Karhyde upholstery.

1933 Saloon. Colours should read:
  Hobar blue, lined cream, brown Karhyde upholstery.
  Willow green, lined cream, green Karhyde upholstery.
  Black, lined white, green Karhyde upholstery.

1934 Saloon. Colours should read:
  Green cellulose with green leather upholstery.
  Blue cellulose with blue leather upholstery.
  Black cellulose with brown leather upholstery.

Page 192
1934 Saloon. Colours should read:
  Green cellulose with green leather upholstery.
  Blue cellulose with blue leather upholstery.
  Black cellulose with brown leather upholstery.

1934 Special sports, four-seater. Colours should read:
  Red cellulose, black wings, red leather upholstery.
  Green cellulose, black wings, green leather upholstery.
  Saxe blue cellulose, dark blue wings and wheels, blue leather upholstery.
  Cream cellulose, green wings, green leather upholstery.
  All black with red leather upholstery.

Page 203 – Line 23
'February 1937' should read 'January 1937'.

Page 205
Delete 'Chieftain tanks' and 'until the seventies' and add 'still in service'.

Page 206
For chassis numbers 901-48612 and 48613-16500, add note that these are the Morris
Motors Ltd. listed change-over numbers. Research suggests that the change-over
from 35/E to S1/E would be in the region of 45424.

Page 207
Note that after September 1935 body colours for fixed-head and sliding-head
saloons were the same.

1935 Saloon, sliding-head. Morris literature of the period also quoted an all black
version with red leather upholstery.

Page 218 – Line 3
Alexander Duckhams Series II saloon, for 'Twenty-five', should read 'sixteen'.

Page 233
Add to chassis number table:
   'Series III Ten-Four. S3/TN108711 - S3/TN122430'.

Page 234
Twelve-Four Saloon colours:
   Maroon with red leather upholstery replaced earlier models in grey with blue upholstery.

Page 235
Twenty-Five Six Saloon colours:
   Grey with blue upholstery only applicable to 1937/8 models.
   Beige/Brown with brown upholstery was added for 1939 models.

Page 256
Series E Eight:
   Pre-war saloon and tourers should read chassis numbers 'SE/E101 - SE/E54675'.
   Post-war saloons should read 'SE/E54676 - SE/E122209'.

Series M Ten:
   Chassis numbers for pre-war saloons should read 'SM/TN101 - SM/TN27120'.

Page 307 – Line 34
'two stroke unit fitted east-to-east' should read 'east to west'.

Index